Laboratory Manual
for Olmsted/Williams

D0166002

Chemistry
THE MOLECULAR SCIENCE

Chemical Education Resources, Inc.
Wellesley, MA

 Mosby

St. Louis Baltimore Berlin Boston Carlsbad Chicago London Madrid
Naples New York Philadelphia Sydney Tokyo Toronto

Mosby
Dedicated to Publishing Excellence

Copyright © 1994 by Mosby–Year Book, Inc.

All rights reserved. Except in classes in which *Chemistry: The Molecular Science*
is used, no part of this publication may be reproduced, stored in a retrieval system,
or transmitted in any form or by any means, electronic, mechanical, photocopying,
recording, or otherwise, without prior written permission from the publisher.

Printed in the United States of America

Mosby–Year Book, Inc.
11830 Westline Industrial Drive
St. Louis, Missouri 63146

QD33.046

ISBN 0-8016-5072-0

CONTENTS

Laboratory Techniques: Safety Precautions

prepared by **Norman E. Griswold**, Nebraska Wesleyan University

As every chemist knows, there are many potential hazards in chemistry laboratories, some of which can be serious. However, chemists can avoid accidents by being thoroughly familiar with appropriate experimental techniques, by knowing the hazards associated with the chemicals they use, and by taking common sense safety precautions. Unlike most research and production laboratories, a school laboratory contains a relatively large number of students working, many without extensive knowledge of laboratory techniques, equipment, chemical hazards, or safety precautions. Consequently, students usually need some initial instruction in order to help them develop the necessary knowledge and awareness to avoid accidents.

By studying this module, you will learn basic safety rules and procedures applicable to chemistry laboratories. There is also a section describing what you should do if an accident occurs. On the last page, there is a Laboratory Safety Agreement that you must sign before you perform any experiments in the laboratory.

I. General Safety Rules

A. For Personal Protection

1. Wear eye protection at all times in the laboratory.

All staff, students, and visitors in the laboratory must wear proper eye-protection glasses or goggles that are approved by the appropriate authorities. Many states have eye-protection laws and require use of splashproof goggles. Your eye protection should protect both the front and sides of your eyes, should meet Occupational Safety and Health Administration (OSHA) standards, and should fit over your prescription glasses, if you wear them.

Normally, you should not wear contact lenses in the laboratory. If they are permitted and you do wear them, you must wear *fitted* goggles at all times. Contact lenses are hazardous, because they concentrate vapors under the lens, trap foreign matter, and interfere with the effectiveness of eyewash fountains. Soft contact lenses are particularly hazardous, because they can absorb and retain chemical vapors. Ordinary eyeglasses or sunglasses are not adequate protection in the laboratory.

2. Wear sensible clothing and tie back long hair.

In the chemistry laboratory, "sensible" clothing means old clothes that are fire resistant and not too loose, especially the sleeves. You should remove neckties, scarves, and jewelry. It is a good idea to wear a laboratory coat or apron.

"Sensible" clothing also includes shoes that cover your entire foot. Sandals and high heels are not safe. Shoes provide initial foot protection from dropped containers and spilled chemicals.

Keep all extra clothing, such as coats, hats, and scarves, off your laboratory bench and, if possible, out of the laboratory altogether.

If you have long hair, it can easily catch fire when you are close to an open flame. Usually this happens when you bend forward and your hair falls in front of your shoulders. Hair sprays only make matters worse. Consequently, you should fasten your long hair back.

3. Avoid absorbing, inhaling, or ingesting chemicals while you are in the laboratory.

There are three main routes for entry of chemicals into the body.

(1) The respiratory tract (lungs): Vapors, mists, smoke, and even dust particles can carry lethal

Copyright © 1993 by Chemical Education Resources, Inc., P.O. Box 357, 220 S. Railroad, Palmyra, Pennsylvania 17078
No part of this laboratory program may be reproduced or transmitted in any form or by any means, electronic or mechanical, including photocopying, recording, or any information storage and retrieval system, without permission in writing from the publisher. Printed in the United States of America

amounts of many substances into your body through your lungs. You can detect the presence of some toxic vapors by their noticeable odors, but many have no odor at all. For example, mercury metal has no odor, but its vapor pressure at room temperature allows mercury to enter the atmosphere at about one hundred times its threshold toxicity limit. To prevent inhaling toxic fumes, use the fume hood as your laboratory instructor directs.

If you are directed to detect odorous fumes by smell, do so with great care. Be certain that the reaction has ceased before you remove the container holding the reaction mixture from the fume hood. Place your hand near the container and gently fan some of the vapor toward you. Sniff carefully; do not inhale deeply. Return the container to the fume hood.

(2) The digestive tract: Never put anything in your mouth while you are in the laboratory. This precaution applies not only to chemicals, your fingers, and pipets, but also to beverages, food, and cigarettes. Do not bring food or drink into the laboratory, because it might become contaminated. Never put glassware in your mouth. Never use mouth suction to fill a pipet. Instead, your laboratory instructor will demonstrate the proper procedure for filling a pipet, using a rubber bulb.

(3) The skin: Some toxic substances can enter your body through your skin. Such substances include phenol, nitrobenzene, carbon tetrachloride, and solutions of cyanides. Some liquids, such as acids and bromine, can burn or otherwise damage your skin. Because many of these potentially hazardous liquids look like water, it is easy to forget their toxic or corrosive properties. Your cleanliness habits in the laboratory are an important part of skin-related accident prevention. Clean up spills immediately. Thoroughly wash your skin or any clothing that contacts a chemical. Periodically wash your hands and arms while working in the laboratory. Wash your hands and face with soap or detergent at the end of the laboratory period.

B. Working in the Laboratory

1. Do not attempt any unauthorized or altered experiments.

Consult your laboratory instructor before you make any change in an experimental procedure. Based on experience and knowledge, your instructor may be able to foresee possible hazardous results due to the altered procedure.

2. Know where safety equipment is located and how to use it.

Safety equipment includes fire extinguishers, safety showers, eyewash fountains, fire blankets, and an alarm or emergency telephone system. Your laboratory instructor will identify and describe the use of all these items during your first laboratory period. In addition, your instructor will clearly indicate the locations of all laboratory exits and describe an emergency evacuation plan. If such an introduction is not presented during your first laboratory period, ask your laboratory instructor to do so.

3. Never work alone in the laboratory.

Your laboratory instructor must be within your sight and hearing range at all times while you are working in the laboratory. If you encounter difficulties, your instructor must be able to assist you.

4. Use the fume hood when necessary.

Always perform experiments utilizing either toxic substances or substances with strong, irritating odors in a fume hood with proper airflow. Be certain that the hood is turned on. Do not place any portion of your body except your hands and lower arms into the hood.

Among the substances that you should use exclusively in a fume hood are chlorine, bromine, formaldehyde, phenol, sulfur dioxide, hydrogen sulfide, carbon disulfide, glacial acetic acid, and concentrated aqueous ammonia. If you are using highly toxic or carcinogenic substances such as cyanides, benzene, or chloroform, use a fume hood with an airflow of 60–100 linear feet per minute.

There are many other toxic substances. Therefore, it is imperative that you know the hazardous properties of *all* the substances you will use in an experiment. A major source of such information is the **Material Safety Data Sheet (MSDS)**, which must be available in the laboratory for every hazardous chemical you use. The MSDS sections on reactivity and health hazard data are of special interest for this purpose.

5. Dispose of waste and excess materials as directed by your laboratory instructor.

Each person is responsible for handling waste and excess materials in ways that will minimize environmental contamination and personal hazard. You should be given clear procedures to follow for disposing of waste materials. Some general guidelines are:

(1) Dispose of chemicals as directed by your laboratory instructor. Most reaction by-products and surplus chemicals must be neutralized or deactivated before disposal.

(2) Promptly dispose of waste materials into the proper labeled container.

(3) Place broken glass and porcelainware in separate containers specifically designated for this purpose. Ask for assistance before disposing of a

broken thermometer, because special handling is often required due to the mercury present in many thermometers.

(4) Place used matches and paper in a trash container, not in the sink. Dampen matches before disposing of them. Ask your laboratory instructor for disposal directions for any paper contaminated with a chemical.

II. General Rules for Handling Equipment

1. Only use equipment that is in good condition.

Defective equipment is a major accident source. Avoid using:

(1) beakers, flasks, funnels, graduated cylinders, or test tubes with chipped or broken rims;

(2) cracked beakers, flasks, graduated cylinders, test tubes, or crucibles;

(3) test tubes, beakers, or flasks with star-shaped cracks at or near the bottom;

(4) beakers, flasks, or test tubes with severely scratched bases;

(5) burets, pipets, or funnels with chipped tips;

(6) glass tubing or glass rods with sharp edges;

(7) hardened rubber stoppers;

(8) screw clamps, buret clamps, or support rings with nonfunctioning parts.

2. Assemble your experimental apparatus carefully.

Keep your work space uncluttered by extra equipment and chemicals. Use only equipment that is dry, at least on the outside. Wet glassware is slippery and can easily be dropped and broken.

When you use a ring stand with a platform base, assemble the attached apparatus directly over the base, and not to one side. Be sure to keep the assembly back from the edge of the laboratory bench, and tighten the clamps firmly. When you use a Bunsen burner, be sure you can quickly move it away from the ring stand assembly. Place a wire gauze under any glassware that you are going to heat with a Bunsen burner.

3. Avoid touching hot objects.

Burns are among the most common accidents in chemistry laboratories. Use the following precautions to avoid burns.

Do not hold reaction vessels in your hand: many chemical reactions generate large amounts of heat. Place flat-bottomed reaction vessels on the laboratory bench before you mix chemicals in them. If you cannot place the vessel on a laboratory bench, attach a clamp to the vessel, and clamp it to a ring stand.

When you heat or mix chemicals in a test tube, do not hold the test tube in your hand. Hold it with a test-tube clamp, or place it in a beaker.

When you heat chemicals in a container, remember that you are applying heat to not only the chemical but also to the container and any clamp holding the container. Be very careful when you touch clamps or containers that have been heated or been near a heat source.

Heated glass seems to stay hot forever. Therefore, after you work with glass in or near a flame, lay the glass aside to cool. Use great caution when touching the glass after cooling. Protect your hands with a towel or heat-absorbing gloves if you must move hot equipment.

Do not lay hot glassware directly on the laboratory bench. The hot object may cause charring or other damage. Instead, place hot glass on a ceramic-centered wire gauze, altered as follows. Bend each corner of the gauze downward to form a 90° angle. Place the gauze on the bench so that the folded corners act as legs. This should keep the center of the gauze about 1 cm above the bench surface. Assume that anything on this gauze is hot. In this way, you can avoid many painful finger burns.

4. Use extreme caution when you insert glass into stoppers.

Occasionally, you must insert a piece of glass into a rubber stopper. If you force the glass into the stopper, the pressure can cause the glass to suddenly break and seriously injure your hand.

Use the following procedure when inserting glass tubing, glass rods, thermometers, funnels, thistle tubes, or other objects with small diameters into rubber stoppers.

Before beginning to insert glass tubing, check to be sure that the ends of the tubing are smooth. If the ends are rough, fire polish the tubing, and set it aside to cool before you try to insert it into a stopper. Next, check to see that the stopper is pliable when rolling it in your palm or squeezing it with your fingers. If your stopper is hard and inflexible, use a different one. Be certain that the stopper properly fits the intended opening. You will feel frustrated if you successfully insert a glass tube into a stopper, only to find that you used the wrong-size stopper. Also, make sure that the stopper hole is approximately the same diameter as the tubing you are inserting.

Using water or glycerine, lubricate the stopper hole and about 2 or 3 cm at the end of the tubing. Keep the rest of the tubing dry so that you can easily grip it with your hand. Wrap your hands with the opposite ends of

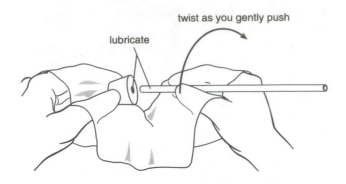

Figure 1   *Inserting glass tubing into a stopper*

a cloth. Hold the stopper in one wrapped hand. With the other wrapped hand, grasp the tubing near the lubricated end, as shown in Figure 1. Carefully insert the tubing into the stopper hole, using gentle pressure and a twisting motion. If the tubing does not slide through the hole when you exert slight pressure, apply additional lubricant to the tubing. ***Never use force***. Usually, once you introduce the tubing into the hole, it will slide easily through the hole.

Adjusting the tubing to the proper insertion depth is not difficult. Grasp the tubing as close to the stopper as possible. Continue to protect your hands with the cloth while you make the adjustment.

If you are inserting bent tubing, you may be inclined to grasp the bend in order to twist the tubing. You should avoid this, because the bend is generally the weakest part of a piece of bent tubing. The tubing will break if you put too much pressure on this point. Similarly, do not hold a thistle tube or a funnel by its top when inserting its stem into a rubber stopper.

III.   Handling Chemicals

A.   General Rules

1.   Read labels carefully.

(1)   Be sure that you select the specified substances. Different chemicals can have similar names. Compare, for example, sodium nitrate and sodium nitrite, or stannic chloride and stannous chloride. The names of these compounds look quite similar, but their chemical behaviors are very different.

(2)   Use the specified concentration or form of all substances. If you use the wrong concentration of a solution, you may get unexpected or undesired results. Some solids are available as strips, wire, granules, or powder. In some experiments, it is vital that you use the correct solid reagent form.

(3)   Know the hazards associated with the substances you are using. To help warn users of hazards associated with specific chemicals, the National Fire Protection Agency (NFPA) has developed a diamond-shaped symbol for chemical labels that rates the hazard level for many chemicals. An example of the NFPA symbol is shown in Figure 2. The NFPA symbol has a red segment at the top that indicates flammability hazard. The left segment is blue and shows the health hazard. The right segment is yellow and indicates reactivity. The bottom white segment is used to show special hazards, such as radioactivity, with a symbol of the hazard. If no specific hazard applies, the segment is left

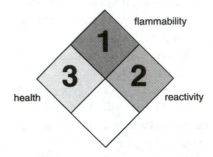

Figure 2   *An example of the NFPA symbol for chemical hazards*

blank. Within each segment, there is a boldface black number that indicates the degree of hazard. The numerical ratings are:

4 = extreme hazard (use goggles, gloves, protective
      clothing, and fume hood)
3 = severe hazard
2 = moderate hazard
1 = slight hazard
0 = no special hazard

2.   Obtain only the specified amount of each substance.

Safety regulations strictly forbid your returning unused chemicals to their original containers. If you accidently returned a chemical to the wrong container, you might contaminate the entire contents of the container. Therefore, estimate the amount of each chemical you will need, and obtain only that amount from the reagent bottle. If it is difficult for you to estimate the proper amount, be conservative. You can always return for more if you need it.

If you obtain too much of a chemical, you must properly dispose of the excess. Consult your laboratory instructor for specific disposal instructions.

3.   Leave chemicals in their proper places.

Bring your own container to the dispensing bottle on the reagent shelf. Do not take the dispensing bottle to your laboratory bench. It is not wise to carry large amounts of chemicals around the laboratory.

4. Clean up all spills immediately.

If you spill a chemical, it is your responsibility to immediately consult your laboratory instructor about the proper cleanup and disposal procedure. You are the only one who knows specifically what was spilled and where. The cleanup only takes a short time, and quick action can prevent further accidents.

If you spill a chemical on a balance, clean it up carefully and quickly. Otherwise, this delicate and expensive instrument may become corroded and permanently damaged.

Take care to remove any drops clinging to the outside of a reagent bottle after you have withdrawn reagent.

5. Label all containers to identify their contents.

When you obtain chemicals from the reagent shelf, be certain to label the containers into which you dispense the chemicals. Even if you plan to use the chemical immediately, you may have an excess that you must dispose of later. Don't trust your memory to keep track of such chemicals—use labels. You can use a pencil to write directly on the etched spot on many flasks and beakers. If you use pencil, you can easily erase these labels later.

B. Handling Liquids

1. Use proper technique when you obtain a liquid.

Take a labeled container to the reagent shelf. Transfer the liquid to your container at a nearby sink, in order to avoid spills on the bench, shelf, or floor.

From a dropper bottle: Be sure that you never touch your container or its contents with the dropper. Proper technique is shown in Figure 3. Never lay the dropper on any surface. Be careful to avoid contaminating the dropper bottle contents.

From a stoppered bottle: When you transfer the liquid, hold the stopper in your hand or place it upside down on a flat surface. Do not lay the stopper on its side. Proper technique is shown in Figure 4. If you follow this procedure, you will avoid contaminating the contents of the reagent bottle.

When you return the stopper to the reagent bottle, do not interchange stoppers. You may contaminate the reagent if you insert a stopper intended for another bottle.

When using a pipet to transfer liquid, pour some of the liquid into your own labeled container and pipet

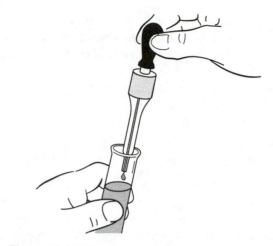

Figure 3 *Transfer of a liquid using a dropper*

from that container into another. **Do not pipet directly from a reagent bottle**.

From automatic dispensers: If you use an automatic dispensing buret assembly, you will fill the buret by using a rubber bulb to apply pressure to a reservoir containing the solution. The apparatus has an automatic zero adjustment. You will dispense the liquid through a stopcock. Ask your laboratory instructor for specific directions for using this apparatus.

Another dispensing device you may use is the Repipet®. This device consists of a syringe pump mounted on top of a bottle. You can set the stop on the syringe to deliver a specified volume of liquid. To draw liquid into the pipet, lift the syringe plunger as far as the stop. Then, fully depress the plunger to deliver the specified amount of liquid from the delivery tube.

2. Be careful when you mix liquids.

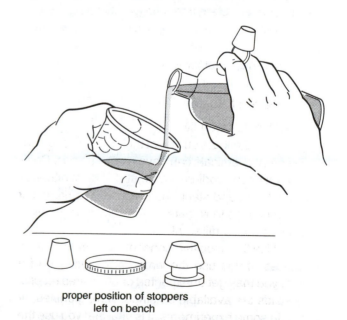

proper position of stoppers
left on bench

Figure 4 *Pouring liquids from a bottle*

When you mix liquid chemicals with water, always add the chemicals to the water, rather than vice versa. If you do this, the resulting solution will always be the more dilute and less hazardous solution. An accidental spattering from more dilute solution will be less dangerous. Always add chemicals slowly, in small amounts. It is especially important to use this procedure when mixing an acid and water, so that the water can absorb the heat generated by the dilution process.

3. Use a rubber bulb when you pipet liquids.

Never use your mouth to provide suction for filling a pipet. Instead, use a small rubber bulb at the top of the pipet to create suction to draw liquids into the pipet. The proper technique for filling a pipet is shown in Figure 5 and will be demonstrated by your laboratory instructor. Details of this technique are given in Experiment 4, *Laboratory Techniques: Volume Measurement of Liquids*.

4. Heat liquids cautiously.

Use a beaker or flask on a ceramic-centered wire gauze on a ring stand when heating a larger amount of liquid. Secure the container with a clamp.

Heat small amounts of liquids in test tubes. Attach a test tube clamp to the upper part of the test tube. Place the lower part of the test tube in a beaker of boiling water.

If you need to heat a liquid in a test tube to a temperature higher than 100 °C, you may heat the test tube and its contents directly in a burner flame, but you must be extremely cautious. The liquid at the bottom of the test tube may boil more quickly than the rest, and the resulting expansion may cause hot liquid to spurt out the top of the test tube. Therefore, you should never look into a test tube or point the open end of the test tube toward anyone while you are heating a test tube in this manner. Heat the test tube very gradually by placing the side of the tube, rather than the bottom, in the flame. Constantly move the test tube into and out of the flame. Proper technique is shown in Figure 6.

Be extremely cautious whenever you heat **flammable** liquids. You should only heat flammable liquids in a steam bath or on a special hot plate; either way, your laboratory instructor should carefully supervise the procedure.

Never heat a liquid in a graduated cylinder or in any other volumetric glassware.

5. Dispose of liquids properly.

Dispose of liquids as specified by your laboratory instructor. Never pour flammable or water-immiscible liquids into the sink or other outlet that drains into the sewer. Transfer these liquids to appropriately labeled

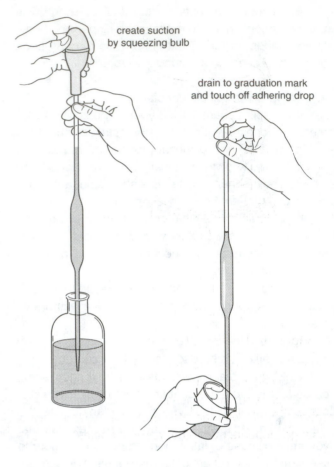

create suction by squeezing bulb

drain to graduation mark and touch off adhering drop

Figure 5 *Filling a pipet*

containers for disposal. Among the substances you must dispose of in this way are solvents such as acetone, ethyl alcohol, hexane, kerosene, methyl alcohol, and toluene.

C. Handling Solids

1. Transfer solids into widemouthed containers.

Solids are somewhat more difficult to transfer than are liquids. Take a labeled widemouthed container, such as a beaker, to the reagent shelf to obtain the solid you need. Make the transfer at or near the reagent shelf.

2. Transfer solids by rotating the reagent bottle.

During the transfer, hold the reagent bottle stopper in your hand or lay it on the bench, as shown in Figure 4, in order to prevent contamination.

Never insert your spatula directly into the reagent bottle. Instead, pour the chemical from the reagent bottle into your container. While pouring, slowly rotate the tipped bottle back and forth about an imaginary axis that passes through the top and bottom of the bottle, as shown in Figure 7. If you tip the bottle too far in an attempt to transfer the solid without rotating the bottle,

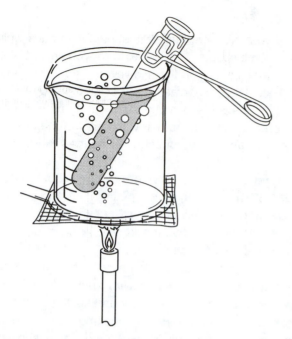

Figure 6 *Heating a liquid in a test tube*

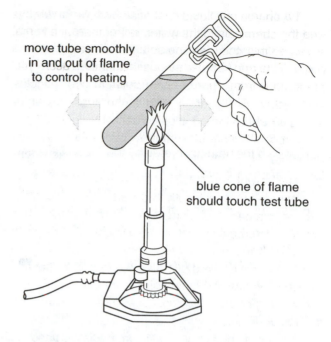

move tube smoothly
in and out of flame
to control heating

blue cone of flame
should touch test tube

large chunks may suddenly fall into your container, possibly causing a spill.

When you are finished, be sure to close the reagent bottle with the proper stopper. Do not interchange stoppers from other reagent bottles.

3. When you mix chemicals, add solids to liquids.

When mixing a solid with a liquid, add solid to liquid, rather than vice versa, with continuous stirring. Add solids in small amounts unless you are instructed otherwise.

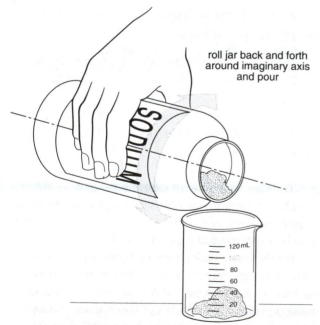

roll jar back and forth
around imaginary axis
and pour

SODIUM

120 mL
80
60
40
20

Figure 7 *Rotating a reagent bottle in order to pour a solid safely*

4. Dispose of solids properly.

Dispose of solids as specified by your laboratory instructor. Place the solids in appropriately labeled containers for disposal.

IV. When an Accident Occurs

Immediately attend to all physical and chemical injuries. Ask another student to report the accident to your laboratory instructor.

A. Burns

If you burn your skin with a hot object, flames, or chemicals, immediately flush the affected area with cold water. Continue flushing for 20 min.

If a large area of skin is burned and must be flushed, be careful not to get chilled.

Tell your laboratory instructor about the accident, even if the burn does not seem serious. Your instructor will decide whether or not you should seek medical attention.

In general, physicians prefer that you do not apply ointment to burns, especially serious ones. Often, they must remove the ointment because it can inhibit healing.

B. Splattered Chemicals

1. In your eyes.

You must wear eye protection at all times in the laboratory. Therefore, there should be no opportunity for chemicals to splatter in your eyes.

If a chemical splashes on your face while you are wearing splashproof goggles or glasses, ***do not remove your goggles***. Instead, drench your face and goggles at the nearest eyewash fountain. After you have removed all of the chemical from your goggles and face, take off your goggles, and drench the part of your skin where the goggles contact your face.

If a chemical does somehow get in your eye, immediately *YELL* for help in getting to an eyewash fountain. Drench your eye with water. Force your eye open, if necessary. Hold your eyelid away from your eyeball, and roll your eyeball so that the water flushes the entire eyeball. Continue to flush your eye with water for at least 15 min.

Ask your laboratory instructor whether or not you should seek medical attention.

2. On your skin.

For small spills on your skin, immediately rinse the affected area with large amounts of water. Remove any jewelry which might interfere with a thorough rinsing. Continue rinsing the area for 5–10 min.

If the spilled chemical is an acid, your laboratory instructor should follow the first water rinsing with a rinse of the sodium hydrogen carbonate solution kept in the laboratory for such an emergency. Resume rinsing with water for 5–10 min longer.

For spills of a base, your laboratory instructor should follow the first water rinsing with a rinse of boric acid solution, also found in the laboratory for emergency use. Resume rinsing with water for 5–10 min longer.

If a large area of your skin is affected, use the safety shower.

Ask your laboratory instructor whether or not you should seek medical attention.

3. On your clothing.

If possible, immediately remove all affected clothing. ***Seconds count***, so don't hesitate because of modesty. Drench the skin contacted by the contaminated clothing with plenty of water. For large spills, immediately use the safety shower, regardless of any initial cold or discomfort. However, avoid becoming too chilled. Your laboratory instructor will supervise your use of the shower and tell you when to leave it.

Discard all clothing that is chemically contaminated, including shoes and belts, following the directions of your laboratory instructor. Change into clean, dry clothes as soon as possible.

Ask your laboratory instructor whether or not you should seek medical attention.

C. Fire

If you have long hair, you should tie it back while you are in the laboratory. If you do so, you are not likely to catch your hair on fire. However, if your hair does catch on fire, immediately *YELL* for help. Use your hands to keep the burning hair away from your face and drench the burning hair with water at the nearest sink. Your laboratory instructor and nearby students will help.

If your clothing is on fire, use the ***STOP-DROP-ROLL-YELL*** technique for putting out the flames. ***STOP*** what you are doing, ***DROP*** to the floor, and ***ROLL*** to try and smother the flames. *YELL* for others to help quickly extinguish the fire.

Whether it's your clothing or your hair on fire, **do not run** to a safety shower or a fire blanket. Running increases the burning rate and therefore the possibility that hot, toxic fumes will enter your lungs. Let someone else run and get the fire blanket. Also, safety showers are for chemical spills, not extinguishing fires!

If a fire blanket is used, your laboratory instructor will have it removed as soon as possible, in order to minimize skin burns from fused clothing. After the blanket has been removed, your laboratory instructor will supervise drenching you in the safety shower.

While you are receiving aid, another student should try to shut off or reduce the fuel supply to the original fire. If necessary, she/he should direct the spray from a fire extinguisher at the base of the fire. If the fire cannot be quickly extinguished, everyone should leave the laboratory. Then activate the building fire alarm, call the fire department, and notify the campus emergency services.

After the fire has been extinguished, tell your laboratory instructor which extinguishers were used, so that they can be tagged as empty and replaced with filled ones.

D. Injury

Thoroughly wash minor cuts, making certain to remove all foreign materials from the wound, such as traces of chemicals and broken pieces of glass. Apply a bandage to keep the wound clean and to avoid further irritation.

For major cuts and severe bleeding, fast action is critical. Have someone call for emergency medical aid. Your laboratory instructor will quickly apply direct pressure to the open wound with a clean compress or cloth, until a tight bandage can be applied to control bleeding. When medical personnel arrive, they will decide what action to take. Only a medical professional is trained to treat serious injury.

Laboratory Safety Quiz

1. What part of your body requires special protection in the laboratory? Briefly describe the required protection.

3. Briefly describe how you can avoid ingesting chemicals into your digestive tract?

2. In your laboratory, where is the nearest:
   (1)   fire extinguisher?

4. What does MSDS mean?

   (2)   safety shower?

5. Briefly explain what you must do before you dispose of waste or excess materials.

   (3)   eyewash fountain?

   (4)   fire alarm?

6. List the defects that make glassware unusable.

   (5)   telephone?

   (6)   fire blanket?

7. Briefly describe the precautions you should take when you mix chemicals in a test tube.

8. Briefly describe how you should protect your hands while you are inserting glass tubing into a stopper.

9. What does the upper segment of the NFPA symbol indicate?

10. Briefly describe the proper technique for heating liquids in test tubes to temperatures below 100 °C.

11. Briefly explain what you should do *first* if you burn a small area on your skin in the laboratory.

12. What should you do immediately if your clothing starts burning?

name _____ date _____ room number _____

course _____ section _____ locker number _____

Before you begin working in the chemistry laboratory for the first time, read the safety precautions and techniques described in this module for the careful handling of chemicals and equipment, take the Laboratory Safety Quiz to check your understanding, sign and date this agreement, and give it to your laboratory instructor.

Chemistry Laboratory Safety Agreement

When I am in the laboratory, I will:

1. wear approved eye protection at all times;

2. wear sensible clothing and tie back long hair;

3. avoid absorbing chemicals into my body by
   (1) using great care in detecting odors,
   (2) never putting anything in my mouth,
   (3) washing skin and clothing that contact chemicals;

4. not attempt any unauthorized experiments;

5. know where safety equipment is and how to use it;

6. never work alone;

7. use the fume hood when necessary or so directed;

8. dispose of waste and excess materials according to instructions;

9. use only equipment that is in good condition;

10. assemble apparatus carefully;

11. avoid touching hot objects;

12. use extreme caution when inserting glass tubing into stoppers;

13. keep the laboratory clean;

14. handle chemicals with caution by
    (1) reading labels carefully,
    (2) using only the amount required,
    (3) leaving chemicals in their proper places,
    (4) cleaning up all spills immediately with supervision,
    (5) labeling all containers to identify their contents;

15. thoroughly wash my hands and face at the end of each laboratory period;

16. take immediate action as needed in response to burns, splattered chemicals, fire, or injuries;

17. report all accidents and injuries, no matter how minor, to my laboratory instructor.

I have carefully read and fully understand all the safety precautions summarized above. I recognize that it is my responsibility to observe the precautions throughout my chemistry course.

_____ _____

signature   date

_____ _____

laboratory instructor's signature   date

The Gas Burner and Glass Working

prepared by **H. A. Neidig,** Lebanon Valley College
and **J. N. Spencer**, Franklin and Marshall College

Purpose of the Experiment

Study the operation of a gas burner. Cut, firepolish, and bend glass tubing.
Draw dropper tips and insert glass tubing into rubber stoppers.

I. Using a Gas Burner

A. Background Information

We use a burning mixture of gas and air as a source of energy in homes, hospital laboratories, and industry. We use a device called a **gas burner** to mix the combustible gas with air containing oxygen. When we ignite the gas–air mixture, it burns, releasing heat and light. In the chemistry laboratory, we often use gas burners to provide heat to soften or melt substances or to increase the temperature of a chemical system.

Most gas burners in use today are modifications of the one invented by the German chemist, Robert W. Bunsen, in the mid-nineteenth century. A Bunsen burner (Tirrill type) is used in many chemistry laboratories.

A typical laboratory gas burner has a metal mixing tube, or **barrel**, attached to the **base** of the burner, as shown in Figure 1. The combustible gas flows through a piece of rubber or plastic tubing from a **gas cock** on the laboratory bench to the **gas inlet** of the burner. We make coarse adjustments in the gas flow by turning the handle on the gas cock. We regulate the gas–air mixture precisely by careful adjustment of the screw, located at the base of the burner, that controls the **needle**

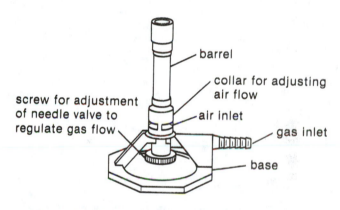

Figure 1 *Components of a gas burner*

valve. As the gas passes up through the burner barrel, it mixes with air being drawn into the barrel through the **air inlets**. We adjust the air flow by rotating the **collar**, which covers the air inlet holes to a greater or lesser extent, depending on its position.

The mixture is ignited by placing a burning match just below the top of the burner barrel, as shown in Figure 2(a). A striker may also be used to ignite the mixture, as shown in Figure 2(b).

We control the type of flame obtained from a burning gas–air mixture by varying the relative amounts of gas and air entering the burner. The different flames have different temperatures.

Copyright © 1990 by Chemical Education Resources, Inc., P.O. Box 357, 220 S. Railroad, Palmyra, Pennsylvania 17078 No part of this laboratory program may be reproduced or transmitted in any form or by any means, electronic or mechanical, including photocopying, recording, or information storage and retrieval system, without permission in writing from the publisher. Printed in the United States of America

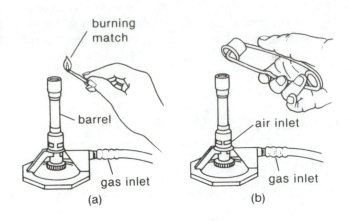

Figure 2 *Igniting the gas–air mixture flowing from a burner:* (a) *using a match;* (b) *using a striker*

When the gas–air mixture contains predominantly gas, the flame, shown in Figure 3(a), is yellow. We refer to this type of flame as a **luminous**, or **reducing**, **flame**.

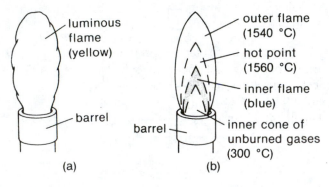

Figure 3 *Types of flames:* (a) *a luminous (or reducing) flame;* (b) *a nonluminous (or oxidizing) flame*

The yellow color of the flame is the result of incomplete burning of the gas, which is caused by an insufficient amount of oxygen (air) in the mixture. We also call this type of flame a **cool flame**. At times, the luminous flame will have a smoky appearance.

As we mix more oxygen with the gas, the temperature of the flame increases. When the gas is being efficiently burned, the hot flame that results is characterized by three distinct cones, as shown in Figure 3(b). We call this flame an **oxidizing**, or **nonluminous**, **flame**, because it does not have a yellow cone. The inner flame is bright blue, and the outer flame, or **envelope**, is almost colorless. Frequently, a colorless inner cone of unburned gas can also be seen. The hottest portion of this flame is immediately above the peak of the blue cone.

In this experiment, you will work with a gas burner, making various adjustments in the relative amounts of air and gas being burned.

B. Procedure

> *Caution:* Wear departmentally approved eye protection while doing this experiment.

> *Note:* Throughout this procedure section and that in Part II, you will be alerted to precautions you must take when working with a gas burner and with glass tubing. If the gas burner you are using is different from the one described in this module, your laboratory instructor will give you additional directions for using it.

> *Caution:* Tie back, or pin down tightly, all long hair. Do not hold your head directly above the top of the burner. Be sure no loose clothing comes near the flame.

1. Attach one end of a piece of rubber or plastic tubing to the gas inlet on your burner. Attach the other end of the tubing to the gas cock on the bench or to the gas supply.

2. Turn the collar over the air inlets so that no air can enter the burner barrel.

3. Rotate the adjustment screw, located at the base of the burner, until it is in the highest possible position (see Figure 1). This closes the needle valve, preventing any gas from entering the burner barrel.

> *Note:* Do Steps 4 and 5 as rapidly as possible.

4. Turn on the gas flow from your gas cock so that the maximum amount of gas can pass from the gas cock to your burner.

5. Immediately strike a match. Hold the burning match at a point just below the top of the burner barrel and at the angle indicated in Figure 2(a). Slowly open the needle valve to allow gas to flow through the burner barrel, and adjust the air inlet collar until you see a flame. Run water on the burnt match and discard it in the trash container specified by your laboratory instructor.

If you use a striker for the ignition, strike the spark at the top of the burner, as shown in Figure 2b.

> **Caution:** If the gas begins to burn inside the barrel, turn off the gas supply immediately. Do not touch the barrel. Allow the barrel to cool. Decrease the amount of air entering the burner. Re-ignite the gas–air mixture and make further adjustments to obtain a proper mixture of gas and air.

> **Note:** The numbers appearing in parentheses indicate the lines on your Data Sheet on which data should be entered.

6. Change the position of the needle valve by turning the adjustment screw. Note any changes in the flame. Record your observations on your Data Sheet (1).

7. Use a test tube clamp or crucible tongs to hold a 4-cm piece of iron wire in the flame for 1 min. Move the wire to another portion of the flame for 1 min. Place the hot wire on a ceramic-centered wire gauze or insulated surface. Record your observations on your Data Sheet (2).

8. Adjust the collar over the air inlets until an oxidizing flame is produced. Record your observations of the flame on your Data Sheet (3).

9. Use a test tube clamp to hold a 4-cm piece of iron wire in the flame for 1 min. Move the wire to another cone of the flame for 1 min. Place the hot wire on a ceramic-centered wire gauze or insulated surface. Record your observations on your Data Sheet (4).

10. Turn off the gas flow to your burner by closing the gas cock on your bench to extinguish the burner flame.

II. Glass Working

A. Background Information

The apparatus required for some experiments will include flasks connected by means of glass tubing and rubber stoppers. Because glass tubing is generally available in 60- and 120-cm lengths, you will have to cut the tubing into shorter, more useful lengths. In some cases, you will need to bend the tubing to obtain a desired angle. Frequently, you will need to insert a

piece of glass tubing into a rubber stopper. Occasionally, you will need to prepare a piece of tubing so that the constricted end can serve as a dropper tip. You will need to develop good technique for glass working to ensure the construction of acceptable apparatus and to minimize the chance of being cut or burned.

In this experiment, you will cut glass tubing, fire-polish the ends of glass tubing, and bend glass tubing. Then you will make droppers or pipets from glass tubing. Finally, you will insert glass tubing into corks or rubber stoppers. By following the steps in this procedure, you will obtain desirable results for the designated operations.

B. Procedure

1. Cutting glass tubing

For this part of the experiment, you will need three 20-cm pieces of glass tubing. You will cut one 20-cm piece into a 15-cm length and a 5-cm length. You will use the 15-cm length to prepare a dropper. One of the remaining 20-cm lengths will be used to prepare a 90° bend, and the other, a 135° bend.

11. Select a 20-cm piece of glass tubing. Make a mark with a pencil or ball-point pen or make a light scratch with a file or a glass scorer, shown in Figure 4, at the point where the tubing is to be cut to yield a 15-cm length. Your laboratory instructor will provide you with a metric ruler to use for these measurements.

> **Caution:** Too much pressure on the file will break the glass tubing instead of scratching it.

12. Place the tubing on the laboratory bench top. Use a firm, single stroke of the file away from you to make a scratch on the tubing at the desired point. See Figure 5(a).

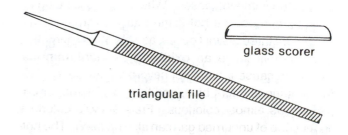

Figure 4 *Tools for scratching glass*

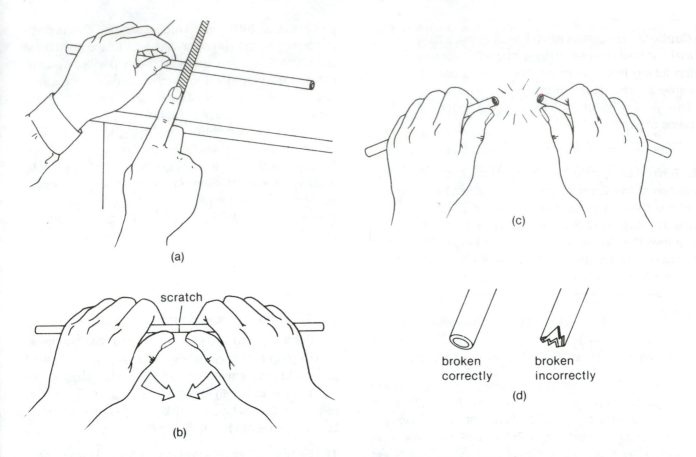

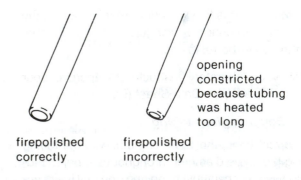

Figure 5 *Cutting or breaking glass tubing: (a) scratching a piece of glass tubing; (b) positioning your hands on the glass tubing; (c) breaking the glass tubing; (d) correct and incorrect breaks*

13. Hold the tubing in your hands so that the scratch is on the side of the tubing away from your body. See Figure 5(b).

Caution: Be careful when breaking glass tubing. If the tubing has been scratched properly, very little pressure will be required to break the tubing.

Handle the tubing with care, in case the end of the tubing is jagged. Avoid cutting yourself with the jagged end of glass tubing.

---

14. Apply light outward pressure to the tubing (away from your body), with your thumbs close to each other on the side of the tubing directly opposite the scratch. Simultaneously, with your fingers pull the ends of the glass tubing toward your body and away from the scratch, while pushing forward with a rapid motion. At the same time, bring your thumbs and wrists together, moving your elbows in against the sides of your body. See Figure 5(c). Figure 5(d) shows the end of a piece of glass tubing broken correctly and the end of another piece broken incorrectly.

2. Firepolishing glass tubing

To make the ends of the tubing smooth and safe to handle, you must firepolish them. You accomplish this operation by heating and rotating each rough end in the hottest part of a nonluminous flame until the end becomes smooth.

Firepolish both ends of the 15-cm tubing, following the directions below.

Figure 6 *Correctly and incorrectly firepolished glass tubing*

Caution: Be careful when handling glass tubing that has been heated. Always place the portion of the tubing that was in the flame on a ceramic-centered wire gauze or on an insulated surface. This glass is hot enough to burn your fingers or a piece of paper.

15. Turn on the gas flow and re-ignite the gas–air mixture. Adjust the flame to give an intense blue cone. Hold the piece of glass tubing at about a 75° angle. Place the rough end of the tubing in the burner flame just above the blue cone, as shown in Figure 7. Continuously rotate the tubing while it is in the flame. Do not allow the tubing to become so hot that its walls collapse. The melted glass will flow together and close the opening.

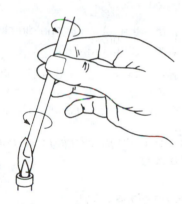

Figure 7 *Firepolishing a piece of glass tubing*

16. When the glass has a smooth edge, remove the tubing from the flame. Place the hot tubing on a ceramic-centered wire gauze or an insulated surface to cool. Do not touch the hot tubing or allow it to touch the bench top.

17. Firepolish the ends of your two 10-cm pieces of glass tubing. Repeat Step 15.

18. Turn off the gas flow to your burner by closing the gas cock on your bench to extinguish the burner flame. Allow the burner barrel to cool.

19. Ask your laboratory instructor to approve your work and to initial your Data Sheet (5).

3. Bending glass tubing

A correct bend should be smooth, with the tubing retaining the same diameter throughout the bend. Figure 8 shows an acceptable bend and one that is not acceptable. Take a 20-cm length of tubing and make a 90° bend like that shown in Figure 9, following the directions below.

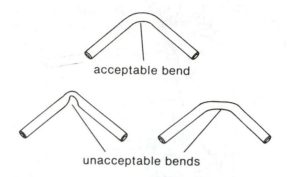

acceptable bend

unacceptable bends

Figure 8 *Examples of acceptable and unacceptable bends in glass tubing*

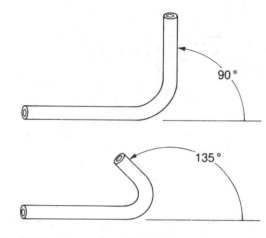

90°

135°

Figure 9 *Examples of a 90° bend and a 135° bend*

20. Place a wing top, or flame spreader, on top of your cooled gas burner, as shown in Figure 10(a).

21. Turn on the gas flow and re-ignite the gas–air mixture. Adjust the flame to give a sharply defined region of intense blue color that spreads uniformly above the wing top, as shown in Figure 10(a).

22. Select one of the two pieces of ***cooled*** glass tubing that has both ends firepolished.

23. Grasp the glass tubing with both hands, one hand on either side of the location where the bend is to be made.

24. Place the tubing just above, and parallel to, the top of the bright blue portion of the flame.

25. Slowly rotate the tubing uniformly and continuously as shown in Figure 10(b).

26. When the glass begins to sag from its own mass, remove the tubing from the flame. Rapidly bend the tubing to the desired angle by quickly lifting the ends of the tubing in an upward and inward direction with your hands, as shown in Figure 10(c).

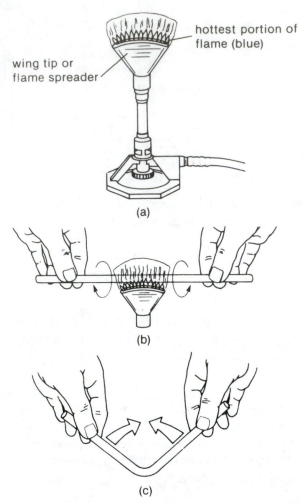

(a)

hottest portion of flame (blue)

wing tip or flame spreader

(b)

(c)

Figure 10 *Bending glass tubing: (a) flame for bending glass tubing; (b) heating glass tubing to soften it; (c) bending glass tubing*

Caution: Handle the glass tubing with care to avoid burning your fingers or hands. Do not place the hot glass directly on the laboratory bench.

---

27. Place the bent glass on a ceramic-centered wire gauze or an insulated surface to cool.

28. Ask your laboratory instructor to examine your work and initial your Data Sheet (6).

29. Take the other 20-cm length of tubing with fire-polished ends and make a 135° bend. Repeat Steps 23–28.

30. Ask your laboratory instructor to examine your work and initial your Data Sheet (7).

31. Turn off the gas flow to your burner by closing the gas cock on your bench to extinguish the burner flame. Allow the burner barrel to cool before removing wing top from your barrel for use in Part 4.

4.   *Preparing droppers or pipets*

Use the 15-cm length of tubing with firepolished ends in the following procedure to prepare droppers or pipets.

32. Remove the wing top from the *cooled* burner. Turn on the gas and re-ignite the gas–air mixture. Adjust the collar to produce a hot, nonluminous flame.

Note: In the following steps, keep your hands parallel to each other when heating and pulling the glass.

---

33. Hold the piece of glass tubing in a horizontal position between your fingers and thumbs, as shown in Figure 11(a). Keep the center of the tubing just above the bright blue cone of the flame.

34. Continuously rotate the tubing in one direction in the flame, as shown in Figure 11(a).

35. When the tubing becomes soft and begins to sag, slowly pull the ends of the tubing away from each other. See Figure 11(b). Keep your hands parallel. Do not pull your hands apart too rapidly.

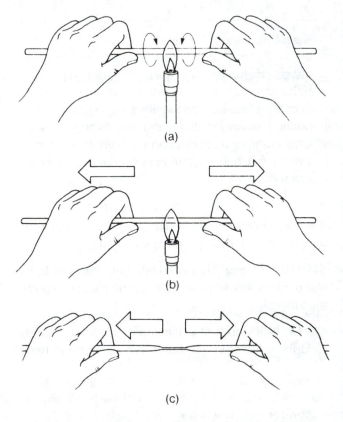

(a)

(b)

(c)

Figure 11 *Preparing droppers: (a) rotating the tubing in the flame; (b) pulling tubing in the flame; (c) pulling tubing out of the flame*

36. Quickly remove the tubing from the flame. Continue to pull the tubing, keeping your hands parallel. See Figure 11(c).

37. Hold the tubing until it hardens. Cut the tips to the desired size using a file or glass scorer.

38. Firepolish the new ends of the tubing. Be careful to heat the constricted tips only briefly to avoid further constriction or closing of the end of the tubing. Place the firepolished droppers on a ceramic-centered wire gauze or an insulated surface.

39. Ask your laboratory instructor to examine your work and to initial your Data Sheet (8).

40. Turn off the gas flow to your burner by closing the gas cock on your bench to extinguish the burner flame.

> *Caution:* Hold the burner by its base when removing the rubber tubing from the burner. Do not grasp the hot burner barrel with your hands.

5. Inserting glass tubing into corks or rubber stoppers

> *Caution:* Special precautions must be taken when inserting glass tubing, glass rods, thermometers, funnels, and thistle tubes into a cork or rubber stopper. Injury to your hands can result if excessive pressure on the glass piece causes it to break suddenly. Good safety measures are to lubricate the glass and stopper, cover both hands with a cloth, and avoid forcing the piece of glass if it does not enter the stopper easily.

The proper procedure for inserting glass tubing into a cork or stopper involves the following steps.

41. Use water or glycerol to lubricate the hole in the stopper and about 2 or 3 cm of the outside of one end of the tubing.

42. Drape the ends of a cloth towel across both hands. Hold the stopper in one hand. With your other hand,

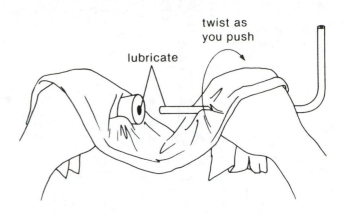

Figure 12 *Inserting a piece of glass tubing into a rubber stopper*

grasp the firepolished tubing with the 90° bend immediately behind the end of the tubing to be inserted into the stopper. See Figure 12.

> *Caution:* Be very careful when inserting a bent piece of glass tubing into a rubber stopper. Do not exert pressure on the bend itself. Grasp the tubing as close to the point of insertion as possible, and apply pressure there.

43. Carefully insert the end of the tubing into the stopper with gentle pressure and a twisting motion.

44. Slowly work the tubing through the stopper. As soon as enough tubing extends through the stopper so that you can grasp it with your fingers, *pull* the exposed end of the tubing, using the hand that was holding the stopper.

45. Ask your laboratory instructor to examine your work and initial your Data Sheet (9).

46. Discard all scrap glass tubing as directed by your laboratory instructor.

> *Caution:* Wash your hands thoroughly with soap or detergent before leaving the laboratory.

name section date

Data Sheet

I. Using a Gas Burner

	nature of flame	*observations with iron wire in flame*
initial flame	(1)	(2)
final flame	(3)	(4)

II. Glass working

	process	*remarks*
(5) firepolishing		

initials

(6) 90° bend

initials

(7) 135° bend

(8) dropper tip

(9) inserting tubing in a rubber stopper

Pre-Laboratory Assignment

1. List safety precautions that should be observed while

    (1) Lighting the gas and adjusting the burner (two precautions).

    (2) Cutting and bending glass tubing (two precautions).

    (3) Inserting glass tubing into a rubber stopper (two precautions).

2. What causes a cool flame? How can a cool flame be recognized?

3. What characterizes efficient burning of a gas–air mixture?

4. What is the purpose of inserting an iron wire into the different cones of the flame?

5. Briefly outline the procedure used to cut tubing.

6. After tubing has been cut, what further procedure should be followed before the cut tubing is used?

7. (1) What color should the flame just above the top of the wing top be when it is used to soften glass?

9. What type of flame is used to heat a piece of glass tubing when making a dropper? How is the dropper tip cut to proper length?

(2) Why is a wing top used to heat tubing for bending?

8. What constitutes an acceptable bend in glass tubing?

Laboratory Techniques: Mass Measurements

prepared by **Norman E. Griswold**, Nebraska Wesleyan University

Background Information

An important part of laboratory technique is the accurate determination of quantities of matter. Sometimes we accomplish this by measuring the space occupied by matter; that is, by finding the volume of matter. However, the use of volume to determine the quantities of solids is not always reliable. For example, a solid in the form of lumps would appear to occupy a different volume from the same sample after being finely ground. Thus, using volume as a reliable measure of a quantity of solid matter has certain limitations.

Apparently, these limitations on the use of volume for measuring solid matter have been known for thousands of years, because archaeologists have found that prehistoric humans developed another method for determining amounts of solid materials. Perhaps as early as 5000 B.C., ancient Egyptians used small stone balances to compare the mass of gold in Egyptian temple treasuries to some standard defined mass.

Mass is a somewhat difficult term to define in words that have physical significance. The mass of an object is defined most often as the quantity of matter it contains, although we sometimes define mass as the property of a body that gives the body inertia. **Inertia** is the property of matter that represents its resistance to being accelerated. Even though these definitions are not very helpful in terms of visualizing what "mass" means, we usually have an intuitive understanding of its meaning, just as we know what is meant by the term "energy," for example.

Even though chemists often use the term "weight" in referring to mass, mass and weight are actually different properties. **Weight** is the force on an object due to the pull of Earth's gravity. Thus, the farther an object gets from the center of the Earth, the less it weighs. When astronauts get beyond Earth's gravitational pull, they experience weightlessness. However, the **mass** of an object remains constant no matter where it is, because mass is determined by comparison to a standard defined mass and the amount of matter in the object is constant. Nevertheless, scientists still use the terms "mass" and "weight" interchangeably, and we will follow that common practice.

The current scientific unit of mass is the **kilogram**, a unit in the metric system. One kilogram is equal to the mass of a platinum-iridium alloy cylinder that is kept at the International Bureau of Weights and Measures at Sevres, France. In the laboratory, chemists find it more convenient to use a smaller unit of mass called a **gram**. A gram is one one-thousandth of a kilogram. The gram is also very nearly equal to the mass of one cubic centimeter of water. Another comparison is that the gram is equal to approximately 0.035 ounce, which is about one-thirtieth of an ounce.

The instrument used for determining the mass of an object is called a **balance**. In general, a balance consists of a beam supported on a sharp knife-edge made from a hard substance. A pan and some weights are attached to the beam. A counter-weight is attached to the other end of the beam. After an object has been placed on the pan, some of the small weights are moved until the beam returns to its balanced condition.

Copyright © 1987 by Chemical Education Resources, Inc., P.O. Box 357, 220 S. Railroad, Palmyra, Pennsylvania 17078
No part of this laboratory program may be reproduced or transmitted in any form or by any means, electronic or mechanical, including photocopying, recording, or any information storage and retrieval system, without permission in writing from the publisher. Printed in the United States of America

The number and mass of weights moved can be used to calculate the mass of the object in the pan. Some laboratory balances are capable of determining mass to the nearest 0.0001 gram and must be operated with great care.

General Weighing Procedures

There are several basic weighing procedures. The one you choose for a particular measurement often depends on what must be accomplished and what is most convenient and least time consuming. Directions for many experiments specify the appropriate weighing procedure. Three common procedures are described below.

1. Weighing a solid object or empty container. A metal or other solid object may be weighed by placing it directly on the balance pan if it is a solid bar or chunk. Granulated or powdered solids should be weighed in a container (see Procedure 2). To avoid any reaction between the solid and the surface of the pan, reactive metals such as the alkali metals should be weighed in containers. When weighing empty containers, you should make sure they are at room temperature and dry.

2. Weighing a container, adding chemical, and then weighing chemical and container together. This procedure should only be used for chemicals that do not absorb water from, or otherwise react with, air. In this procedure, the "container" may be a piece of weighing paper (slick paper, not a paper towel). If the masses of container and chemical must be known to the nearest milligram or better, it is most efficient to first obtain the approximate mass of chemical needed by weighing it alone on a less sensitive balance (for example, a triple-beam or top-loading balance). Then weigh the empty container on the more sensitive analytical balance. Finally, add the chemical to the container and weigh the two together.

3. Weighing a container and a chemical, removing some chemical, and then weighing the container and remaining chemical. In this procedure, you weigh a container, such as a weighing bottle, and chemical together as though they were a single object. Then you carefully remove the desired amount of chemical. It is difficult to estimate exactly how much chemical to remove. You may have to remove some chemical, take a balance reading, and then repeat these two steps several times until you have removed the desired amount. Never return unneeded chemical to the weighing bottle. Finally, you weigh the weighing

bottle and remaining chemical as though they were a single object.

The above procedure, with the following additional steps, is used when a **hygroscopic** chemical (one that absorbs water from the air) is to be weighed. Before it is weighed, the chemical is placed in a container, such as an uncovered weighing bottle, and dried in an oven. After a designated period of time, the bottle and chemical are removed from the oven, and the cover is replaced on the bottle. Then, the covered bottle and dried chemical are cooled to room temperature prior to the weighing procedure. Because the weighing bottle is uncovered for only a short time while chemical is removed during weighing, most chemicals will not absorb significant amounts of water. Frequently, the uncovered weighing bottle and dried chemical are stored in a desiccator until another sample is needed.

Using a Balance

I. Selecting an Instrument

Some masses need to be measured only to the nearest gram, while others may need to be determined to the nearest 0.0001 gram. An unnecessarily exact measurement requires precious time that might be devoted to other work; on the other hand, a rough measurement may lead to faulty conclusions. Therefore, the first decision to be made concerning a mass measurement is the selection of the appropriate balance.

Mass measurements to the nearest gram or tenth of a gram are often made using either a **triple-beam balance** or a **centigram balance.** A triple-beam balance, shown in Figure 1, usually has a capacity of about 600 grams. A centigram balance, shown in Figure 2, is normally limited to about a 110-gram capacity. However, most models have provision for the use of auxiliary masses, which makes it possible to double or even triple their capacities.

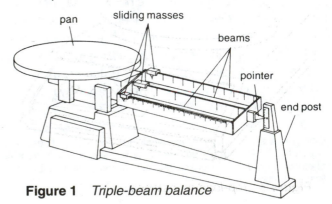

Figure 1   *Triple-beam balance*

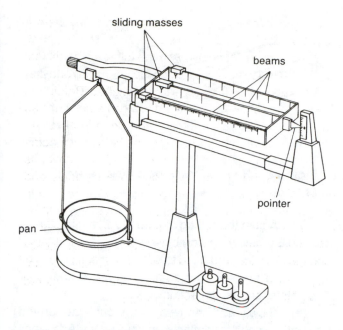

Figure 2 *Centigram balance*

A **top-loading balance** (Figure 3), usually called a "top loader," is used for fast weighing of masses to the nearest 0.1 to 0.001 gram. Many models have capacities from about 100 grams up to about 8000 grams.

An **analytical balance**, shown in Figure 4, is used for careful, exact measurements. These balances are delicate instruments, costing more than $1000 each, and must be used with considerable care. Normally, they have a capacity of between 100 and 2000 grams and are capable of detecting masses to the nearest 0.1 milligram, or 0.0001 gram.

II. General Precautions for the Use of Any Balance

1. Take extreme care to keep chemicals off the pans. Only hard solid objects, such as containers or metal bars, may be placed directly on the pans. All powdered,

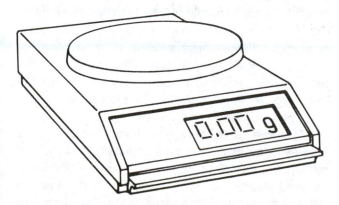

Figure 3 *Top-loading balance*

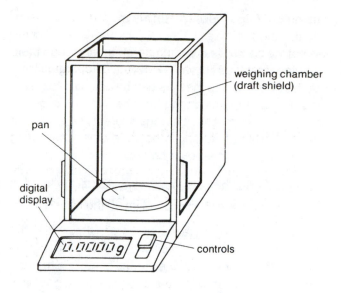

Figure 4 *Analytical balance*

granulated, or pelleted solids—that is, nearly all chemicals—must be weighed in a dry container. Liquids should be weighed in a closed dry container to avoid loss of mass by evaporation or fuming. No wet containers should be placed on the pan. If the balance pan is wet or dirty, consult your instructor.

2. Before measuring the mass of an object, check that the balance is level and that it indicates zero mass when the pan is empty and all standard masses are adjusted to zero. The level of the balance should be checked routinely before each mass measurement. Most balances have a **spirit level**, which is a small window containing a liquid with a bubble. When the balance is level, the bubble is centered in the window. Many balances have leveling disks on their feet that can be turned to adjust the level of the balance.

3. Place objects gently on the pan. Failure to do this can damage the balance, which will certainly affect its accuracy.

4. All objects placed on the pan should be at room temperature. A hot object may cause not only damage to the pan but also convection currents in the air around it. These currents make the object appear to weigh less than its true mass. As the object cools, the convection currents diminish, and the object appears to increase in mass. A cold object also creates a convection current and, in addition, may condense moisture from the air onto its surface. These effects lead to a recorded mass that is too large.

III. The Triple-Beam Balance

As the name implies, a triple-beam balance has three parallel beams, as shown in Figure 1. Each beam has a sliding mass. These masses can be adjusted to grams and tenths of grams on the front beam, tens of grams on the rear beam, and hundreds of grams on the center beam. Use the following procedure when measuring masses with a triple-beam balance.

1. Check that the balance is level. Adjust the leveling disk, located on one foot of the balance, if necessary.

2. Set all sliding masses at zero. Check that the balance indicates zero mass with the pan empty. The pointer at the right end of the beams should point to the zero mark of the scale on the end post. If the pointer is moving, it should swing an equal distance above and below the center mark; do not wait for it to stop swinging. A recessed balance adjustment, which is usually under the pan, may need to be adjusted. If this does not improve the situation, consult your instructor for assistance.

3. Gently place the object to be weighed on the pan. The pointer at the end of the beams will deflect upward.

4. On the beam with the largest sliding mass, slide the mass toward the right from notch to notch until the pointer at the end of the beams deflects downward.

5. Move the sliding mass back one notch to the left. The pointer should deflect upward again.

6. On the beam with the intermediate sliding mass, slide the mass toward the right from notch to notch until the pointer deflects downward.

7. Move the intermediate sliding mass back one notch to the left. The pointer should deflect upward again.

8. Move the smallest of the sliding masses along its beam (there are no notches on this beam) until the pointer swings an equal distance above and below the center mark on the scale. Do not wait for the pointer to stop at the center mark; this takes too long.

9. The mass of the object is equal to the sum of the values indicated by the sliding masses on all three beams. Figure 5 shows a mass of 498.5 grams. Record the mass of your object.

10 If you have weighed a container and wish to add chemical to it and reweigh, follow this step next. Otherwise go to Step 11.

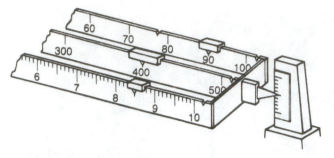

Figure 5   *Reading mass on a triple-beam balance*

(a)   Adjust the sliding masses to a reading that is about 0.5 g below the combined mass of the container and the chemical you need to add. The mass is set a bit low to help prevent your adding too much chemical. The pointer should deflect downward.

(b)   Slowly add chemical to the container on the pan until the pointer deflects upward again. Add small amounts so as not to get more than you want in the container. Additions may be made by spatula or scoopula. If the chemical is granular with no lumps, it is possible to pour it slowly into the container, but this takes practice.

(c)   Adjust the smallest sliding mass until the pointer is centered on the scale as before.

(d)   The mass of the container and chemical is the sum of the values indicated by the sliding masses on all three beams. Record the new mass.

11. Remove the object from the pan and slide all masses back to the left end of the beams.

IV. The Centigram Balance

The centigram balance, shown in Figure 2, is very similar to the triple-beam balance. There are two apparent differences. One is that the capacity of the centigram balance is usually smaller than that of the triple-beam balance. The other difference is that some centigram balances have a **beam-support lever**, located just below the knife-edge. This lever causes the beam to be supported by something other than the knife-edge while objects are being added to the pan and while the balance is not in use.

To weigh an object on a centigram balance with a beam-support lever, place the object on the pan, then rotate the lever 180 degrees. Use the same procedure as used for the triple-beam balance to determine a mass. After recording the mass, return the beam support lever to its original position.

V.   The Top-Loading Balance

Top loaders consist of a box with a pan on the top and with controls and a read-out at the front, as shown in Figure 3. Electronics within the box give fast, reliable results that are displayed digitally. Different manufacturers design their top loaders slightly differently. For example, one manufacturer's top loaders have a single control bar at their front edge. Another manufacturer provides individual touch buttons for the various controls. These differences make it difficult to give specific directions for use of the top loader in your laboratory, but the following general procedure should help.

1.   Press the button or control bar that turns the top loader on. The display should indicate zero grams. If it does not, ask your instructor for assistance.

2.   Gently place the object on the pan. The mass of the object will appear on the display. Record this value.

3.   If the object you are weighing is a container and you wish to add chemicals to it and reweigh, follow this step next. Otherwise go to to Step 4.
   (a)   Press the tare (T) button or control bar to **tare** the container. The tare function resets the balance to 0.00 grams with the container on the pan. From this point on, the display will show only the mass of chemical added to the container.
   (b)   Slowly and carefully add chemical to the tared container until the desired amount has been added. The digital read-out will indicate your progress. Some top loaders have graphic displays to indicate how much of the weighing range you have used and how much is still available. The pan moves as chemical is added, and the reading may fluctuate for a few seconds after you stop adding chemicals. The letter "g" will appear when the reading is stable.
   (c)   Record the mass of added chemical from the digital read-out.

4.   Remove the object or the container and chemical from the pan.

5.   Press the button or control bar that turns the top loader off. This is the same control that turned it on.
   Some top loaders have data interfaces built into them for connection to computers and printers. Operation of such interfaces involves the use of a keyboard and provides a number of useful applications.

VI.   The Analytical Balance

Analytical balances are used for very accurate, quantitative measurements of mass to the nearest 0.0001 gram. These balances are more delicate than those described earlier. The precautions given in Part II must be observed scrupulously to avoid damage. Analytical balances look like a top loader with a glass-sided box over the pan, as shown in Figure 4. The box acts as a draft shield so that very accurate weighings will be unaffected by any air currents in the laboratory. The draft shield (or weighing chamber) opens to allow access to the pan from either side and from the top.

Analytical balances come in a variety of models, with several optional accessories. Although the controls may vary a bit from model to model, the following general procedure for using analytical balances should help you in using your balance.

1.   Check the balance to see that:
   (a)   The balance is level. Leveling disks are normally located on two feet of the balance.
   (b)   The pan is clean and the weighing-chamber doors are closed.
   If either of these conditions does not apply, consult your instructor.

2.   If you wish to simply weigh a solid object, use the procedure in this step. Otherwise go to Step 3.
   (a)   Press the button or control bar that turns the balance on. The read-out should indicate zero grams. If it does not, ask your instructor for assistance.
   (b)   Open an appropriate weighing-chamber door.
   (c)   Gently place the object in the center of the pan.
   (d)   Close the weighing-chamber door. The mass of the object will appear on the read-out.
   (e)   Record the mass of the object.
   (f)   Go to Step 4.

3.   Use this procedure if you wish to weigh a container and then add a certain mass of chemical to it.
   (a)   Open an appropriate weighing-chamber door.
   (b)   Gently place the container in the center of the pan.
   (c)   Close the weighing-chamber door.
   (d)   Press the button or control bar that turns the balance on. The mass of the container will appear on the read-out.
   (e)   Record the mass of the container.
   (f)   Press the tare (T) button or control bar. The tare function resets the balance display to 0.0000 grams with the container on the pan. From this point on,

the read-out will show only the mass of chemical added to the container.

(g) Open an appropriate weighing-chamber door.

(h) Slowly and carefully add chemical to the tared container until the desired amount has been added. The read-out will indicate your progress. The pan moves as chemical is added, and the reading may fluctuate for a few seconds after you stop adding chemical. The letter "g" will appear when the reading is stable.

(i) Close the weighing-chamber door.

(j) Record the mass of chemical from the read-out.

4. Press the button or control bar that turns the balance off.

5. Open an appropriate weighing-chamber door.

6. Remove everything from the pan.

7. Close the weighing-chamber door.

Procedure

To learn how to use a balance efficiently, practice measuring the mass of a variety of objects. The reliability of your determinations may be checked by undertaking the following exercise.

Determine the mass of two objects placed on the pan together. Use such objects as a small flask and a stopper, a small metal bar and a watch glass, or a key and a coin. Your instructor may give you specific assignments. Record the assignment number in your laboratory notebook. Then determine the mass of each object individually. Record each mass in your laboratory notebook. Finally, add the individual masses and compare the result to the first (combined) mass measurement. If they are not the same, practice weighing some more; the success of many experiments depends on the ability to make accurate mass measurements.

Laboratory Techniques: Volume Measurement of Liquids

prepared by **Norman E. Griswold**, Nebraska Wesleyan University

Liquids take the shapes of their containers, so volume measurements are particularly convenient for determining amounts of liquids. The unit of measurement for volume is based on an imaginary cube that measures exactly one meter on each side. The volume contained by this cube is called a cubic meter (m^3). Smaller units of volume are somewhat more convenient and include such units as cubic feet, cubic inches, and cubic centimeters (cm^3). Other units, some not so obviously related to a cube, are also employed. These include pints, quarts, gallons, and liters.

Scientists normally use the metric system for measurements, so chemists usually express volumes in metric units of cubic centimeters or liters. The liter (L) is defined as 0.001 m^3 and is equivalent to the volume of 1000 grams of water. One one-thousandth of a liter is called a milliliter (mL). One milliliter is almost equal to a cubic centimeter, so for most practical purposes, the terms milliliter and cubic centimeter are used interchangeably.

Calibrated containers, called volumetric ware, have one or more volume marks and are used for volume measurements. Several types of volumetric ware are shown in Figure 1. **Graduated cylinders** are used for measuring approximate volumes of liquids. **Pipets** and **burets** are used for accurately measuring

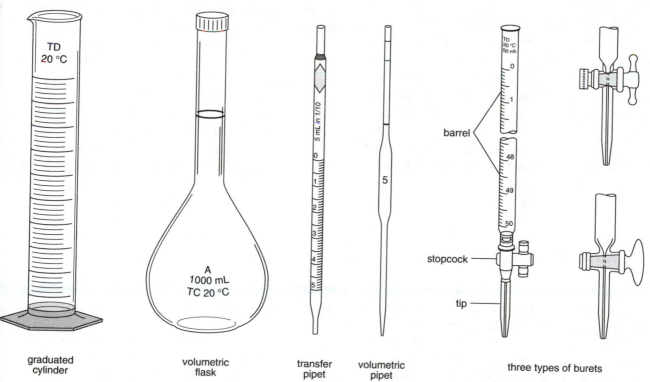

graduated
cylinder

volumetric
flask

transfer
pipet

volumetric
pipet

barrel

stopcock

tip

three types of burets

Figure 1   *Several types of volumetric ware*

Copyright © 1986 by Chemical Education Resources, Inc., P.O. Box 357, 220 S. Railroad, Palmyra, Pennsylvania 17078 No part of this laboratory program may be reproduced or transmitted in any form or by any means, electronic or mechanical, including photocopying, recording, or any information storage and retrieval system, without permission in writing from the publisher. Printed in the United States of America

liquid volumes that are transferred or delivered. **Volumetric flasks** are used for preparing solutions of accurately known concentration.

Volumetric ware can be calibrated and marked in terms of either what they contain or what they can deliver. Manufacturers usually label their volumetric ware to indicate the type of calibration, using TD for "to deliver" and TC for "to contain." For example, a 25-mL pipet is calibrated *to deliver* 25.00 mL at a specified temperature when the liquid level falls from the calibration mark through the orifice. A 100-mL volumetric flask is calibrated *to contain* 100.00 mL at a specified temperature when the flask is filled to the calibration mark.

Another aspect of volume measurement of liquids is that the volume of a given mass of liquid varies with temperature. In fact, the volume of the container itself may also vary with temperature. As a consequence, volume measurement of liquids must be referred to some standard temperature. The National Bureau of Standards has adopted 20.0 °C as the temperature for calibration. Most manufacturers label their volumetric ware with the temperature at which the calibration is valid.

Cleaning Volumetric Ware

Volumetric ware must be thoroughly clean before being used. Even slight amounts of grease or other contaminants can affect the accuracy of a measurement. Any foreign matter may react with the contents of a solution, altering its concentration. In some cases, foreign material might dissolve in the solution and then contaminate any experiment in which the solution is used. Chemists normally empty, clean, and thoroughly rinse each item of volumetric ware immediately after use.

I. Graduated Cylinders and Volumetric Flasks

Use the following procedure to clean graduated cylinders and volumetric flasks. Empty any solution from the container. Rinse the container two or three times with tap water. Add a dilute detergent or soap solution and use an appropriate brush, if necessary. Pour out the detergent solution and rinse the container two times with tap water. Rinse three times with small amounts of distilled water.

After the final rinsing, a uniform, often nearly invisible, film of water adheres to the inner surface of the container. The appearance of uneven wetting or droplets is a sign that the container needs further cleaning. Check with your laboratory instructor for directions on further cleaning your volumetric ware.

If the container must be dried, it is usually left standing in air at room temperature. Never apply direct heat from a flame to volumetric ware. The container will expand on heating and, after cooling, will be distorted from its original form, causing an error in the original calibration. Such distorted volumetric ware will not deliver the designated volume. For the same reason, drying volumetric ware in an oven is not normally recommended.

II. Pipets

> *Caution:* Always use a small rubber suction bulb to draw a liquid into a pipet. Never use your mouth to suck liquid into a pipet!

Pipets may be cleaned with a warm solution of detergent. Draw in detergent solution using a rubber suction bulb until the pipet is about one-third full. Remove the bulb from the pipet and quickly place your forefinger over the top opening of the pipet. Hold the pipet nearly horizontal and carefully rotate it so that all interior surfaces come in contact with the detergent solution. Remove your finger to drain the solution through the tip of the pipet. Rinse the pipet several times with tap water and then with distilled water.

The proper techinque for rinsing a pipet is to draw a small portion of water through the tip into the pipet. Rotate the pipet as described above and drain the rinse water through the tip. *Do not* rinse the pipet by holding it under a faucet.

Note that it is generally not necessary to be concerned about thorough cleaning in areas above the calibration mark(s). If warm detergent solution does not completely clean your pipet, see your laboratory instructor.

III. Burets

A buret must be very clean before it is used. The following special techniques are used to properly clean a buret. Fill the buret with water, and allow it to drain through the tip. Only a uniform, nearly invisible, film of water should remain on the interior surfaces. If any water droplets are visible on the inside wall of the buret, the buret is still dirty. Burets with glass stopcocks are especially likely to become dirty from the spreading of stopcock grease over the inside wall. In the following procedure, note that the buret brush is not pushed into the buret past the bottom calibration mark. The buret is always rinsed so that the rinse water is discharged through the stopcock. These practices reduce the possibility of transferring stopcock grease to the interior surfaces of the buret.

Dip a buret brush into warm detergent solution. Carefully insert the brush into the top of the buret. Scrub the inner surface of the buret wall, being careful not to push the brush past the bottom graduation mark. Remove the brush and add small amounts of tap water to the buret. Drain the water through the stopcock after each addition. Repeat the rinsing until the buret is free of detergent solution.

Add a few milliliters of distilled water to the buret. Hold the buret in an almost horizontal position and rotate it so that the water rinses the entire inside surface of the buret. Completely drain the buret through the stopcock before adding another few milliliters of distilled water and draining. Three small portions of distilled water added and drained in this manner quickly and completely remove all the tap water and replace it with a film of distilled water.

Using Volumetric Glassware

I. Graduated Cylinders

Graduated cylinders, shown in Figure 1, are made of either plastic or glass and vary in capacity from 5 mL to 2000 mL. The subdivisions marked on these cylinders are usually in units of about 1% of the total capacity; that is, 0.1 mL on a 10-mL cylinder, 1 mL on a 100-mL cylinder, and 5 mL on a 500-mL cylinder. Most cylinders are calibrated "to deliver" (TD) and are equipped with a pouring spout. Many cylinders have a hexagonal base to prevent rolling if the cylinder is upset. The spout is positioned above a corner of the base to avoid contact between the spout and the table. An optional accessory for glass cylinders is a plastic cylinder guard, which fits around the upper part of the cylinder to absorb the impact if the cylinder is tipped over.

When liquids are contained in plastic tubes, the liquid surface is flat and the volume is easily read. However, when a liquid is in a glass tube, such as a graduated cylinder, the liquid surface exhibits a definite curvature, called a **meniscus**. For most liquids, this curvature is concave (downward), and the bottom of the meniscus is used for reading the volume. However, the surfaces of certain liquids, such as mercury, form convex (upward) curves. In these cases, you must read the top of the meniscus.

Before attempting to read the volume of a liquid in a piece of volumetric glassware, be certain to observe the scale used on the glassware.   When reading volumes, your eye must be level with the bottom of the meniscus in order to observe the correct volume (see Figure 2). For example, when reading a graduated cylinder, if your eye is above the meniscus, you will observe too large a reading. If your eye is below the meniscus, you will observe too small a reading. This apparent variation in volume caused by viewing from different points is known as **parallax**. Proper reading of the meniscus and avoiding error due to parallax are important in the use of all volumetric glassware.

Keep the cylinder out of the way when it is not in use. A cylinder is relatively easy to tip over if bumped with an elbow or a hand. When you are not using your cylinder, place it on its side on the laboratory bench or in your locker.

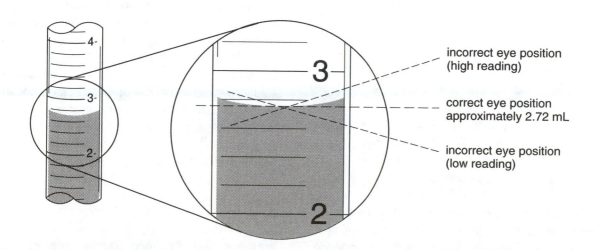

incorrect eye position (high reading)

correct eye position approximately 2.72 mL

incorrect eye position (low reading)

Figure 2   *Reading a meniscus*

II. Pipets

Two basic types of pipets are shown in Figure 1, each having a number of modifications. **Graduated** or **measuring pipets** are calibrated in convenient units so that any volume up to the maximum capacity of the pipet can be delivered. Measuring pipets are available with capacities from 0.1 mL to 50 mL and are calibrated "to deliver." They are drained from one calibration mark to a lower calibration mark on the pipet.

A specific kind of measuring pipet is the **serological pipet**, which has calibration marks extending to its tip. Serological pipets are available with capacities ranging from 0.1 mL to 10 mL. A ground-glass band near the top of the pipet indicates that, after free drainage has ceased, the small amount remaining in the tip must be blown out, using a rubber bulb, to obtain the total rated capacity. Serological pipets are used for delivering exact volumes of liquid samples to tubes or slides for medical tests.

The other basic type of pipet is the **volumetric** or **transfer pipet**, which is used for delivering a single, fixed volume of liquid. Volumetric pipets are calibrated with only one mark and are available with capacities ranging from 1 mL to 200 mL. Manufacturers now color-code both volumetric and measuring pipets with a band of color near the top to aid in easy identification. For example, one manufacturer places a blue band on 1-mL pipets, an orange band on 2-mL pipets, a white band on 10-mL pipets, and a red band on 50-mL pipets.

The following instructions are specifically for use with volumetric pipets. With minor modifications, the instructions may be applied to graduated pipets. Solutions and liquids are drawn into a pipet by applying a slight vacuum at the top, using a rubber suction bulb.

First, use the following steps to rinse the pipet with a small portion of the solution being measured to avoid diluting the sample with any distilled water left in the pipet. Pour a small amount of the solution to be measured into a clean, dry beaker.

Attach a rubber suction bulb to the pipet. If you are right-handed, hold the suction bulb in your left hand. Partially squeeze the bulb and slip it onto the end of the pipet, as shown in Figure 3(a). Make certain the connection is secure enough to prevent air leaks.

Draw some of the solution into the pipet by inserting the pipet tip into the solution. Keep the tip well below the surface of the solution so that no air is drawn into the pipet along with the solution. Release the pressure on the bulb slowly, to avoid drawing solution into the bulb. Quickly disconnect the bulb. Place your right forefinger (or your left forefinger if you are left-handed)

on the top of the pipet to prevent the solution from draining out. Hold the pipet in a nearly horizontal position. Rotate the pipet to allow the solution to contact all interior surfaces. Remove your finger briefly during this process to allow the solution to enter the upper stem of the pipet. Allow the solution to drain out through the pipet tip. Discard this solution as directed by your laboratory instructor. Repeat this step with at least two more portions of the solution.

Next, use the pipet to transfer a measured volume of the solution from the beaker to a receiving vessel. Use the same procedure as you used for drawing solution into the pipet when you rinsed it. Draw the solution up to a level that is above the calibration mark on the pipet, but be careful not to draw solution into the bulb. Lightly rest the tip of the pipet against the bottom of the beaker, and gently remove the bulb. Quickly press your forefinger on top of the pipet, as before. Do not allow the liquid level to fall below the calibration mark. Your index finger on top of the pipet gives maximum flow control, especially if your finger is slightly moist. However, too much moisture should be avoided, because your finger will slip off the end of the pipet, and the liquid level will drop below the calibration mark.

After removing the pipet from the solution, wipe the outside of the pipet tip with a clean tissue. Slowly and carefully drain some of the solution back into the beaker, until the bottom of the meniscus aligns exactly with the calibration mark. Rotating the pipet against your forefinger may help control the liquid level. When the bottom of the meniscus is at the calibration mark, touch the tip of the pipet against the side of the beaker so that the hanging drop is transferred to the beaker [Figure 3(b)].

Slowly move the pipet to the vessel to which you wish to transfer the measured solution. Hold the tip of the pipet against the inside surface of the vessel to avoid splatter. Lift your finger to allow the solution to drain from the pipet. When the flow stops, hold the pipet vertically for 15 seconds more to allow for complete draining. Do not attempt to remove the small amount of solution remaining in the tip of the pipet, because the pipet was calibrated to deliver a specified volume of liquid *excluding* the liquid remaining in its tip. Touch the tip to the inside of the vessel so that the hanging drop is transferred to the vessel.

III. Buret

A buret consists of a long, narrow, calibrated tube with a mechanism at one end to control the flow of liquid. The most common sizes of burets are 10-mL, 25-mL,

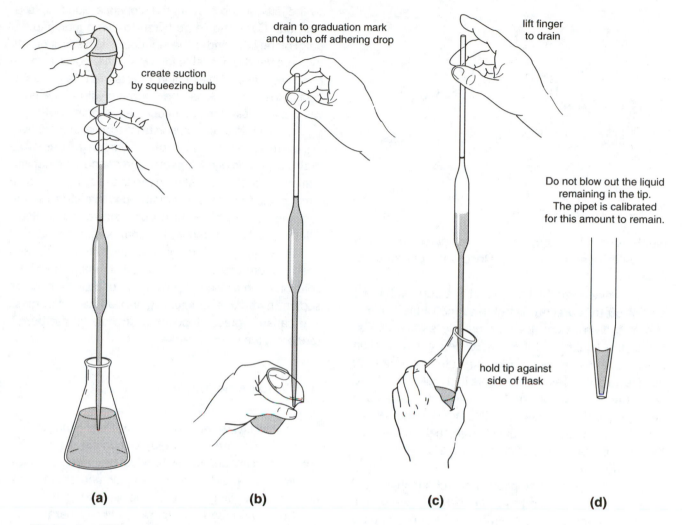

create suction
by squeezing bulb

drain to graduation mark
and touch off adhering drop

lift finger
to drain

Do not blow out the liquid
remaining in the tip.
The pipet is calibrated
for this amount to remain.

hold tip against
side of flask

(a)   (b)   (c)   (d)

Figure 3   *Filling a pipet*

50-mL, and 100-mL. The precision attainable with a buret is somewhat better than that attainable with a measuring pipet. Burets differ chiefly in the type of flow-control mechanism, or valve, used. Three common buret valves are: a tightly fitting glass bead in a short length of rubber tubing; a tapered glass stopcock in a glass barrel; and a Teflon stopcock in a glass or Teflon barrel.

Teflon stopcocks do not need lubrication, but glass stopcocks do need continuous lubrication, using a thin film of stopcock grease. To lubricate a stopcock, carefully remove the tapered glass stopcock plug from the barrel, and wipe all the old grease from the surfaces of both the stopcock and the barrel. Spread a small amount of stopcock grease lightly over the stopcock surface, being careful to avoid the area near the hole. Insert the stopcock into the barrel. Rotate the stopcock in one direction several times. If the proper amount of grease has been used, the stopcock will appear nearly transparent where the barrel contacts the plug.

Before using the buret, determine if the stopcock is liquid tight by filling the buret with water. Let the buret stand for a short time. Consult your laboratory instructor if it leaks.

When the buret is clean and ready to use, rinse it three or more times with small portions of the solution to be dispensed from the buret. Close the stopcock. Place a short-stem funnel in the buret. Slowly pour solution through the funnel into the buret. Do not allow the solution to overflow the funnel. Fill the buret to above the zero calibration mark at the top. Remove the funnel to prevent dripping from it during use of the buret.

Eliminate air bubbles in the tip of the buret by opening the stopcock for a second or two. Lower the level of the solution so that the bottom of the meniscus is at, or slightly below, the zero calibration mark. It is an unnecessary waste of time to exactly align the liquid level with the zero mark, as long as the liquid level is read and recorded before proceeding further in the experi-

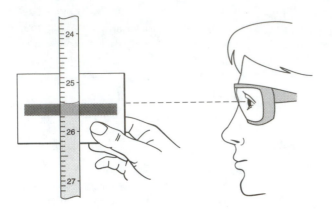

Figure 4   *Reading a buret*

ment. Touch the buret tip to a wet glass surface to remove the hanging droplet. Record your initial buret reading.

You must read the liquid level in a buret with great care. A 50-mL buret is calibrated every 0.1 mL, but one fifth of a division can be reproducibly estimated. Estimate the liquid level if it is between the calibration marks, and record every reading to the nearest 0.02 mL. Several devices are available that make it easier to read the level of the liquid more exactly. One such device consists of a dark, usually black, strip on a white card. Hold the card immediately behind the buret, with the top of the black strip slightly below the meniscus, as shown in Figure 4.

Burets with glass stopcocks must be manipulated carefully to avoid pulling the stopcock from the barrel. Using the right hand to operate the stopcock usually leads to leaking of the solution at the stopcock. Figure 5 shows the preferred method. Note that, although the stopcock valve handle is to the right, the stopcock is operated with the **left** hand. Light pressure is applied

by the left hand to prevent the stopcock plug from slipping out. Care should be taken to avoid touching the palm of the left hand to the stopcock. In this procedure, the right hand is available for swirling the solution in the receiving flask. When using burets with Teflon stopcocks, the stopcock can be operated with either hand without the risk of the stopcock plug slipping out.

You should drain solution from a buret at a relatively slow rate, so that the film of solution on the buret wall drains reproducibly. After you have drained the desired volume, touch a glass stirring rod to the buret tip. Use a wash bottle to rinse the stirring rod with distilled water, allowing the washings to go into the receiving flask. After a brief pause, record the final buret reading.

When you are through with the buret, drain and rinse it thoroughly. This is especially important if the solution used in the buret was a strong base, such as sodium hydroxide. Sodium hydroxide can etch glass and cause a glass stopcock to "freeze" in the barrel, rendering the buret useless.

IV.   Volumetric Flasks

Volumetric flasks vary in capacity from 5 mL to 5 L. These flasks are calibrated to contain a specified volume and are commonly used for preparing chemical solutions. The calibration mark is located on a narrow neck to allow filling to a reproducible volume.

Solutions may be prepared in two general ways. The first procedure involves the use of a solid solute. Usually you would weigh the solute and dissolve it in a solvent in a clean beaker. Transfer the resulting solution to a volumetric flask, using a funnel. Rinse the

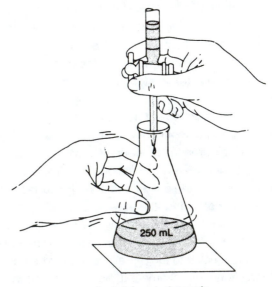

for right-handed people   for left-handed people

Figure 5   *Manipulating a stopcock*

beaker with a small amount of solvent and transfer the rinse through the funnel into the flask. Repeat this procedure several times with additional portions of solvent to thoroughly rinse the beaker and the funnel. Weighing the solute directly into a volumetric flask is also possible, but not as practical.

The second procedure involves the careful dilution of a more concentrated solution. In this case, you would transfer an accurately measured sample of the solution to be diluted to the volumetric flask, using a pipet. Techniques for using the pipet are given in Section II.

Add solvent while constantly agitating the solution, until the flask is about two-thirds full. Avoid unnecessary handling of the body of the flask, because heat from your hand may cause the liquid to expand and result in an incorrect dilution. Hold the flask by the neck only, above the calibration mark. Add more solvent until the liquid level is a few milliliters below the calibration mark. Agitate the solution thoroughly. Final alignment of the solution level with the calibration mark may be done in a dropwise manner with a small medicine dropper or a Pasteur pipet. Be sure to avoid parallax error.

After you have aligned the solution level with the calibration mark, secure the stopper or the plastic cap firmly on top of the flask. Hold the stopper firmly in place with your forefinger and slowly invert the flask many times, so that the solution becomes completely homogeneous.

Solutions are generally not stored for any length of time in volumetric flasks. Transfer the solution to a clean, dry bottle for storage. Then rinse the volumetric flask thoroughly and invert it until dry. Store the flask with the stopper in place. If the stopper is glass, insert a small strip of paper between the stopper and the flask to avoid having the stopper stick.

Laboratory Techniques: Filtration

prepared by **Norman E. Griswold**, Nebraska Wesleyan University

---

Purpose of the Experiment

Provide instruction in filtering techniques. Form and filter a barium sulfate precipitate in a conical funnel. Form and filter a calcium carbonate precipitate in a Büchner funnel.

Background Information

One of the basic laboratory manipulations of chemists is the separation of precipitates from the liquids in which they are formed. Although such separations are given the general term **filtration**, there are a variety of conditions involved in the efficient filtration of certain precipitates. For example, precipitates vary in particle size and in crystallinity; precipitate solubilities are dependent on temperature; and some precipitates are filtered to be discarded, while others must be carefully dried and accurately weighed. Thus, there are a variety of filtration media and several filtration techniques.

Several types of filtration media are shown in Figure 1. The **conical funnel** shown in Figure 1(a) is fitted with filter paper and is used quite commonly when the precipitate is crystalline and does not need to be accurately weighed. This filtering medium is the only one that is satisfactory for gelatinous precipitates. The funnels can be obtained in a variety of sizes, from 25 mm top diameter to as large as 250 mm, and with stem lengths that vary from 40 mm to 150 mm. The technique for using this type of funnel is described in detail in the section "Filtering with the Conical Funnel."

Filter paper is available in sizes ranging from diameters of 5.5. cm to about 18.5 cm; it has a variety of characteristics such as different filtering rates, ranging from 40 to 1000 mL min^{-1}, and different retentivities to retain precipitates of varying particle size. In addition, some commercially available filter paper is treated with

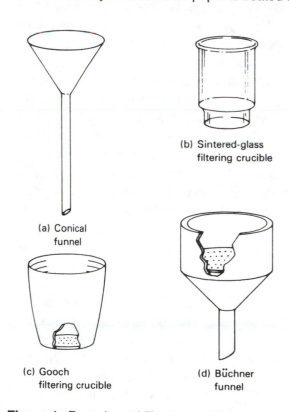

(a) Conical funnel

(b) Sintered-glass filtering crucible

(c) Gooch filtering crucible

(d) Büchner funnel

Figure 1: *Funnels and filtering crucibles*

Copyright © 1984 by Chemical Education Resources, Inc., P.O. Box 357, 220 S. Railroad, Palmyra, Pennsylvania 17078
No part of this laboratory program may be reproduced or transmitted in any form or by any means, electronic or mechanical, including photocopying, recording, or any information storage and retrieval system, without permission in writing from the publisher. Printed in the United States of America

hydrochloric and hydrofluoric acids to eliminate inorganic matter from the paper, so that, after careful ingnition, virtually no ash remains. For example, a 9-cm circle of ashless paper typically leaves less than 0.1 mg of ash. The selection of the proper grade of filter paper for a given filtration is very important. Manufacturers' literature can be consulted for complete specifications of various types of filter paper.

The **sintered-glass filtering crucible** shown in Figure 1(b) is a glass-walled crucible with a porous glass disk sealed into the bottom. If the crucible has porcelain walls, it is known as a Selas filtering crucible. It is used for filtering crystalline precipitates that can be dried at temperatures no higher than about 200 °C. This type of crucible has the advantage of rapid filtration rate, inert filtering medium, and the availability of several different porosities of the glass disk. Disadvantages include the high cost and the difficulty in cleaning the filtering crucible. These crucibles are available with capacities of 15, 30, and 50 mL. The technique for using this crucible is described in the section "Suction Filtering."

The **Gooch filtering crucible** shown in Figure 1(c) is made of porcelain and contains a perforated porcelain bottom. A mat of long-fibered asbestos can be formed over the small holes in the bottom of the crucible; this must be done before each filtration. Although this filtering device is somewhat inconvenient to prepare, and the porosity of the asbestos made is not easily controlled, the device has the advantage that it can be used for precipitates that must be ignited at temperatures as high as 1200 °C. As an alternative to the asbestos mat, glass fiber disks with diameters ranging from 2 to 12.5 cm can be used. When placed in a Gooch crucible, these disks retain very fine precipitates and allow more rapid filtration than filter paper. However, these disks can only be heated to 500°C.

There are several other types of filtering media. Among these are **Büchner funnels** made of porcelain, glass, or polyethylene. A typical Büchner funnel is shown in Figure 1(d). Filter paper or a glass fiber disk is placed over the perforations in this funnel and suction is used to accelerate the filtration. Membrane filters made of cellulose esters, containing many tiny uniform holes that occupy up to 80% of the filter volume, can be used with Büchner funnels.

In addition, filter cones, filter disks, filter tubes and filter sticks are available for special filtration applications.

Precipitate Preparation

Precipitates are generally formed in beakers in order to facilitate the transfer and the washing of the precipitate. Very often, the precipitation is carried out using hot solutions. The precipitating agent is usually contained in a dilute solution and is added slowly to a solution of the other reactant with constant stirring. The quantity of precipitating agent added should be calculated in advance if the approximate composition of the sample solution is known. This procedure avoids incomplete precipitations in the case where too little precipitating agent is added, and extreme waste or increased solubility of the precipitate if a very large excess of precipitating agent is added. To ensure complete precipitation, at least a 10% excess of precipitant is desirable, but it is always advisable to test for completeness of precipitation by adding an additional small portion of the precipitant after the precipitate has settled. The appearance of additional precipitate indicates that more precipitant must be added.

A precipitate is generally not ready for filtration immediately after it is formed. In a process known as **digestion**, the precipitate is allowed to stand for a period of time in the reaction mixture to allow the particles to increase in size so that the precipitate is more efficiently filtered. The increase in particle size has the added effect that there is a smaller total surface area available for the adsorption of impurities on the precipitate, and thus increases the purity of the precipitate. In some cases, the digestion process is carried out at elevated temperatures.

The length of time needed for digestion varies from a few minutes for gelatinous precipitates to as much as a day or two, with an hour or so being very common. For digestions longer than a few minutes, the beaker should be covered with a watch glass to prevent dust and other foreign particles from contaminating the precipitate. During the digestion process, the filtering apparatus can be prepared.

Funnel Preparation

I. The Conical Funnel

The conical funnel is placed in a special filtering rack to keep it stable. After the proper grade of filter paper is carefully selected, the proper radius must be chosen. Usually one selects a filter paper circle with a radius slightly less than the distance from the top edge of the funnel to the top of the stem as shown in Figure 2(a).

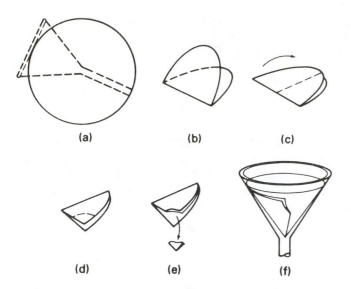

Figure 2: *Folding filter paper*

The circle of filter paper is folded exactly in half as shown in Figure 2(b). A second fold is made as shown is Figure 2(c), but this time the edges of the paper should not quite match, as shown in Figure 2(d). The angle formed by the two edges of the paper should be about five to ten degrees. Then a corner of the smaller section of the paper is torn off, as shown in Figure 2(e). If the mass of the precipitate is to be determined, the torn corner should be placed in the filter paper cone, since the original mass of the filter paper included this torn corner. This procedure will allow the filter paper to fit more tightly in the funnel and will therefore increase the rate of the filtration. The larger section of the paper is opened and the resulting cone is inserted into the funnel. This provides a cone with one thickness halfway around the inside surface of the funnel and three thicknesses around the other half, as shown in Figure 2(f). This funnel is now ready for a filtration.

II.   The Büchner Funnel

Filtration often proceeds quite slowly but the rate can be increased by using a Büchner funnel and applying suction.

> **Caution:** Both the filter flask and the trap should be wrapped with tape for safety reasons. If an implosion occurs while the taped flask and trap are under reduced pressure, the risk of injury from flying glass will be decreased.

The stem of the Büchner funnel is inserted into a rubber stopper of proper size to fit the filter flask in the apparatus for suction filtration, shown in Figure 3. The tubing used on the filter flask is either rubber or plastic and is heavy-walled to prevent collapse when suction is applied; the filter flask and the reap are also heavy-walled to prevent implosion. The source of suction is a water aspirator. The trap prevents accidental backflow of tap water from the aspirator into the filtering flask. The filter flask should be very clean in case the filtration must be repeated or the filtrate is to be saved. A filter paper circle with a diameter that allows the circle to be placed flat and cover the perforations in the funnel is selected. The funnel is now ready for filtration.

III.   The Sintered-Glass and Gooch Crucibles

Sintered-glass and Gooch crucibles also require suction for filtration. A crucible and crucible-holder assembly are substituted for the Büchner funnel and stopper shown in Figure 3. The apparatus employs heavy-walled tubing, rubber stoppers, and heavy-walled flasks as described for Büchner funnels. Before assembling the apparatus, the empty Gooch crucible should be heated, cooled, and weighed until constant mass is obtained.

The glass filtering medium in sintered-glass crucibles is permanently sealed into the crucible. Filter porosities of coarse (C), medium (M), and fine (F) are available. These crucibles should not be heated much above 200 °C, so the heating preparation should be done in an oven.

If a Gooch crucible is to be used with an asbestos mat, it must be freshly prepared prior to each filtration. Although asbestos has been used as a filtering medium for many years, there are disadvantages to its use. Asbestos is on the OSHA list of carcinogens, so appropriate precautions must be taken when working

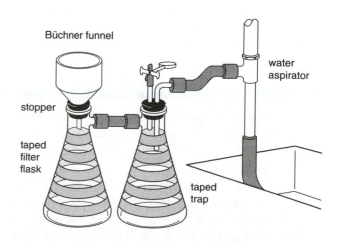

Figure 3: *Apparatus for suction filtration*

with it. Asbestos is also hygroscopic, so the initial heating to constant weight before filtration must be done under exactly the same conditions as the precipitate heating. Another disadvantage of using asbestos is that a mat made of it can rupture; special techniques are necessary to help avoid this problem. Small circles of borosilicate glass fiber are available commercially in diameters ranging from 2 to 12.5 cm; these can be substituted for asbestos mats. Glass mats can tolerated temperatures up to 500°C and are much less hygroscopic than asbestos. These disks are ready-made but are used in pairs to avoid loss if one disintegrates accidentally when liquid is added. The disks are placed in the Gooch crucible, which is then dried to constant mass in an oven, prior to use.

Filtering with the Conical Funnel

Before prodeeding with the actual filtration, it should be ascertained that everything is ready. A container must be placed under the funnel in such a way that the stem of the funnel touches the side of the beaker to avoid splattering the filtrate. A filled wash bottle and a long glass stirring rod with fire-polished ends should be available for use. Before the filtraiton is begun, the precipitate should have settled to the bottom of the beaker.

The paper in the funnel should be moistened with distilled water from a wash bottle, and the edges of the paper should be firmly pressed to the funnel to assure a tight fit. A little extra attention to this adjustment is time well spent, because considerable filtration time may be saved. Sufficient distilled water is added to the funnel so that the stem of the funnel is filled with water. If the paper fits tightly, the stem will remain filled after the water flows through the filter paper in the funnel. This column of water in the stem exerts a pull on the liquid to be filtered and hastens filtration considerably.

The filtration is begun by decating as much of the liquid as possible from the precipitate into the funnel. By filtering the bulk of the liquid first, the filter paper does not become clogged with precipitate, and the filtration process is hastened. The flow of liquid can be guided with the glass rod to avoid splattering, as shown in Figure 4. The paper cone should never be more than about three-fourths full.

Usually the precipitate must be washed. The fastest way to wash the precipitate is to direct a stream of wash solution from the wash bottle directly on the remaining precipitate in the precipitation beaker. After waiting a short time for settling, the wash solution is decanted through the filter paper using the same tech-

Figure 4: *Proper filtering technique*

nique as described previously. The use of several small portions of wash solution are more efficient than one washing with a large volume of water.

If washing the precipitate is critical, a drop or two of the filtered wash solution may be collected in a small test tube and subjected to the appropriate qualitative test to see if a particular ion is still passing through the filter paper. For example, the presence of chloride ion in the wash solution is determined by adding silver ion, while sulfate ion is determined by adding barium ion. If the ion is still present in the wash solution, the washing of the precipitate should be continued.

After the precipitate is washed thoroughly, the precipitate may be transferred from the beaker to the filter funnel with the aid of a stream of wash solution from the wash bottle, as shown in Figure 5. After washing as much of the precipitate as possible out of the

Figure 5: *Transferring a precipitate to a funnel*

precipitation beaker, the remaining portions of precipitate may be transferred to the funnel by using a rubber policeman fitted over the end of a glass rod. A **rubber policeman** is a flexible piece of gum rubber with a flat end cut diagonally. After using the rubber policeman, wash the policeman thoroughly just above the funnel so that the washings enter the funnel. Finally, by using a gentle stream of wash solution from the wash bottle, the precipitate is washed down from the sides of the filter paper cone into the bottom of the cone. The filter assembly should be allowed to stand until all the liquid has passed through the filter paper.

The next step depends on the nature of the separation. For example, if the filtered solution, the **filtrate**, is of major interest, the precipitate might be discarded here. Another possibility is that as much precipitate is to be collected as possible without accurate weighings being necessary. This situation arises when new compounds or intermediates are being synthesized. Finally, a quantitative analysis procedure might be used in which the precipitate will be accurately weighed. The procedure might involve the ignition of the precipitate. In this case, if filter paper is used for the filtration, an ashless filter paper should be used. Methods of drying and of igniting precipitates are described later.

Suction Filtering

Just before starting the filtration, suction is begun by turning on the tap water at the water aspirator. Thereafter, throughout the entire process, the suction is controlled by the pinch clamp on the trap shown in Figure 3. The water tap is not turned off until filtration and washing are complete and the pinch clamp has been removed.

When the suction is on, the filtration process itself is very similar to that used for the conical funnel. That is, the bulk of the liquids is filtered first, followed by the wash solutions, and finally by the bulk of the precipitate itself. Refer to "Filtering with the Conical Funnel" for details on the procedure. Some liquid should be present in the crucible during the entire filtration. The flow of liquid through the crucible can be regulated by the pinch clamp on the trap. After the filtration and washings are complete, preliminary drying of the precipitate may often be accomplished by continuing to pull air through the precipitate for a short time. However, with some kinds of precipitates, this practice is not always acceptable because some substances will react with carbon dioxide in the air.

Drying Precipitates

In the event that the precipitate is to be collected and weighed without ignition, the filter paper and the precipitate may be carefully removed from the funnel, the filter paper opened out flat, and the paper containing the precipitate placed on a watch glass for drying. Great care must be exercised in handling the filter paper because the paper is wet and therefore tears very easily. If the solution remaining in the precipitate consists of a low-boiling liquid, the watch glass and filter paper plus precipitate can be allowed to stand in air for drying. If the residual liquid in the precipitate consists of a small quantity of water, the watch glass can be placed on top of a beaker of gently boiling water. The steam from the boiling water causes the water in the precipitate to evaporate, and the precipitate is dried. If a very dry precipitate is desired, the watch glass and precipitate may be placed in an oven for drying for at least one-half hour.

> *Caution:* Be careful handling the hot watch glass when you remove it from the beaker of boiling water and the oven.

Ignition of Precipitates

I. Precipitates Collected in Filter Paper

Ignition of precipitates collected in filter paper is carried out in a crucible made of porcelain, silica, or platinum. The first step in the procedure is to transfer the paper and precipitate to a preweighed crucible. This transfer must be done very carefully since the paper will be wet and easy to tear. Usually the cone of filter paper is carefully flattened somewhat. Then the two corners at the top of the paper cone are folded over, followed by folding the top of the paper cone down as shown in Figure 6. The precipitate is now totally enclosed in filter paper and can be placed into the preweighed crucible. The crucible is placed on a triangle situated above a burner as shown in Figure 7.

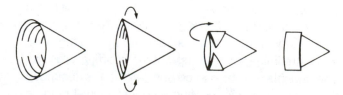

Figure 6: *Folding filter paper before ignition.*

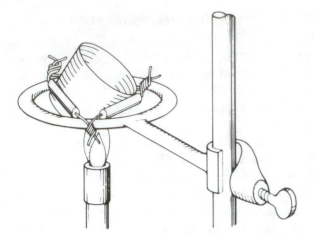

Figure 7 *First phase of ignition of a precipitate.*

The ignition must be carried out in two steps. The first step of the ignition is the slow charring of the paper at a relatively low temperature with the crucible cover off. This step requires constant attention and care. The paper should never burst into flame, but, if it does, the flame should be extinguished immediately by replacing the crucible cover using the crucible tongs. The rate of heating should be diminished, and then, when the cover is removed again, the charring should be continued until the paper appears to have been removed completely.

In the second phase of ignition, the cover is placed loosely on the crucible, the flame temperature is increased, and the crucible and its contents are heated at the full ignition temperature. The crucible must be placed in the hottest part of the flame so that no soot is deposited on the outside of the crucible. When the ignition is ended, the crucible is allowed to cool in air for a few minutes until the crucible seems cool when a hand is held *near* the crucible. Then, the crucible is placed in a desiccator for at least one half-hour before weighing. The ignition, cooling, and weighing sequence is repeated until successive weighings agree.

II. Precipitates Collected in Sintered-Glass or Gooch Crucibles

If the precipitate to be subjected to ignition was filtered through a sintered-glass or Gooch crucible, the crucible should be placed in a second solid-bottom crucible so that the gases from the burner do not come into direct contact with the precipitate. The full ignition temperature can be reached directly without the need for a preliminary low-temperature charring process. The crucible may be cooled and weighed as described above. The ignition, cooling, and weighing should be repeated until constant mass is attained.

Procedure

Caution: Wear departmentally approved eye-protection while doing this experiment.

I. Filtration with a Conical Funnel

Measure in a clean graduated cylinder 45 mL of 0.2*M* barium chloride solution. Transfer this solution to a clean 250-mL beaker. In a second clean graduated cylinder, measure 50 mL of 0.2*M* sodium sulfate solution.

With continuous stirring using a glass stirring rod, slowly pour the sodium sulfate solution into the barium chloride solution. Continue to stir the mixture for a few minutes.

Weigh a clean, dry watch glass and a piece of filter paper to the nearest centigram. Record the total mass of the watch glass and paper on the Data Sheet.

Place a clean filtering funnel in the funnel stand. Fold the piece of filter paper as shown in Figure 2 and described in the section "Funnel Preparation: The Conical Funnel". Place the torn corner of the filter paper in the paper cone, since the mass of this piece of filter paper was included when the watch glass and filter paper were weighed.

Place a 400-mL beaker under the funnel so that the stem of the funnel touches the side of the beaker. Moisten the filter paper in the funnel with distilled water from a wash bottle. Firmly press the edges of the filter paper against the funnel. Add sufficient distilled water to the funnel to fill the stem of the funnel.

Filter the precipitate by decanting as much as possible of the **supernatant**, the liquid portion of the reaction mixture, from the precipitate into the funnel. With a glass stirring rod guide the supernatant from the beaker into the funnel, as shown in Figure 4. After the majority of the supernatant has been filtered, add about 20 mL of distilled water to the precipitate in the beaker by directing a stream of distilled water from a wash bottle on the remaining precipitate in the beaker. Allow the precipitate to settle, and decant the wash solution through the funnel using the procedure previously described. Wash the precipitate with two additional 20-mL portions of distilled water, and each time transfer the wash solution to the funnel.

After the precipitate is thoroughly washed, transfer the precipitate from the beaker to the funnel with the aid of a stream of distilled water from the wash bottle, as shown in Figure 5. After washing as much of the precipitate as possible out of the beaker into the funnel, transfer the remaining portions of the precipitate from the beaker to the funnel by using a stirring rod fitted with a rubber policeman. After the transfer of the precipitate is completed, direct a stream of distilled water from the wash bottle on the rubber policeman, holding the attached stirring rod so that the washings go into the funnel.

By using a gentle stream of distilled water from the wash bottle, wash the precipitate down the sides of the paper cone into the bottom of the cone. Collect 1 mL of filtrate in a test tube. To the filtrate in the test tube add 1 mL of 0.2M barium chloride solution to test for sulfate ion in the filtrate. If a precipitate forms, wash the precipitate on the filter paper by carefully adding distilled water from the wash bottle to the funnel. Retest a second sample of filtrate for sulfate ion. If sulfate ion is still present in the filtrate, wash the precipitate again. Allow the funnel assembly to stand until all the liquid has passed through the filter paper.

Carefully remove the filter paper and precipitate from the funnel and place the paper and precipitate on the preweighed watch glass. Carefully open the filter paper on the watch glass.

Place the watch glass and contents in an oven at 100°C for one-half hour. By using crucible tongs, remove the watch glass and contents from the oven and allow them to cool. When cool, weigh the watch glass and contents to the nearest centigram. Enter this mass on the Data Sheet.

Replace the watch glass and contents in the oven for an additional 15 minutes. By using crucible tongs, remove the watch glass and contents from the oven and allow them to cool. When cool, weigh the watch glass and contents to the nearest centigram. Enter this mass on the Data Sheet. Repeat this procedure until the mass of the precipitate remains constant.

Note: Place barium sulfate in container designated by the laboratory instructor.

II. Filtration with a Büchner Funnel

Prepare the reaction mixture consisting of 45 mL of 0.01M calcium nitrate and 50 mL of 0.01M sodium carbonate exactly as described under Part I for barium chloride and sodium sulfate. Allow this mixture to stand while preparing the filtration assembly as described under "Funnel Preparation: The Büchner Funnel". Weigh a clean, dry watch glass and a piece of filter paper to the nearest centigram. Record the total mass of the watch glass and paper on the Data Sheet.

Begin the suction by turning on the water tap at the water aspirator. Moisten the filter paper in the funnel with distilled water from a wash bottle. Filter the reaction mixture through the Büchner funnel as described in Part I, including decanting the supernatant, washing the precipitate, and transferring the precipitate to the funnel. Thoughout the filtration, keep some liquid in the crucible. When all contents of the beaker have been transferred to the Büchner funnel, wash the precipitate with two 20-mL portions of distilled water. Allow the suction to continue for several minutes. Air is pulled through the precipitate to initiate drying. Remove the pinch clamp from the trap before turning off the water tap at the water aspirator.

Carefully remove the filter paper and precipitate from the funnel, and place the paper and precipitate on a preweighed watch glass. Dry and weigh the watch glass and contents in an oven as described in the last two paragraphs of Part I.

Note: Place calcium carbonate in a container designated by the laboratory instructor.

Data Sheet

I. Filtration with a Conical Funnel

| | weighings | | |
	first	second	third
mass of precipitate, watch glass, and filter paper, g	_____	_____	_____
mass of watch glass and filter paper, g	_____	_____	_____
mass of precipitate, g	_____	_____	_____

II. Filtration with a Büchner Funnel

	first	second	third
mass of precipitate, watch glass, and filter paper, g	_____	_____	_____
mass of watch glass and filter paper, g	_____	_____	_____
mass of precipitate, g	_____	_____	_____

Laboratory Techniques:
Visible Absorption Measurements

prepared by **Norman E. Griswold**, Nebraska Wesleyan University

Background Information

Scientists use a variety of methods to study chemical substances. One broad group of methods consists of **spectroscopic techniques**. In these techniques, electromagnetic radiation is passed through a sample. Some of the radiation is absorbed or reflected by the substance, while the remainder passes straight through or is transmitted. Spectroscopic methods measure the radiation that is transmitted or is reflected by a substance. Each substance transmits and/or reflects a unique pattern of wavelengths, so spectroscopy can be used to determine composition.

The wavelengths of electromagnetic radiation vary over an extremely wide range (10^3 to 10^{-10} cm). The various spectroscopic methods are classified by the portion of the radiation spectrum utilized for the specific technique. Three common methods are **infrared (IR) spectroscopy** (using wavelengths from 2.5×10^{-4} to 5×10^{-3} cm), **visible (VIS) spectroscopy** (4×10^{-5} to 7×10^{-5} cm), and **ultraviolet (UV) spectroscopy** (1×10^{-5} cm to 4×10^{-5} cm). These methods are used to identify substances, but they have other uses as well. Infrared spectroscopy is used to study rotations and vibrations within molecules. Ultraviolet and visible spectroscopy are used to study energy levels of outer electrons, to distinguish geometric isomers, and to establish the presence of alternate double and single bonds in sequence.

Visible spectroscopy is often used to determine the concentrations of solutions. As the name implies, the *visible* portion of the electromagnetic spectrum corresponds to the range of light wavelengths visible to the human eye. Only solutions with color can be studied using this method. This module explains the determination of concentrations of colored solutions of known substances using a Bausch & Lomb Spectronic 20 spectrophotometer.

I. The Spectrophotometer

A **spectrophotometer** is an instrument that measures the effects of samples on electromagnetic radiation. The instrument consists of a radiation (light) source, a grating to separate the light into its component wavelengths, and lenses and slits to focus the light. A photoelectric device detects the level of light that passes through the sample and displays it on a meter. There are also lenses and slits to focus the light. Several manufacturers produce spectrophotometers for visible spectroscopy. One that is commonly used for an introduction to visible spectroscopy is the Bausch & Lomb (B & L) Spectronic 20 shown in Figure 1.

The sample holder has a hinged cover that opens to permit insertion of a sample tube. Sample tubes are special test tubes called **cuvettes**. Cuvettes are made carefully to achieve consistent size and excellent optical properties. The sample holder cover must be closed while the meter is being read.

The instrument is turned on by turning the **power switch/zero knob** clockwise until it clicks. When the instrument is ready, the same knob is used to set the meter at 0% *T* with the sample holder empty. When the sample holder is empty, a solid block, called an oc-

Copyright © 1987 by Chemical Education Resources, Inc., P.O. Box 357, 220 S. Railroad, Palmyra, Pennsylvania 17078
No part of this laboratory program may be reproduced or transmitted in any form or by any means, electronic or mechanical, including photocopying, recording, or any information storage and retrieval system, without permission in writing from the publisher. Printed in the United States of America

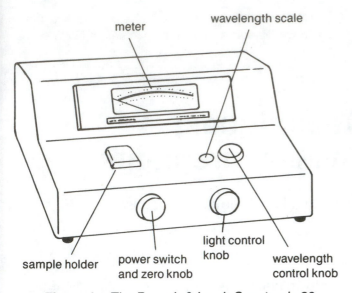

Figure 1 *The Bausch & Lomb Spectronic 20*

cluder, drops into the light path and prevents any light from reaching the detector. Under this condition, 0% of the light is transmitted. A schematic diagram of the optical system of a B & L Spectronic 20 is shown in Figure 2.

In the Spectronic 20, white light (light emanating from a tungsten lamp) passes through an entrance slit and lenses to a diffraction grating. A **diffraction grating** is a sheet of plastic-coated glass ruled with many fine, accurately spaced parallel grooves (600 per mm). The grating acts like a prism: white light shining on the grating is separated into wavelength components. These components are reflected by the grating in a fan-shaped pattern, with the shortest wavelengths (violet) at one end and the longest wavelengths (red) at the other. The grating is rotated using the **wavelength control knob** on top of the instrument. This rotation makes it possible to focus a very narrow wavelength

band on a second slit, so that only light within the selected wavelength band will strike the sample. Light that passes through the sample falls upon the **photo-tube detector**, which electronically measures the intensity of the transmitted light.

The **meter** on the Spectronic 20 includes two scales for reading the output of the instrument. One is a nonlinear scale of **absorbance (A)** units, with values from zero to infinity. The other is a linear scale with units of **percent transmittance ($\%T$)**, ranging from zero to 100. The linear nature and the smaller range of possible values of $\%T$ make it more convenient to read these values. On the other hand, absorbance values are often more useful in analysis. For the most precise results, it is best to read $\%T$ values and convert them to A values mathematically. The conversion is derived in the following way. Percent transmittance is defined as

$$\%T = \left(\frac{I_t}{I_o}\right)(100) \qquad \text{(Eq. 1)}$$

where

I_t = intensity of the beam transmitted by the sample

I_o = intensity of the beam directed at the sample

Absorbance is defined as

$$A = \log \frac{I_o}{I_t} \qquad \text{(Eq. 2)}$$

The logarithm of Equation 1 is taken.

$$\log (\%T) = \log \frac{I_t}{I_o} + \log 100 \qquad \text{(Eq.3)}$$

Two simple rules of logarithms [$\log (a/b) = -\log (b/a)$ and $\log 10^n = n$] allow Equation 3 to be rewritten as

$$\log (\%T) = -\log \frac{I_o}{I_t} + 2.00 \qquad \text{(Eq. 4)}$$

Equation 2 is substituted into Equation 4.

$$\log (\%T) = -A + 2.00 \qquad \text{(Eq. 5)}$$

Equation 5 is rearranged to give Equation 6.

$$A = 2.00 - \log (\%T) \qquad \text{(Eq. 6)}$$

This useful conversion equation can be used as follows.

1. Read each $\%T$ value from the meter on the Spectronic 20.

2. Determine the logarithm of the $\%T$ value, using a calculator.

3. Subtract the result from 2.00 to get the equivalent A value.

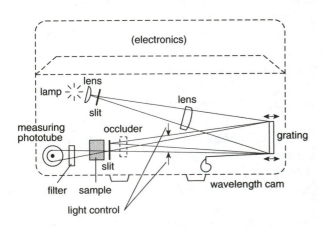

Figure 2 *Schematic of the Spectronic 20 optical system*

The **light control knob** on the front of the instrument permits the use of the entire meter scale while keeping the meter readings from going off the scale. After the wavelength control knob is adjusted to the correct value, a cuvette containing the solvent (or all components of the solution except the substance being studied) is placed in the sample holder. Then, the light control knob is adjusted until the meter shows 100% T. The light control knob must be adjusted for each wavelength setting.

II. Choosing a Wavelength

The absorbance of a substance could be determined at many wavelengths, but one wavelength setting gives the best results. The best wavelength is selected from the absorption spectrum of the substance. The **absorption spectrum** is a graph of absorbance (ordinate) versus wavelength (abscissa). Figure 3(a) shows a simple spectrum with one peak. Figure 3(b) shows the additive nature of the spectra of substances A and B in the same solution. The solid curve is the spectrum for the mixture. The dashed curves are the spectra for the individual substances at the concentrations present in the mixture.

Use the following considerations in selecting the proper wavelength for concentration determinations.

1. A slight shift in wavelength should not result in an appreciable change in absorbance. For the spectrum in Figure 3(a), the wavelength at Q is better than the wavelength at P, because the line is more nearly horizontal at Q than at P.

2. The absorbance should change significantly when the concentration changes. In Figure 3(a), if the concentration of the solution were cut in half, the absorbance would be at R and S for the two wavelengths shown. The difference in absorption, $Q - R$, is greater than the difference between P and S. Therefore, the wavelength for Q and R, is preferable. This consideration favors the choice of a high point on the curve.

3. The absorbing substance being studied should be the only absorbing substance at the wavelength chosen. In solutions containing more than one absorbing substance, a wavelength should be chosen at which only the substance being studied has significant absorption. In Figure 3(b), the wavelength at X is preferable to the wavelength at Y for determining concentrations of substance A. At X, substance B has very little absorption and does not affect the measurement.

III. The Beer-Lambert Law

It is reasonable to expect that the intensity of light passing through a solution would be influenced by the distance the light must travel through the solution and by the concentration of the absorbing solute. The mathematical relationship for these influences is known as the **Beer-Lambert law** shown in Equation (7).

$$A = \varepsilon bc \qquad \text{(Eq. 7)}$$

where

A = absorbance (defined in Equation 2)

ε = molar absorptivity

b = thickness of the absorbing solution (in centimeters)

c = concentration of the absorbing substance (in moles of solute per liter of solution)

Molar absorptivity is a proportionality constant. It has a specific value for each absorbing substance at each wavelength. In most experimental work, cuvettes with a uniform diameter are used for the entire determination, so that b is constant. When doing an experiment using a fixed wavelength and cuvettes of uniform size, the absorbance, A, is directly proportional to the concentration, c. A graph of A versus c should be a straight line.

(a)

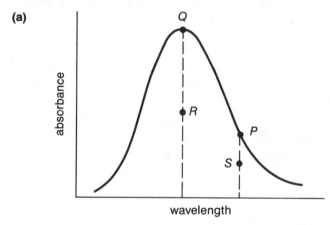

(b)

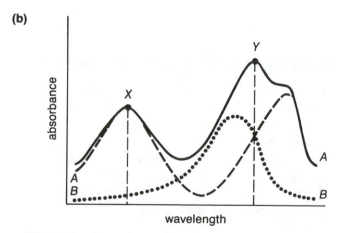

Figure 3   *Two absorption spectra*

Deviations from the Beer-Lambert law can occur. Two reasons for deviation are:

1. A solution is so concentrated that solute particles interact.

2. The absorbing substance interacts with the solvent.

Thus, when beginning an experiment, it is necessary to experimentally establish that the Beer-Lambert law is applicable to the absorbing substance being studied. Several solutions of different known concentrations of the substance are carefully prepared. These solutions are called **standard solutions**. The absorbances of the standard solutions are measured. A graph of A (ordinate) versus c (abscissa) is prepared. If a straight-line graph is obtained, then the Beer-Lambert law is applicable. Such a graph is called a **Beer's law calibration graph** or, often, a **Beer's law plot**, shown in Figure 4.

A Beer's law calibration graph must be plotted for each absorbing substance and each wavelength used. The graph is used to determine concentrations of solutions of the absorbing substance, using a fixed wavelength. A sample of a solution of unknown concentration is placed in the sample holder, and the absorbance is determined from the spectrophotometer. This absorption is located on the line of the calibration graph. In Figure 4, the point X has the absorbance value 0.420. The concentration of this solution is the c value of this point. In Figure 4, the concentration that corresponds to $A = 0.420$ is indicated by point Y and has the value $20 \times 10^{-5} M$ (or $2.0 \times 10^{-4} M$).

Procedure

In the procedure that follows, it will be helpful to refer to Figure 1 to review the locations of various components of the B & L Spectronic 20.

1. Preparing solutions.

Prepare several standard solutions of known concentrations for use in plotting a Beer's law calibration graph for the substance being studied. The range of concentrations should extend beyond all expected unknown concentrations. Your laboratory instructor will inform you of the expected range and will give you specific directions for preparing the solution. Concentrations of 10^{-4} to 10^{-6} moles L^{-1} are common.

2. Preparing your laboratory record book for data.

Record the identity of the absorbing substance. Record the wavelength to be used. For the standard solutions, prepare a data table with three columns with

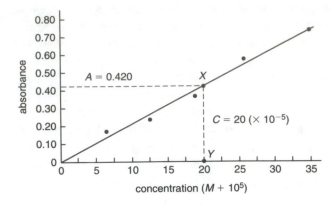

Figure 4   A Beer's law calibration graph

the headings $\%T$, A, and c (mol L^{-1}). For the solutions of unknown concentration, pre-pare a similar data table, but also include a column for the code number or letter used to identify each solution.

3. Turning on the power to the instrument.

Turn the power switch/zero knob clockwise until it clicks. The pilot lamp should light. Allow an older instrument to warm up for at least five minutes. A new or reconditioned instrument with solid-state electronics can be used immediately.

4. Selecting the wavelength.

Adjust the wavelength control to the desired value on the wavelength scale. The wavelength is given in many experiments. If it is not given, you must choose it using an absorbance spectrum for the absorbing substance, as described previously. Record the selected wavelength in your laboratory record book.

5. Zeroing the instrument.

Make sure there is no cuvette in the sample holder. Close the cover on the sample holder. Adjust the zero knob clockwise until the meter indicates a reading of $0\%T$.

6. Obtaining cuvettes.

Handle the cuvettes only by the upper rim. Finger smudges on the lower part of the tube affect the passage of light through that part of the tube. Before making a measurement, rinse the cuvette with three small portions of the solution to be used. After pouring the solution into the cuvette, inspect the outside of the cuvette for stray drops or smudges. Wipe the surface of the cuvette with a special tissue such as lens paper before placing it in the sample holder.

7. Setting the light control knob.

Pour the cuvette about half full of the **reference solution**. The reference solution consists of all components of the solutions being studied except the substance whose concentration is being determined. If the solutions being studied consist of a single substance dissolved in distilled water, then the reference "solution" is distilled water. Insert the half-filled cuvette into the sample holder, pushing down firmly. If the cuvette has a vertical line etched on it, the line should be toward the front of the instrument. Close the cover on the sample holder. Adjust the light control until the meter indicates a reading of 100% T.

Remove the cuvette containing the reference solution from the sample holder. Save it for periodic checking of the light control setting. The instrument is now ready to use.

8. Obtaining data for a calibration graph.

Rinse a clean cuvette three times with small portions of the first standard solution. Add the first standard solution to a cuvette until it is half full. Wipe the cuvette with a tissue. Place the cuvette in the sample holder and close the cover. Read the %T value indicated on the meter. Record the value in your laboratory record book. Remove the cuvette from the sample holder. Discard the solution, following your laboratory instructor's directions. Rinse the cuvette three times with small portions of distilled water. Rinse the cuvette three times with the second standard solution, discarding the rinses. Add the second standard solution until the cuvette is half full. Read the %T value for this solution. Continue this cycle with all of the remaining standard solutions. Periodically check the 100% T reading using the reference solution. When you are finished, remove the cuvette and rinse it three times with distilled water.

9. Obtaining data for solutions of unknown concentrations.

Rinse the cuvette three times with small portions of the solution of unknown concentration. Discard the rinses. Add solution to the cuvette until it is half full. Determine the %T value for this solution. Record the value in your laboratory record book. Empty the cuvette and obtain %T readings for two more samples of the same solution. Record all data in your laboratory record book. If concentrations of other solutions of the same substance need to be determined, check the wavelength and light control settings, and start over at Step 7. When you are finished, remove the cuvette from the sample holder.

10. Turning off the instrument (if appropriate).

When you have completed all determinations, turn off the instrument *if no one else is waiting to use it*. Turn the power switch/zero knob counterclockwise until it clicks. The pilot lamp will no longer be lighted. Place a dust cover over the instrument, if one is available.

11. Cleaning the cuvettes.

Rinse the cuvettes several times with tap water, then with distilled water. Be careful not to scratch them. Cuvettes can be dried in an oven at 110 °C. They should be stored in a special place to avoid confusing them with regular test tubes.

12. Processing the data and preparing a calibration graph.

Use a calculator to convert all %T values to A values, using Equation 6. Record the results in the data tables in your laboratory record book. Prepare a calibration graph, using the following procedure.

(a) Allow plenty of space for plotting data points.

(b) Draw the axes using a straightedge.

(c) Label each axis clearly. Use the ordinate for absorbance and the abscissa for concentration.

(d) Determine the scale to use on each axis. Your scales do not have to start at zero. Scale units on the two axes do not have to be the same size. Choose scale-unit sizes that will produce a graph with easy point plotting and interpretation.

(e) Plot data and draw the most representative straight line.

(f) Write a title on your graph.

13. Interpreting the data.

For your first solution of unknown concentration, locate the absorbance on the ordinate of your calibration graph. Use a straightedge to find the point on the straight line that corresponds to this absorbance, and mark it. Now use your straightedge to find the position on the abscissa that is directly beneath the mark you have just made. Make a second mark where the straightedge crosses the abscissa. Read the concentration value at this mark. Record the value in your data table beside the corresponding absorbance value. You have now determined the concentration of your unknown solution from its absorbance value.

Repeat this procedure using absorbance values for other solution of unknown concentrations.

Questions

1. When iron(III) chloride is added to a solution of salicylic acid, a purple complex forms. Different amounts of iron(III) chloride were added to standard solutions of salicylic acid. The percent transmittance of each solution was determined with a Spectronic 20. Convert these %T values to absorbance values.

2. Prepare a calibration graph from the data in Question 1, using the graph paper on the next page of this module.

3. Using the calibration graph prepared in Question 2, determine the concentration corresponding to an absorbance of 0.55.

_____
answer

standard solution code letter	FeCl$_3$ concentration, M	%T	A
A	40.0×10^{-5}	17.9	_____
B	32.0×10^{-5}	25.0	_____
C	24.0×10^{-5}	35.7	_____
D	16.0×10^{-5}	50.2	_____
E	8.0×10^{-5}	70.8	_____

Density of Liquids and Solids

prepared by **H. A. Neidig**, Lebanon Valley College
and **J. N. Spencer**, Franklin and Marshall College

Purpose of the Experiment

Determine the density of an unknown sample of rubbing alcohol, a rubber stopper, and an unknown metal, using mass and volume measurements.

Background Information

One of the physical properties of matter is **density**. This property is dependent on the volume and the mass of a sample of matter. The relationship between density, volume, and mass is shown in Equation 1.

$$\text{density} = \frac{\text{mass}}{\text{volume}}, \quad \text{or} \quad d = \frac{m}{V} \quad \text{(Eq. 1)}$$

The density of a liquid or of a solution is usually reported in units of grams per milliliter ($g\ mL^{-1}$). The density of a solid is reported in units of grams per cubic centimeter ($g\ cm^{-3}$). Because 1 mL is equivalent to 1 cm^3, these units are interchangeable.

The experimental procedure for obtaining laboratory data in order to calculate the density of an unknown sample of rubbing alcohol involves two steps. One step is to measure the mass of the sample. The second step is to measure its volume.

The mass of a sample of a substance is measured using a balance. If the substance is a liquid, the volume of the sample can be measured using a piece of calibrated glassware. For solid substances, the volume can be found by measuring the volume of liquid displaced by the sample. If the solid has a regular shape, such as a cube, its volume can be calculated using Equation 2. For other geometric shapes, appropriate equations are used to calculate their volumes.

$$\text{volume} = \text{length} \times \text{width} \times \text{thickness} \quad \text{(Eq. 2)}$$

$$= \text{cm} \times \text{cm} \times \text{cm}$$

$$= cm^3 = mL$$

When measuring volume by displacement, begin by pouring a liquid, such as water, into a graduated cylinder. Measure and record the volume of water. Add the weighed sample of the solid to the water in the cylinder. Measure the combined volume of water and the submerged solid. The difference between these two volumes is the volume of the solid.

In this experiment, you will use the experimentally determined mass and volume of the sample to calculate the density of the substance.

Procedure

Caution: Wear departmentally approved eye protection while doing this experiment.

Copyright © 1990 by Chemical Education Resources, Inc., P.O. Box 357, 220 S. Railroad, Palmyra, Pennsylvania 17078
No part of this laboratory program may be reproduced or transmitted in any form or by any means, electronic or mechanical, including photocopying, recording, or any information storage and retrieval system, without permission in writing from the publisher. Printed in the United States of America

I. Determining the Density of an Unknown Rubbing Alcohol Solution

> *Note:* Your laboratory instructor will give you directions for using your balance and will inform you as to the number of significant digits to the right of the decimal point to use when recording your data.

> *Note:* The following instructions pertain to a top-loading balance.

1. Turn the balance on by pressing the control bar. The display should show zero grams. If it does not, consult your laboratory instructor.

2. Use crucible tongs to place a container or a piece of weighing paper on the center of the balance pan.

3. Press the control bar or the tare bar to display the container mass. Press the tare bar to return the balance display to zero.

   The mass of the container subtracted from the total mass of the container and its contents is called the **tare**. The tare function on your balance resets the display to zero grams with the empty container on the pan. The mass of the container is retained in the memory of the balance. This amount will be subtracted when the container and contents are weighed, to display only the mass of the contents.

4. Use tongs to remove the container from the balance. Place the substance or object to be weighed in the container.

5. Use tongs to replace the container and its contents on the center of the balance pan. The mass of the substance or object alone will appear on the display.

6. Read the mass of the substance or object, using the number of significant digits to the right of the decimal point specified by your laboratory instructor.

7. Use tongs to remove the container and its contents from the balance pan.

8. Place the balance in the rest position.

> *Note:* The following instructions pertain to a centigram balance. If your centigram balance has an arrest lever, your laboratory instructor will give you additional directions regarding its use.

1. With all three sliding masses set to zero, check the balance to see whether the pointer freely swings an equal distance on either side of the zero mark. If it does not, consult your laboratory instructor.

2. Use crucible tongs to place the container on the center of the balance pan. The pointer on the center beam will deflect upward.

3. Move the appropriate sliding masses on the beams to balance the mass of the container. See Experiment 3 for additional information on the use of a centigram balance.

4. Read the positions of the masses on the beams. Calculate the mass of the container, using the number of significant digits to the right of the decimal point specified by your laboratory instructor.

5. Use tongs to remove the container from the balance pan. Add the substance or object to be weighed.

6. Use tongs to replace the container and its contents on the center of the balance pan.

7. Move the appropriate sliding masses on the beams to balance the mass of the container and its contents.

8. Read the positions of the masses on the beams. Calculate the combined mass of the container and its contents.

9. Return the sliding masses to their zero positions on the beams.

10. Use tongs to remove the container and its contents from the pan.

> *Note:* The numbers appearing in parentheses indicate the lines on your Data Sheet on which data should be entered.

1. Weigh a clean, dry, 10-mL graduated cylinder. Record this mass on your Data Sheet (3).

> **Caution:** Isopropyl alcohol is flammable and toxic if ingested. Keep isopropyl alcohol away from open flame and heat sources.

2. Use crucible tongs to remove the graduated cylinder from the balance. Obtain from your laboratory instructor a bottle containing your unknown rubbing alcohol solution. Record the code number of your unknown on your Data Sheet (1).

3. Pour 5 to 6 mL of rubbing alcohol solution from the bottle into the graduated cylinder. This amount of alcohol is Sample #1.

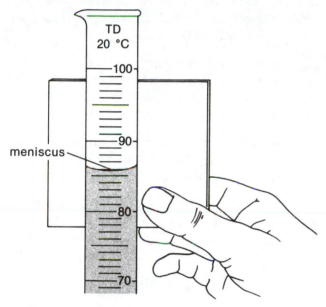

Figure 1 *Finding the meniscus*

> **Note:** The following instructions pertain to reading a volume in a graduated cylinder.
>
> **1.** Place a piece of white paper or card directly behind the cylinder at the meniscus. See Figure 1.
>
> **2.** Position your head so that your eye is at the same height as the level of the liquid. If you are holding the glassware, be sure that the glassware is exactly vertical.
>
> **3.** Look straight at the meniscus through the glassware so that you see only a concave line, not a concave surface.
>
> **4.** Read the level of the liquid at the bottom of the meniscus, the curved surface of the liquid. See Figure 2.

4. Read the volume of Sample #1 in the cylinder to the nearest one-tenth of a milliliter, 0.1 mL. Record this volume on your Data Sheet (4).

5. Weigh Sample #1 and the graduated cylinder, using the number of significant digits to the right of the decimal point specified by your laboratory instructor. Record this combined mass on your Data Sheet (2). If you are using a top-loading balance, you will not record the combined mass but will record only the mass of Sample #1 on your Data Sheet (7).

6. Transfer the alcohol sample to a test tube.

> **Note:** In Steps 7–10, you will do a determination of the volume and mass of a second sample of the alcohol solution, using Sample #2.

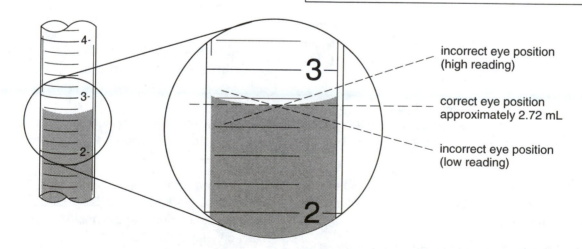

Figure 2 *Using line of sight to read a meniscus*

7. Pour 4 to 5 mL of alcohol solution from the test tube into the graduated cylinder. This amount of alcohol solution is Sample #2.

8. Read the volume of Sample #2 in the graduated cylinder to the nearest 0.1 mL. Record this volume on your Data Sheet (6).

9. Weigh Sample #2 and the graduated cylinder, using the number of significant digits to the right of the decimal point specified by your laboratory instructor. Record this combined mass on your Data Sheet (5). If you are using a top-loading balance, you will not record the combined mass but will record only the mass of Sample #2 on your Data Sheet (9).

10. Pour Sample #2 and the alcohol solution remaining in the test tube into the container specified by your laboratory instructor and labeled, "Discarded Unknown Rubbing Alcohol Solutions."

II. Determining the Density of a Rubber Stopper

11. Obtain a rubber stopper from your laboratory instructor and record its identifying number on your Data Sheet (12).

> **Caution:** Use crucible tongs to handle the rubber stopper.

12. Weigh the rubber stopper, using the number of significant digits to the right of the decimal point specified indicated by your laboratory instructor. Record this mass on your Data Sheet (13).

13. Add approximately 50 to 60 mL of water to your 100-mL graduated cylinder. Read the volume of liquid in the cylinder to the nearest milliliter. Record this volume on your Data Sheet (14).

14. Slightly tilt the graduated cylinder and carefully slide the rubber stopper down the inside surface of the cylinder. Avoid splashing any of the water out of the cylinder.

15. Read the volume of liquid in the graduated cylinder to the nearest milliliter. Record this volume on your Data Sheet (15).

16. Drain the water from the graduated cylinder. Dry the stopper. Return the stopper to your laboratory instructor.

III. Determining the Density of an Unknown Metal

17. Obtain an unknown metal sample from your laboratory instructor and record its identifying number on your Data Sheet (18).

> **Caution:** Use crucible tongs to handle the metal sample.

18. Wipe the sample carefully with a dampened cloth, dry thoroughly, and weigh. Record the mass on your Data Sheet (19).

19. Add 50 to 60 mL of water to your 100-mL graduated cylinder. Read the volume of liquid in the graduated cylinder to the nearest milliliter. Record this volume on your Data Sheet (20).

20. Slightly tilt the graduated cylinder and carefully slide the metal sample down the inside surface of the cylinder. Avoid splashing any of the water out of the cylinder.

21. Read the volume of the liquid in the graduated cylinder to the nearest milliliter. Record this volume on your Data Sheet (21).

22. Drain the water from the graduated cylinder. Dry the metal sample. Return the sample to your laboratory instructor.

Calculations

Do the following calculations and record the results on your Data Sheet.

I. Determining the Density of an Unknown Rubbing Alcohol Solution

> **Note:** If you used a top-loading balance for this experiment, you will not need to do Steps 1 and 3.

1. Calculate the mass of Sample #1. Subtract the mass of the graduated cylinder (3) from the mass of the graduated cylinder and Sample #1 (2). Record this mass on your Data Sheet (7).

2. Calculate the density of Sample #1. Divide the mass of Sample #1 (7) by the volume of Sample #1

(4). Record the density of Sample #1 on your Data Sheet (8).

3. Calculate the mass of Sample #2. Subtract the mass of the graduated cylinder (3) from the mass of the graduated cylinder and Sample #2 (5). Record this mass on your Data Sheet (9).

4. Calculate the density of Sample #2. Divide the mass of Sample #2 (9) by the volume of Sample #2 (6). Record the density of Sample #2 on your Data Sheet (10).

5. Calculate the mean density of your unknown rubbing alcohol solution. Add the density of Sample #1 (8) and of Sample #2 (10), and divide by two. Record this mean density on your Data Sheet (11).

II. Determining the Density of a Rubber Stopper

6. Calculate the volume of the rubber stopper. Subtract the volume of the water (14) from the volume of the water and stopper (15). Record the volume of the stopper on your Data Sheet (16).

7. Calculate the density of the rubber stopper. Divide the mass of the rubber stopper (13) by the volume of the rubber stopper (16). Record this density on your Data Sheet (17).

III. Determining the Density of an Unknown Metal

8. Calculate the volume of the metal sample. Subtract the volume of the water (20) from the volume of the water and metal sample (21). Record the volume of the unknown metal on your Data Sheet (22).

9. Calculate the density of the unknown metal. Divide the mass of the unknown metal (19) by the volume of the metal (22). Record the density on your Data Sheet (23).

10. Identify the unknown metal. Use the list of metals and their densities supplied by your laboratory instructor to determine the identity of your unknown metal. Record its identity on your Data Sheet (24).

Post-Laboratory Questions

(Use the spaces provided for the answers and additional paper if necessary.)

1. The density of ice at 0 °C is 0.9168 g mL^{-1}, and that of liquid water at 0 °C is 0.9999 g mL^{-1}.

(1)   What are the volumes of 1.000 g of ice and of 1.000 g of water at 0 °C?

(2)   A sealed glass container with a capacity of exactly 100. mL contains 96.0 mL of liquid water at 0 °C. If the water freezes, will the container rupture?

2. The volume of the nucleus of a carbon atom is about 9.9 × 10^{-39} mL. The molar mass of carbon is 12.00 g mol^{-1}. What is the density of the carbon nucleus?

3. Liquid mercury has a density of 13.6 g mL^{-1}. Which of the following substances will float on mercury, and which will sink?

	density, g mL^{-1}
neptunium	20.4
nickel	8.9
osmium	22.6
zinc	7.1
lead	11.4

4. A perfect cube of aluminum metal was found to weigh 20.00 g. The density of aluminum is 2.70 g mL^{-1}. What are the dimensions of the cube?

Data Sheet

I. Determining the Density of an Unknown Rubbing Alcohol Solution

(1) code number of alcohol solution   _____

(2) mass of Sample #1 and graduated cylinder, g   _____

(3) mass of graduated cylinder, g   _____

(4) volume of Sample #1, mL   _____

(5) mass of Sample #2 and graduated cylinder, g   _____

(6) volume of Sample #2, mL   _____

(7) mass of Sample #1, g   _____

(8) density of Sample #1, $g\ mL^{-1}$ _____

(9) mass of Sample #2, g   _____

(10) density of Sample #2, $g\ mL^{-1}$ _____

(11) mean density of the unknown rubbing alcohol solution, $g\ mL^{-1}$ _____

II. Determining the Density of a Rubber Stopper

(12) identifying number of rubber stopper   _____

(13) mass of rubber stopper, g   _____

(14) volume of water in graduated cylinder, mL   _____

(15) volume of water and rubber stopper in graduated cylinder, mL   _____

(16) volume of rubber stopper, mL   _____

(17) density of the rubber stopper, $g\ mL^{-1}$ _____

III. Determining the Density of an Unknown Metal

(18) identifying number of unknown metal   _____

(19) mass of metal, g   _____

(20) volume of water in graduated cylinder, mL   _____

(21) volume of water and metal in graduated cylinder, mL   _____

(22) volume of metal, mL   _____

(23) density of unknown metal, g mL^{-1} _____

(24) identity of unknown metal   _____

Calculations   (Show all your work. Use additional paper if necessary.)

name _______________ section _______ date _______

Pre-Laboratory Assignment

1. At 25 °C, 10.0181 g of an unknown liquid was found to have a volume of 6.75 mL.

(1) Calculate the density of the liquid.

answer _______

(2) Which of the following liquids was the unknown?

	density, g mL^{-1} at 25 °C
water	0.9982
toluene	0.8669
chloroform	1.4832

answer _______

(3) If the unknown liquid had been water, what would the volume have been?

(4) What mass would a 10.00-mL sample of each of the liquids in (2) have?

water _______ answer

toluene _______ answer

chloroform _______ answer

2. A stopper was found to have a mass of 5.06 g. When placed in a graduated cylinder containing 45.2 mL of water, the volume of stopper and water was found to be 49.4 mL. Calculate the density of the stopper.

answer _______

3. A chemist was given four unidentified, water-insoluble, cubes measuring $1 \times 1 \times 1$ cm and asked to arrange these substances in order of their increasing density. These cubes were labeled A, B, C, and D. As a reference, the chemist was also given the following liquids, whose densities in g mL^{-1} at 20 °C are given below.

water	0.9982	nitromethane	1.1371
toluene	0.8669	chloroform	1.4832

The chemist added one of the four substances to one of the liquids and observed whether the substance floated or sank. By repeating this procedure with the other substances and liquids, he was able to make a series of observations about the relative densities of the substances and the liquids. Use the following selected observations to arrange the four unknown substances in order of increasing density. Briefly defend your order.

(1)   Substance A sank in chloroform.

(2)   Substance B floated in water but sank in toluene.

(3)   Substance C sank in water but floated in chloroform and nitromethane.

(4)   Substance D sank in nitromethane but did not sink as rapidly as Substance A did in nitromethane.

_____   _____   _____   _____

least   *most*

dense   *dense*

Detecting Signs of Chemical Change

prepared by **M. L. Gillette**, Indiana University/Kokomo
and **H. A. Neidig**, Lebanon Valley College

Purpose of the Experiment

Mix combinations of liquids, solutions, and solids. Use laboratory observations to determine if the mixing causes a chemical change to occur.

Background Information

The limestone statue of a founding father standing in front of the county courthouse can undergo two basic kinds of change. If a piece of the nose is chipped off, we call the process a **physical change**. The physical appearance of the statue is altered, but the chemical composition of the statue remains unchanged.

Another type of physical change involves phase changes, such as when liquid water freezes or boils. These processes are reversible and do not change the chemical composition of the water. However, if, the statue's surfaces become eroded over time through the action of acid rain, we say that those portions of the statue have undergone a chemical change.

Chemical change occurs when the chemical composition of a substance is altered through chemical reaction. In this case, the limestone is chemically decomposed by the acid rain.

When a chemical reaction occurs, experimentally detectable changes are evident. However, not every detectable change results from a chemical reaction. For example, the appeal of a cup of coffee is improved for some by the addition of milk or cream. The coffee appearance differs after the milk is added, but the addition does not cause a chemical reaction. With the proper equipment, we can separate the coffee from the milk using the different physical properties of the two solutions.

On the other hand, the brown color and rough surface texture of toasted bread do indicate that a chemical reaction has occurred in the toasted portion of the bread. We can scrape the toasted layer off the bread, but we cannot return those crumbs to their original, untoasted condition. Hence, in the laboratory, we must carefully determine which observations are connected with physical changes and which with chemical changes.

Observations indicating a chemical change include:

(1)   **formation of a precipitate**, or solid, when two clear solutions are mixed;

(2)   a **color change**, one that does not simply represent the dilution of either of the solutions, when two solutions are mixed;

(3)   **evolution of a gas** that may or may not have an odor;

(4)   a **temperature change** in the reaction mixture that is not caused by external heating or cooling sources.

We can detect these phenomena using sight, touch, hearing, and smell.

Copyright © 1992 by Chemical Education Resources, Inc., P.O. Box 357, 220 S. Railroad, Palmyra, Pennsylvania 17078
No part of this laboratory program may be reproduced or transmitted in any form or by any means, electronic or mechanical, including photocopying, recording, or any information storage and retrieval system, without permission in writing from the publisher. Printed in the United States of America

In many chemical reactions, a precipitate forms because one or more of the reaction products is insoluble in the reaction mixture. The insoluble product or products usually settle to the bottom of the reaction vessel as a solid, although insoluble liquids may also form. And the insoluble product may also appear as a cloudy suspension or as a collection of particles that only slowly, if ever, settles to the bottom of the vessel.

Color changes that are indicative of a chemical reaction sometimes result when the chemical form of a compound changes. If a drop of clear, colorless hydrochloric acid solution (HCl) is placed on a piece of blue litmus paper, the moistened portion of the paper turns red. This is because the chemical form of litmus changes in the presence of acid, and the two chemical forms of litmus are different colors.

If one or more of the reaction products is insoluble in the solution and is a gas at the reaction temperature, bubbles will appear in the solution. Identification of the gas is possible if the gas has a characteristic odor, color, or recognizable chemical properties.

Every chemical reaction involves a transfer of energy. In many cases, this energy transfer involves easily detectable heat, or thermal energy, changes. A heat change causes an increase or decrease in the reaction mixture temperature. If the reaction mixture temperature increases during a reaction, we say that the reaction is **exothermic**; this means that it gives off heat to the solution. If the reaction mixture temperature decreases during the reaction, we say that the reaction is **endothermic**; this means that it absorbs heat from the solution.

In this experiment you will mix solutions of different substances and record your observations. You will also combine solids with solutions and record your observations. Based on these observations, you will conclude whether or not a chemical change has occurred in each case.

Procedure

> ### Chemical Alert
> 1M ammonia—toxic and corrosive
> ammonium chloride—irritant
> 0.1M cobalt(II) chloride in ethanol—irritant and flammable
> 0.1M copper(II) sulfate—toxic and irritant
> 1M hydrochloric acid—toxic and corrosive
> 1M sodium hydroxide—toxic and corrosive

> *Caution:* Wear departmentally approved eye protection while doing this experiment.

I. Preparing Your Glassware

1. Wash 12 test tubes with soap or detergent solution.

Thoroughly rinse each test tube three times using 5 mL of tap water each time. Then, thoroughly rinse each test tube once with 5 mL of distilled or deionized water.

Dry the test tubes with an absorbent paper towel. Place all test tubes in a test-tube rack.

> *Note:* The laboratory assignments are summarized in Table 1. Your laboratory instructor will tell you which laboratory assignments you will study. If you do not perform all the assignments, you may be told to share data with students who performed those assignments you did not.

2. Label one test tube for each laboratory assignment you will study.

II. General Procedure

> *Caution:* 1M NH$_3$, HCl, and NaOH solutions are toxic and corrosive. Prevent contact with your eyes, skin, and clothing. Do not ingest these solutions.
>
> Ethanol in solution is flammable.
>
> If you spill any solution, immediately notify your laboratory instructor.

3. To obtain solid reagent, first crease a small, clean piece of waxed or weighing paper. Remove the amount of solid reagent that fills the end of a spatula from the reagent container, and place the solid on the creased portion of the paper.

To transfer the solid to a test tube, center one end of the crease over the test tube mouth. Holding the paper in a V-shape, carefully tilt it until the solid slides along the crease, as shown in Figure 1.

> *Note:* Calibrated Pasteur or Beral pipets may be used to directly deliver the liquids into the test tubes. If applicable, your laboratory instructor will describe and demonstrate use of these pipets.

Table 1 *Laboratory assignments*

assignment	chemical names	chemical symbol
A	0.1 M cobalt(II) chloride hexahydrate 95% ethanol	$CoCl_2 \cdot 6\,H_2O$ C_2H_5OH
B	0.1 M cobalt(II) chloride hexahydrate dissolved in 95% ethanol water	$CoCl_2 \cdot 6\,H_2O$ H_2O
C	zinc 1 M hydrochloric acid solution	Zn HCl
D	zinc 0.1 M copper(II) sulfate solution	Zn $CuSO_4$
E	0.1 M copper(II) sulfate solution 1 M hydrochloric acid solution	$CuSO_4$ HCl
F	0.1 M copper(II) sulfate solution 0.5 M sodium hydrogen carbonate solution	$CuSO_4$ $NaHCO_3$
G	0.1 M copper(II) sulfate solution 1 M ammonia solution	$CuSO_4$ NH_3
H	magnesium 0.1 M copper(II) sulfate solution	Mg $CuSO_4$
I	magnesium 1 M hydrochloric acid solution	Mg HCl
J	1 M hydrochloric acid solution 1 M sodium hydroxide solution	HCl NaOH
K	1 M hydrochloric acid solution 0.5 M sodium hydrogen carbonate solution	HCl $NaHCO_3$
L	ammonium chloride water	NH_4Cl H_2O

4. Use a 10-mL graduated cylinder to measure and transfer milliliter volumes of liquids to a test tube. After each use, rinse the cylinder with 5 mL of tap water.

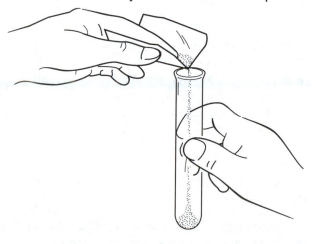

Figure 1 *Transferring a solid to a test tube*

Then, rinse the cylinder three times, using 5 mL portions of distilled water each time. ***Follow individual assignment directions for disposal of rinses.***

Allow the cylinder to drain as completely as possible between uses, in order to avoid diluting solutions with water.

> ***Note:*** Your laboratory instructor will demonstrate satisfactory methods for mixing test tube contents. ***Never*** mix test tube contents by placing your finger or thumb over the mouth of the test tube and inverting the tube.

5. To mix the contents of a test tube, you may be told to hold the test tube near the top, using the finger and thumb of one hand. To mix, you should tap the test

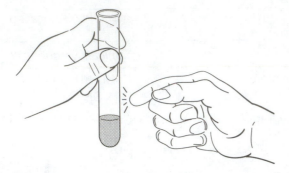

Figure 2 *Mixing the contents of a test tube*

tube gently near the bottom, using the forefinger of your free hand, as shown in Figure 2.

6. Make and record observations of color change, gas evolution, heat change, and precipitate formation.

> ***Note:*** When observing solution color, be sure to carefully distinguish between the terms ***clear*** and ***colorless***. **Clear** means transparent, and **color-less** means without color. You make your observation of color change easier by holding the test tube against a white background for contrast.
>
> If bubbling of a gas is not obvious, evidence of gas formation might be seen on test tube walls or a solid surface.
>
> In order to observe heat changes, hold the lower portion of the test tube so you can feel if the tube warms or cools when you mix the two components.

III. Laboratory Assignments

> ***Note:*** The numbers appearing in parentheses indicate the specific lines on your Data Sheet on which the indicated data should be entered.

A. CoCl₂ · 6 H₂O and 95% C₂H₅OH

$$A. \quad CoCl_2 \cdot 6\ H_2O \text{ and } 95\%\ C_2H_5OH$$

> ***Note:*** The success of this part of the experiment requires that you begin with an ***absolutely dry, clean*** test tube.

7. Transfer the amount of $CoCl_2 \cdot 6\ H_2O$ that fills the end of a microspatula to a ***dry***, clean, appropriately labeled test tube.

8. Add 2 mL of 95% C_2H_5OH to the test tube. Mix well.

9. Record your observations on your Data Sheet (1). Indicate whether or not a chemical change occurred (2).

10. Discard contents of test tube and all rinses in the container provided by your laboratory instructor and labeled "Discarded $CoCl_2 \cdot 6\ H_2O$–C_2H_5OH Mixture."

B. CoCl₂ · 6 H₂O in 95% C₂H₅OH and H₂O

$$B. \quad CoCl_2 \cdot 6\ H_2O \text{ in } 95\%\ C_2H_5OH \text{ and } H_2O$$

> ***Note:*** In order to obtain acceptable results for this part of the experiment, the graduated cylinder and test tube in the next step must be ***absolutely dry and clean***.

11. Using a ***dry*** graduated cylinder, transfer 1 mL of $CoCl_2 \cdot 6\ H_2O$ in 95% C_2H_5OH solution to a ***dry***, ***clean***, appropriately labeled test tube.

12. Add 1 mL of distilled water to the solution in the test tube. Mix well.

13. Record your observations on your Data Sheet (3). Indicate whether or not a chemical change occurred (4).

14. Discard contents of test tube and all rinses in the container provided by your laboratory instructor and labeled "Discarded $CoCl_2 \cdot 6\ H_2O$–C_2H_5OH Mixture."

C. Zn and HCl Solution

15. Transfer a piece of Zn the size of half a pea to the appropriately labeled test tube.

> ***Caution:*** $1M$ HCl solution is a corrosive solution that can cause skin irritation. Prevent contact with your eyes, skin, and clothing. Avoid inhaling the vapors and ingesting the solution.
>
> If you spill any HCl solution, immediately notify your laboratory instructor.

16. Add 2 mL of $1M$ HCl solution to the test tube. Mix well.

17. Record your observations on your Data Sheet (5). Indicate whether or not a chemical change occurred (6).

18. Discard contents of test tube and all rinses in the container provided by your laboratory instructor and labeled "Discarded Zn–HCl Mixture."

D. Zn and CuSO₄ Solution

19. Transfer a piece of Zn the size of half a pea to the appropriately labeled test tube.

20. Add 2 mL of 0.1 M CuSO₄ solution to the test tube. Mix well.

21. Record your observations on your Data Sheet (7). Indicate whether or not a chemical change occurred (8).

22. Discard contents of test tube and all rinses in the container provided by your laboratory instructor and labeled "Discarded Zn–CuSO₄ Mixture."

E. CuSO₄ Solution and HCl Solution

23. Transfer 1 mL of 0.1 M CuSO₄ solution to an appropriately labeled test tube.

24. Add 1 mL of 1 M HCl solution to the solution in the test tube. Mix well.

25. Record your observations on your Data Sheet (9). Indicate whether or not a chemical change occurred (10).

26. Discard contents of test tube and all rinsings in the container provided by your laboratory instructor and labeled "Discarded CuSO₄–HCl Mixture."

F. CuSO₄ Solution and NaHCO₃ Solution

27. Transfer 1 mL of 0.1 M CuSO₄ solution to an appropriately labeled test tube.

28. Add 1 mL of 0.5 M NaHCO₃ solution to the solution in the test tube. Mix well.

29. Record your observations on your Data Sheet (11). Indicate whether or not a chemical change occurred (12).

30. Discard contents of test tube and all rinses in the container provided by your laboratory instructor and labeled "Discarded CuSO₄–NaHCO₃ Mixture."

G. CuSO₄ Solution and NH₃ Solution

31. Transfer 1 mL of 0.1 M CuSO₄ solution to an appropriately labeled test tube.

Caution: 1 M NH₃ solution is corrosive, and irritating. Prevent contact with your eyes, skin, and clothing. Permanent fogging of soft contact len-

ses may result from NH₃ vapors. Avoid inhaling vapors and ingesting the solution.
   If you spill any NH₃ solution, immediately notify your laboratory instructor.

32. Add 1 mL of 1 M NH₃ solution to the solution in the test tube. Mix well.

33. Record your observations on your Data Sheet (13). Indicate whether or not a chemical change occurred (14).

34. Discard contents of test tube and all rinses in the container provided by your laboratory instructor and labeled "Discarded CuSO₄–NH₃ Mixture."

H. Mg and CuSO₄ Solution

Note: If the Mg surface is not shiny, rub the Mg with sandpaper to remove the oxide coating.

35. Transfer a shiny, 1-cm strip of Mg to the appropriately labeled test tube.

36. Add 2 mL of 0.1 M CuSO₄ solution to the test tube. Mix well.

37. Record your observations on your Data Sheet (14). Indicate whether or not a chemical change occurred (15).

38. Discard contents of test tube and all rinses in the container provided by your laboratory instructor and labeled "Discarded Mg–CuSO₄ Mixture."

I. Mg and HCl Solution

39. Transfer a shiny, 1-cm strip of Mg to the appropriately labeled test tube.

40. Add 2 mL of 1 M HCl solution to the test tube. Mix well.

41. Record your observations on your Data Sheet (17). Indicate whether or not a chemical change occurred (18).

42. Discard contents of test tube and all rinses in the container provided by your laboratory instructor and labeled "Discarded Mg–CuSO₄ Mixture."

J. HCl Solution and NaOH Solution

43. Transfer 1 mL of 1 M HCl solution to an appropriately labeled test tube.

> **Caution:** 1*M* NaOH solution is corrosive and toxic and it can cause skin irritation. Prevent contact with your eyes, skin, and clothing. Do not ingest the solution.
>
>   If you spill any NaOH solution, immediately notify your laboratory instructor.

44. Add 1 mL of 1*M* NaOH solution to the solution in the test tube. Mix well.

45. Record your observations on your Data Sheet (19). Indicate whether or not a chemical change occurred (20).

46. Discard contents of test tube and all rinses in the container provided by your laboratory instructor and labeled "Discarded $CuSO_4$–NH_3 Mixture."

K. HCl Solution and NaHCO₃ Solution

47. Transfer 1 mL of 1*M* HCl solution to an appropriately labeled test tube.

48. Add 1 mL of 0.5*M* $NaHCO_3$ solution to the solution in the test tube. Mix well.

49. Record your observations on your Data Sheet (21). Indicate whether or not a chemical change occurred (22).

50. Discard contents of test tube and all rinses into the drain.

L. NH₄Cl and H₂O

51. Transfer the amount of NH_4Cl that fills the end of a microspatula to a *dry*, appropriately labeled test tube.

52. Add 2 mL of room-temperature distilled water to the test tube. Mix well.

53. Record your observations on your Data Sheet (23). Indicate whether or not a chemical change occurred (24).

54. Discard contents of test tube and all rinses in the container provided by your laboratory instructor and labeled "Discarded NH_4Cl Solution."

> **Caution:** Wash your hands thoroughly with soap or detergent before leaving the laboratory.

Post-Laboratory Questions

(Use the spaces provided for the answers and additional paper if necessary.)

1. Based on your experimental data, list the laboratory assignment(s) that:

(1)   did not produce a chemical reaction;

(2)   produced solids in solution;

(3)   produced gaseous products;

(4)   resulted in color changes;

(5)   resulted in heat changes.

3. Based on your observations of the laboratory assignment(s) that produced gases, can you conclude that the same gas was produced in each assignment? Briefly explain.

4. Based on your observations of the laboratory assignment(s) that resulted in a color change, can you conclude that:

(1)   Precipitate formation is always accompanied by a solution color changes? Briefly explain.

(2)   A precipitate must form in order for a color change to occur? Briefly explain.

2. Based on the your observations of the laboratory assignment(s) that produced precipitates, can you conclude that solids form only when two solutions are mixed? Briefly explain.

5. Based on your observations of the laboratory assignment(s) that resulted in heat changes, can you conclude that heat is always released in a chemical reaction? Briefly explain.

name   section   date

Data Sheet

laboratory assignment	observations	Did a chemical change occur?
A (1)	(2)	
B (3)	(4)	
C (5)	(6)	
D (7)	(8)	
E (9)	(10)	
F (11)	(12)	
G (13)	(14)	

laboratory assignment		observations	Did a chemical change occur?
H	(15)	(16)	
I	(17)	(18)	
J	(19)	(20)	
K	(21)	(22)	
L	(23)	(24)	

Pre-Laboratory Assignment

1. Briefly describe the hazards associated with solutions of
  (1)   1*M* HCl

  (2)   1*M* NH$_3$

  (3)   1*M* NaOH

2. Briefly describe the meaning of each of these phrases as it pertains to this experiment.
  (1)   The solution turned blue but remained clear.

  (2)   The reaction was endothermic.

  (3)   A precipitate formed in solution.

  (4)   The change that occurred was physical, and not chemical.

3. Determine whether each of the following observations is evidence of a physical or a chemical change. Briefly explain.
  (1)   On a hot day, drops of water collect on the outside of a glass holding iced tea.

  (2)   Clear, colorless lemon juice is added to a glass of iced tea. The tea color changes from brown to yellow-brown.

  (3)   Water is added to a glass of iced tea. The tea color lightens.

Studying Chemical Reactions and Writing Chemical Equations

prepared by **M. L. Gillette**, Indiana University/Kokomo
and **H. Anthony Neidig**, Lebanon Valley College

Purpose of the Experiment

Describe chemical reactions by writing chemical equations based on laboratory observations and information about reactions of different substances.

Background Information

The human digestive process depends on a high concentration of hydronium ion (H_3O^+) in the stomach. Sometimes the acid concentration becomes too high, which causes discomfort ("acid stomach"). When this condition occurs, many people take antacids to help neutralize the excess acid. The active ingredient in many antacids is magnesium hydroxide, $Mg(OH)_2$. When solid $Mg(OH)_2$ mixes with acidic solution, a chemical reaction occurs. We can describe this reaction as: $Mg(OH)_2$ reacts with H_3O^+ to produce magnesium cations (Mg^{2+}) and water (H_2O).

We can make the same statement more concisely by writing a chemical equation for the reaction. A **chemical equation**, the symbolic description of a chemical process, shows the substances that react, the **reactants**, and the substances that form as a result of a chemical change, the **products**. An arrow indicates that the process occurs and is read "yields." The chemical equation for the reaction of $Mg(OH)_2$ and H_3O^+ is shown in Equation 1.

$$Mg(OH)_2(s) + 2\ H_3O^+(aq) \rightarrow Mg^{2+}(aq) + 4\ H_2O(l) \qquad \text{(Eq. 1)}$$

The abbreviations in parentheses following each of the chemical formulas indicate the physical state of each reactant and product. The meanings of the four abbreviations commonly used in equations are summarized in Table 1.

Table 1 *The meanings of physical-state abbreviations used in chemical equations*

abbreviation	meaning
aq	substance is dissolved in water
g	substance is a gas at reaction temperature and pressure
l	substance is a pure liquid at reaction temperature and pressure
s	substance is a solid that is insoluble in the reaction mixture

---

Copyright © 1993 by Chemical Education Resources, Inc., P.O. Box 357, 220 S. Railroad, Palmyra, Pennsylvania 17078
No part of this laboratory program may be reproduced or transmitted in any form or by any means, electronic or mechanical, including photocopying, recording, or any information storage and retrieval system, without permission in writing from the publisher. Printed in the United States of America

The 2 that appears in front of H_3O^+ and the 4 in front of H_2O are called **coefficients**; they indicate that two H_3O^+ ions are consumed and four H_2O molecules are formed in the reaction. Coefficients are used to **balance** an equation so that the reaction involves only a rearrangement of the reactant atoms to form products. When no coefficient is shown, as is the case with $Mg(OH)_2$ in Equation 1, a coefficient of 1 is understood.

The reaction of iron (Fe) with oxygen (O_2) to form iron(III) oxide (Fe_2O_3, ferric oxide) is another familiar reaction. The common name for Fe_2O_3 is rust. The chemical equation for this reaction is given in Equation 2.

$$4\ Fe(s) + 3\ O_2(g) \rightarrow 2\ Fe_2O_3(s) \qquad \text{(Eq. 2)}$$

Notice an important difference between Equations 1 and 2. In the first reaction, *two* reactants produce *two* products whereas in the second reaction, *two* reactants produce only *one* product.

Classifying Chemical Reactions

Many chemical reactions are conveniently classified as one of four major types. This classification is based on the type of chemical transformation that occurs.

Type I: Combination, Synthesis, or Formation Reactions

As the name suggests, a **combination**, or **synthesis reaction**, occurs when two substances combine to form a compound. The reaction in Equation 2 is a combination reaction. Equation 3 is a generalized combination reaction, while Equations 4–6 are specific examples.

$$A + B \rightarrow AB \qquad \text{(Eq. 3)}$$

$$8\ Mg(s) + S_8(s) \rightarrow 8\ MgS(s) \qquad \text{(Eq. 4)}$$

$$2\ Ca(s) + O_2(g) \rightarrow 2\ CaO(s) \qquad \text{(Eq. 5)}$$

$$H_2(g) + Cl_2(g) \rightarrow 2\ HCl(g) \qquad \text{(Eq. 6)}$$

Type II: Decomposition Reactions

A **decomposition reaction** occurs when a compound breaks apart to form two or more products. Equation 7 shows a generalized example, while Equations 8–10 give specific examples.

$$AB \rightarrow A + B \qquad \text{(Eq. 7)}$$

$$2\ HgO(s) \rightarrow 2\ Hg(l) + O_2(g) \qquad \text{(Eq. 8)}$$

$$BaCO_3(s) \rightarrow BaO(s) + CO_2(g) \qquad \text{(Eq. 9)}$$

$$2\ PbO_2(s) \rightarrow 2\ PbO(s) + O_2(g) \qquad \text{(Eq. 10)}$$

Type III: Single Displacement Reactions

When one element displaces another element from a compound, we call the process a **single displacement reaction**. Equation 11 shows a generalized example, and Equations 12 and 13 show specific examples.

$$A + BC \rightarrow AC + B \qquad \text{(Eq. 11)}$$

$$Sn(s) + 2\ AgNO_3(aq) \rightarrow$$
$$Sn(NO_3)_2(aq) + 2\ Ag(s) \qquad \text{(Eq. 12)}$$

$$2\ Al(s) + 6\ HNO_3(aq) \rightarrow$$
$$2\ Al(NO_3)_3(aq) + 3\ H_2(g) \qquad \text{(Eq. 13)}$$

When we write (aq) after the formula of an ionic compound, we mean that the ions are individually solvated by water. We could rewrite Equation 12 as Equation 14 to show the solvation of the ions involved:

$$Sn(s) + 2\ Ag^+(aq) + 2\ NO_3^-(aq) \rightarrow$$
$$Sn^{2+}(aq) + 2\ NO_3^-(aq) + 2\ Ag(s) \qquad \text{(Eq. 14)}$$

Equation 14 is the **complete ionic equation** for the reaction. Note that the solvated nitrate ion (NO_3^-) appears on both sides of Equation 14. Therefore it is not directly involved in the chemical reaction. We call such ions **spectator ions**. Thus, Equation 14 may be rewritten with the NO_3^- ion omitted, as in Equation 15.

$$Sn(s) + 2\ Ag^+(aq) \rightarrow Sn^{2+}(aq) + 2\ Ag(s) \qquad \text{(Eq. 15)}$$

Equation 15 is the **net ionic equation** for this reaction.

Type IV: Double Displacement, Double Replacement, or Metathesis Reactions

A **double displacement reaction** occurs when atoms or ions in two or more different substances change places to form new compounds. One of the new compounds formed is usually either a solid, called a **precipitate**, a slightly dissociated compound, such as H_2O, or a gas. A double displacement reaction involving the formation of a precipitate is often called a **precipitation reaction**. Equation 16 is a general example; Equation 17 is a specific example.

$$AB + CD \rightarrow AD + CB \qquad \text{(Eq. 16)}$$

$$Pb(NO_3)_2(aq) + 2\ NaCl(aq) \rightarrow$$
$$PbCl_2(s) + 2\ NaNO_3(aq) \qquad \text{(Eq. 17)}$$

The complete and net ionic equations for the reaction in Equation 17 are shown in Equations 18 and 19, respectively.

$$Pb^{2+}(aq) + 2\ NO_3^-(aq) + 2\ Na^+(aq) + 2\ Cl^-(aq) \rightarrow$$
$$PbCl_2(s) + 2\ Na^+(aq) + 2\ NO_3^-(aq) \qquad \text{(Eq. 18)}$$

$$Pb^{2+}(aq) + 2\ Cl^-(aq) \rightarrow PbCl_2(s) \quad (Eq.\ 19)$$

In this experiment, you will mix different elements and/or compounds in solution. In some cases, you will heat the mixtures to promote a reaction. You will observe the appearance of the reactants and of the products, and you will determine which type of reaction occurs in each case. You will also write the chemical equations for the reactions you perform. The chemical symbols for the elements and compounds you will work with are listed in Table 2.

Table 2 *Chemical symbols for the elements and compounds used in this experiment*

name	chemical symbol
ammonia	NH_3
ammonium carbonate	$(NH_4)_2CO_3$
carbon(IV) oxide (carbon dioxide)	CO_2
copper	Cu
copper(II) sulfate (cupric sulfate)	$CuSO_4$
copper(II) sulfate pentahydrate	$CuSO_4 \cdot 5\ H_2O$
hydrochloric acid	HCl
iron	Fe
iron(III) chloride (ferric chloride)	$FeCl_3$
lead(II) nitrate	$Pb(NO_3)_2$
magnesium	Mg
oxygen	O_2
potassium iodide	KI
sodium hydroxide	$NaOH$
sulfur	S_8

Procedure

> ### Chemical Alert
> ammonia—toxic and corrosive
> ammonium carbonate—irritant
> $0.1\,M$ copper(II) sulfate—toxic and irritant
> $0.1\,M$ hydrochloric acid—toxic and corrosive
> $0.1\,M$ iron(III) chloride—corrosive
> $0.1\,M$ lead(II) nitrate—toxic, irritant, and oxidant
> magnesium—flammable
> $0.1\,M$ potassium iodide—toxic and irritant
> $0.1\,M$ sodium hydroxide—toxic and corrosive
> sulfur—irritant

Caution: Wear departmentally approved eye protection while doing this experiment.

If you spill any reagent on your hands or skin, immediately wash with ample amounts of running water.

Note: The numbers appearing in parentheses indicate the specific lines on your Data Sheet on which the indicated data should be entered.

I. Reacting Mg with HCl Solution

1. Transfer a 0.5-cm piece of Mg ribbon to the bottom of a clean test tube.

Record on your Data Sheet your description of the appearance of the Mg (1).

Caution: The gas produced by the reaction of Mg and HCl solution is flammable. Be sure that there are no Bunsen burner flames in the area where you are performing this reaction.

2. Measure 2 mL of $0.1\,M$ HCl solution in a 10-mL graduated cylinder. Record on your Data Sheet your description of the appearance of the HCl solution (2).

Transfer the HCl solution to the test tube containing the Mg. Observe the reaction mixture for evidence of a chemical reaction.

Record on your Data Sheet any evidence that a chemical reaction has occurred (3).

3. Transfer the contents of your test tube to the container specified by your laboratory instructor and labeled "Discarded Mg–HCl Reaction Mixtures."

Note: Your laboratory instructor will tell you whether or not you should complete Step 4 before proceeding to Step 5.

4. Write on your Data Sheet the complete and net ionic equations for the reaction of Mg with HCl solution (4, 5). Indicate which of the four general reaction types is represented by this reaction (6).

II. Reacting Pb(NO₃)₂ Solution with KI Solution

Caution: Lead nitrate and KI solutions are toxic. Immediately notify your laboratory instructor if any solution spills.

If these solutions are not dispensed from a container equipped with a Pasteur pipet or other dropper, your laboratory instructor will describe and demonstrate a satisfactory method for dispensing them.

5. Transfer 10 drops of 0.1 M Pb(NO$_3$)$_2$ solution into a clean test tube.

Record on your Data Sheet your description of the appearance of the Pb(NO$_3$)$_2$ solution (7).

6. Record on your Data Sheet your description of the appearance of the 0.1 M KI solution (8).

7. Using a clean Pasteur or Beral pipet, transfer 2 drops of 0.1 M KI solution into the test tube containing the Pb(NO$_3$)$_2$ solution.

Record on your Data Sheet your description of the appearance of the reaction mixture (9).

8. Transfer the reaction mixture in the test tube into the waste container specified by your laboratory instructor and labeled "Discarded Pb(NO$_3$)$_2$–KI Reaction Mixtures." Rinse the test tube with 5 mL tap water three times, and then once with 5 mL of distilled or deionized water. Transfer rinsings to the waste container.

Note: Your laboratory instructor will tell you whether or not you should complete Step 9 before proceeding to Step 10.

---

9. Write on your Data Sheet the complete and net ionic equations for the reaction of Pb(NO$_3$)$_2$ solution with KI solution (10, 11). Indicate which of the four general reaction types is represented by this reaction (12).

III.   Heating Steel Wool (Fe) with S$_8$

10. Transfer to a porcelain crucible enough flattened steel wool to cover the crucible bottom. Record on your Data Sheet your description of the appearance of the steel wool (Fe) (13).

11. Sprinkle the amount of powdered S$_8$ that fills the end of a microspatula over the Fe (steel wool) in the crucible. Rinse the microspatula with distilled water and dry.

Record on your Data Sheet your description of the appearance of the S$_8$ (14).

Caution: Do Step 12 in a fume hood. The fumes from burning S$_8$ are toxic and irritating. Avoid inhaling the fumes.

---

12. Working inside a fume hood, carefully place the crucible and its contents in a wire triangle resting on a ring attached to a ring stand, as shown in Figure 1. Cover the crucible with the crucible cover.

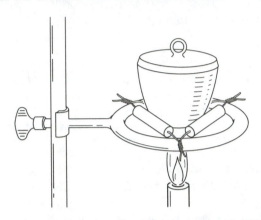

Figure 1   *Positioning a crucible in a wire triangle*

13. Still inside the fume hood, strongly heat the Fe–S$_8$ mixture from beneath with the flame of a Bunsen burner for 5 min. Using crucible tongs, carefully remove the crucible cover and look for evidence of unreacted S$_8$ in the crucible. If you find unreacted S$_8$, reheat the crucible for a minute or two. Check again for unreacted S$_8$. Stop heating when all visible evidence of unreacted S$_8$ has disappeared. Allow the crucible and its contents to cool to room temperature.

14. Using gentle pressure from a glass stirring rod, crush the reaction product. Observe the appearance of the reaction product. Record on your Data Sheet your description of the appearance of the reaction product (15).

15. Transfer the reaction product from the crucible into the container specified by your laboratory instructor and labeled "Discarded Fe–S$_8$ Reaction Products." Rinse the crucible with tap water, and dry the crucible.

Note: Your laboratory instructor will tell you whether or not you should complete Step 16 before proceeding to Step 17.

---

16. Write on your Data Sheet the chemical equation for the reaction of Fe and S$_8$ (16). Indicate which of the four general reaction types is represented by Fe and S reacting (17).

IV.   Heating CuSO$_4$ · 5 H$_2$O

17. Transfer the amount of CuSO$_4$ · 5 H$_2$O that fills the end of a clean, dry microspatula to the bottom of a clean, dry test tube.

Record on your Data Sheet your description of the appearance of the CuSO$_4$ · 5 H$_2$O (18).

18. Grasp the test tube containing CuSO$_4$ · 5 H$_2$O with a test tube holder. Holding the test tube at a 45° angle from the vertical, as shown in Figure 2, strongly heat the bottom of the test tube in a Bunsen burner flame. Carefully observe both the solid and the test tube walls near the open end of the test tube.

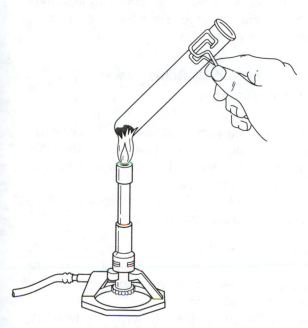

Figure 2   *Heating CuSO$_4$ • 5 H$_2$O in a test tube*

19. Record on your Data Sheet your description of the solid remaining at the bottom of the test tube and of the inside wall of the test tube (19, 20).

20. Transfer the test tube contents into the container specified by your laboratory instructor and labeled "Discarded CuSO$_4$ · 5 H$_2$O Reaction Products." Rinse the test tube twice using 5 mL of tap water each time. Then, rinse the test tube once with 5 mL of distilled water. Transfer the rinses into a 150-mL beaker labeled "Discarded CuSO$_4$ · 5 H$_2$O Rinses."

Transfer the contents of the "Discarded CuSO$_4$ · 5 H$_2$O Rinses" beaker to the container specified by your laboratory instructor and labeled "Discarded CuSO$_4$ · 5 H$_2$O Rinses." Rinse the beaker with 10 mL of tap water and pour these rinses into the drain, diluting with a large amount of running water.

Note: Your laboratory instructor will tell you whether or not you should complete Step 21 before proceeding to Step 22.

---

21. Write on your Data Sheet the chemical equation for the reaction of CuSO$_4$ · 5 H$_2$O when heated (21). Indicate which of the four general reaction types is represented by this reaction (22).

V.   Reacting HCl Solution with NaOH Solution

Caution: Both NaOH and HCl solutions are toxic and corrosive, and they can cause skin burns. Prevent contact with your eyes, skin, and clothing. Do not ingest these solutions. If you spill either solution, immediately notify your laboratory instructor.

Note: Phenolphthalein solution is an acid–base indicator that is red in the presence of excess hydroxide ion, OH$^-$, and colorless in the presence of excess H$^+$ ion.

---

22. Measure 2 mL of 0.1 *M* NaOH solution into the rinsed 10-mL graduated cylinder. Transfer the NaOH solution into a clean test tube. Record on your Data Sheet your description of the appearance of the NaOH solution (23).

Rinse the graduated cylinder twice with 5 mL of tap water each time. Then rinse the graduated cylinder once with 5 mL of distilled water. Transfer the rinses into a 150-mL beaker labeled "Discarded HCl-NaOH Reaction Mixtures."

23. Add one drop of phenolphthalein indicator solution to the NaOH solution in the test tube.

Record on your Data Sheet your description of the appearance of the NaOH solution in the test tube with phenolphthalein added (24).

24. Measure 3 mL of 0.1 *M* HCl solution into the rinsed 10-mL graduated cylinder. Record on your Data Sheet your description of the appearance of the HCl solution (25).

25. Add one drop of phenolphthalein solution to the HCl solution in the graduated cylinder. Record on your Data Sheet your description of the appearance of the solution in the graduated cylinder (26).

26. Carefully pour 1.0 mL of the HCl solution from the graduated cylinder into the test tube containing the NaOH solution. Use a Pasteur or Beral pipet to add the remaining 2.0 mL of HCl solution dropwise from the graduated cylinder until you see a color change in the solution in the test tube.

Record on your Data Sheet the evidence you have that a reaction has occurred (27).

27. Pour the test tube contents into the "Discarded HCl-NaOH Reaction Mixtures" beaker. Rinse the test tube twice with 5 mL of tap water each time. Then, rinse the test tube with 5 mL of distilled water. Pour the rinses into the labeled beaker.

Rinse the graduated cylinder twice with 5 mL of tap water each time. Then rinse the graduated cylinder once with 5 mL of distilled water. Pour the rinses into the "Discarded HCl-NaOH Reaction Mixtures" beaker.

Transfer the contents of the "Discarded HCl-NaOH Reaction Mixtures" beaker to the container specified by your laboratory instructor and labeled "Discarded HCl-NaOH Reaction Mixtures." Rinse the beaker with 10 mL of tap water and pour the rinse into the drain, diluting with a large amount of running water.

> **Note:** Your laboratory instructor will tell you whether or not you should complete Step 28 before proceeding to Step 29.

28. Write on your Data Sheet the complete and net ionic equations for the reaction of NaOH solution with HCl solution (28, 29). Indicate which of the four general reaction types is represented by this reaction (30).

VI. Heating Cu with Atmospheric O_2

> **Note:** Your laboratory instructor will describe and demonstrate the adjustment of a Bunsen burner flame that is necessary to achieve an oxidizing flame.

29. Obtain enough Cu mesh to sparsely cover the bottom of a crucible. Record on your Data Sheet your description of the appearance of the Cu mesh (31).

30. Position the crucible in a wire triangle as shown in Figure 1, but *do not* use a crucible cover. Heat the crucible and its contents until the crucible bottom is glowing red. Remove the heat, and allow the crucible and its contents to cool to room temperature.

31. Record on your Data Sheet your description of the appearance of the Cu mesh after it has been heated and cooled to room temperature (32).

32. Dispose of the Cu mesh in the container specified by your laboratory instructor and labeled "Used Cu Mesh."

> **Note:** Your laboratory instructor will tell you whether or not you should complete Step 33 before proceeding to Step 34.

33. Write on your Data Sheet the chemical equation for the reaction of Cu with atmospheric O_2 (33). Indicate which of the four general reaction types is represented by this reaction (34).

VII. Reacting $CuSO_4$ Solution with Steel Wool (Fe)

34. Obtain an amount of steel wool (Fe) equivalent to the volume of a pencil eraser. Use a clean, dry glass stirring rod to carefully slide the steel wool to the bottom of a clean, dry test tube.

Record on your Data Sheet your description of the appearance of the steel wool (Fe) (35).

35. Measure 2 mL of 0.1M $CuSO_4$ solution into the rinsed 10-mL graduated cylinder. Record on your Data Sheet your description of the appearance of the $CuSO_4$ solution (36).

Transfer the $CuSO_4$ solution into the test tube containing the steel wool (Fe).

36. Observe the appearance of the steel wool (Fe) and the surrounding solution.

Record on your Data Sheet your description of the appearance of the steel wool (Fe) and $CuSO_4$ solution after the reaction has occurred (37, 38).

37. Dispose of the $CuSO_4$ solution in the container specified by your laboratory instructor and labeled "Discarded $CuSO_4$ Solutions." Use your stirring rod to carefully remove the steel wool from the test tube. Place the steel wool in the container specified by your laboratory instructor and labeled "Discarded Steel Wool."

> **Note:** Your laboratory instructor will tell you whether or not you should complete Step 38 before proceeding to Step 39.

38. Write on your Data Sheet the complete and net ionic equations for the reaction of $CuSO_4$ solution with Fe (39, 40). Indicate which of the four general reaction types is represented by this reaction (41).

VIII. Reacting $FeCl_3$ Solution with NaOH Solution

39. Obtain 1 mL of $0.1M$ $FeCl_3$ solution in the rinsed 10-mL graduated cylinder and transfer the solution into a clean test tube. Rinse the graduated cylinder twice, using 5 mL of tap water each time, and then rinse once with 5 mL of distilled water. Transfer the rinses into a 150-mL beaker labeled "Discarded $FeCl_3$–NaOH Reaction Mixtures."

Record on your Data Sheet your description of the appearance of the $FeCl_3$ solution (42).

40. Transfer 1 mL of $0.1M$ NaOH solution into the rinsed 10-mL graduated cylinder. Record on your Data Sheet your description of the appearance of the NaOH solution (43).

41. Transfer the NaOH solution from the graduated cylinder into the test tube containing the $FeCl_3$ solution. Rinse the graduated cylinder twice, using 5 mL of tap water each time, and once with 5 mL of distilled water. Transfer the rinses into the "Discarded $FeCl_3$–NaOH Reaction Mixtures" beaker.

Record on your Data Sheet your description of the appearance of the reaction mixture (44).

42. Transfer the reaction mixture in the test tube into the "Discarded $FeCl_3$–NaOH Reaction Mixtures" beaker. Rinse the test tube three times, using 5 mL of tap water each time and then once with 5 mL of distilled water. Transfer rinses to the discard beaker.

Transfer the contents of the "Discarded $FeCl_3$–NaOH Reaction Mixtures" beaker to the container specified by your laboratory instructor and labeled "Discarded $FeCl_3$–NaOH Reaction Mixtures." Rinse the beaker with 10 mL of tap water and pour the rinse into the drain, diluting with a large amount of running water.

> **Note:** Your laboratory instructor will tell you whether or not you should complete Step 43 before proceeding to Step 44.

43. Write on your Data Sheet the complete and net ionic equations for the reaction of $FeCl_3$ solution with NaOH solution (45, 46). Indicate which of the four

general reaction types is represented by this reaction (47).

IX. Heating $(NH_4)_2CO_3$

44. Transfer the amount of $(NH_4)_2CO_3$ that fills the end of a microspatula to the bottom of a clean, dry test tube.

Record on your Data Sheet your description of the appearance of the $(NH_4)_2CO_3$ (48).

> **Note:** Red litmus paper turns blue in the presence of bases such as ammonia, NH_3.

45. Place a piece of red litmus paper on a small watch glass. Moisten the litmus paper with a drop of distilled water. The moist paper will cling to the watch glass. Record on your Data Sheet your description of the appearance of the litmus paper (49).

> **Caution:** Do Step 46 in a fume hood. When $(NH_4)_2CO_3$ is heated, toxic and irritating fumes are liberated. Avoid inhaling the fumes.

46. Grasp the test tube containing $(NH_4)_2CO_3$ with a test tube holder. Holding the test tube at a $45°$ angle from the vertical, as shown in Figure 2, strongly heat the bottom of the test tube in a Bunsen burner flame. Carefully observe both the solid and the test tube walls near the open end of the test tube.

47. Holding the mouth of the test tube 15 cm or 6 in. from your face, *carefully* fan the fumes coming from the test tube toward your nose, as shown in Figure 3. Note the odor of these vapors.

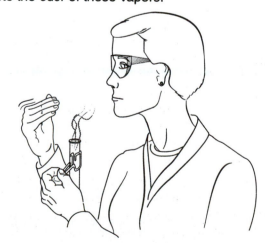

Figure 3   *Detecting odors*

48. Position the watch glass and litmus paper so that the moist paper is next to the test tube mouth. Observe the color of the litmus paper.

49. Record on your Data Sheet your descriptions of: what happened to the solid $(NH_4)_2CO_3$, the inside wall of the test tube, the odor of the vapors from the test tube mouth, and the color of the moist red litmus paper after exposure to these vapors (50, 51, 52, 53).

50. Rinse the test tube twice, using 5 mL of tap water each time. Then rinse the test tube once with 5 mL of distilled water. Transfer the rinses to a 150-mL beaker labeled "Discarded $(NH_4)_2CO_3$ Reaction Rinses."

Transfer the contents of the "Discarded $(NH_4)_2CO_3$ Reaction Rinses" beaker to the container specified by your laboratory instructor and labeled "Discarded $(NH_4)_2CO_3$ Reaction Rinses." Rinse the beaker with 10 mL of tap water and pour the rinse into the drain, diluting with a large amount of running water.

51. Write on your Data Sheet the chemical equation for the reaction of $(NH_4)_2CO_3$ when heated (54). Indicate which of the four general reaction types is represented by this reaction (55).

52. If you have not already done so, complete steps 4, 9, 16, 21, 28, 33, 38, and 43.

Caution: Wash your hands thoroughly with soap or detergent before leaving the laboratory.

Data Sheet

I. Reacting Mg with HCl Solution

(1) appearance of Mg:

(2) appearance of HCl solution:

(3) evidence that a chemical reaction occurred:

(4) complete ionic equation:

(5) net ionic equation:

(6) general reaction type:

II. Reacting $Pb(NO_3)_2$ Solution with KI Solution

(7) appearance of the $Pb(NO_3)_2$ solution:

(8) appearance of the KI solution:

(9) appearance of the reaction mixture:

(10) complete ionic equation:

(11) net ionic equation:

(12) general reaction type:

III. Heating Steel Wool (Fe) with S_8

(13)   appearance of the steel wool (Fe):

(14)   appearance of the S_8:

(15)   appearance of the reaction product:

(16)   chemical equation:

(17)   general reaction type:

IV. Heating $CuSO_4 \cdot 5 H_2O$

(18)   appearance of the $CuSO_4 \cdot 5 H_2O$:

(19)   appearance of solid after heating:

(20)   appearance of the inside wall of the test tube after heating:

(21)   chemical equation:

(22)   general reaction type:

V. Reacting HCl Solution with NaOH Solution

(23)   appearance of the NaOH solution:

(24)   appearance of the NaOH solution with phenolphthalein added:

(25)   appearance of the HCl solution:

(26)   appearance of the solution in the graduated cylinder:

(27)   evidence that a chemical reaction has occurred:

(28)   complete ionic equation:

(29)   net ionic equation:

(30)   general reaction type:

VI.   Heating Cu with Atmospheric O_2

(31)   appearance of the unreacted Cu mesh:

(32)   appearance of the Cu mesh after heating and cooling:

(33)   chemical equation:

(34)   general reaction type:

VII.   Reacting $CuSO_4$ Solution with Steel Wool (Fe)

(35)   appearance of the steel wool (Fe) before reaction:

(36)   appearance of the $CuSO_4$ solution before reaction:

(37)   appearance of the steel wool (Fe) after the reaction has occurred:

(38)   appearance of the $CuSO_4$ solution after the reaction has occurred:

(39)   complete ionic equation:

(40)   net ionic equation:

(41)   general reaction type:

VIII.   Reacting $FeCl_3$ Solution with NaOH Solution

(42)   appearance of the $FeCl_3$ solution:

(43)   appearance of the NaOH solution:

(44)   appearance of the reaction mixture:

(45)   complete ionic equation:

(46)   net ionic equation:

(47)   general reaction type:

IX. Heating $(NH_4)_2CO_3$

(48) appearance of the $(NH_4)_2CO_3$:

(49) initial appearance of the moist red litmus paper:

(50) description of what happened to the solid $(NH_4)_2CO_3$ upon heating:

(51) appearance of the inside wall of the test tube after heating:

(52) description of the odor of the fumes from $(NH_4)_2CO_3$ upon heating:

(53) appearance of the moist red litmus paper after exposure to fumes:

(54) chemical equation:

(55) general reaction type:

Pre-Laboratory Assignment

1. Briefly answer the following questions about some of the procedures you will be using in this experiment.

(1) Why would it be dangerous to perform Part I of the procedure of this experiment close to a lighted Bunsen burner?

(2) Why is it necessary to perform the reaction of Fe with S_8 solution under a fume hood?

(3) What solutions are used in Part II of the Procedure? Briefly describe the hazards associated with each of these solutions.

(4) Although the crucible is covered while heating in Part III of this experiment, why should you heat an *uncovered* crucible in Part VI?

(5) Why is it important to wash your hands before leaving the laboratory?

2. Briefly explain what is meant by each of the following terms as it relates to this experiment.

(1) precipitation reaction

(2) spectator ion

(3)   the designation (aq) following the formula of a compound

(4)   decomposition reaction

(5)   coefficients

3. (1)   Write the complete ionic equation for the double displacement reaction that occurs when aqueous solutions of barium nitrate, $Ba(NO_3)_2$, and sodium chromate (Na_2CrO_4) are mixed.

(2)   Name the spectator ions in this chemical reaction.

(3)   Write the net ionic equation for this reaction.

Determining the Comparative Reactivities of Several Metals

prepared by **M. L. Gillette**, Indiana University/Kokomo
and **H. A. Neidig**, Lebanon Valley College

Purpose of the Experiment

Determine the comparative reactivities of a group of metals from their reactions with 1 M hydrochloric acid solution and with solutions of metallic salts.

Background Information

One characteristic of metal atoms is their tendency to release electrons to another chemical species and form positively charged ions called **cations**. The process of releasing electrons is called **oxidation**. The species accepting the electrons undergoes **reduction**. Some metal atoms are oxidized more easily than others. Hence, we say that some metals are more reactive than others. We can determine the comparative reactivities of several metals by comparing their reactions with reagents that can undergo reduction. A particular metal may or may not react with a specific reagent. In addition, two metals may each react with a specific reagent, but with differing intensities.

In this experiment, you will compare the reactions of several metals with dilute hydrochloric acid solution (HCl) to determine the comparative reactivities of the metals. For example, some metals, here represented generically by "M," react with 1 M HCl solution as shown in Equation 1.

$$M(s) + 2\ HCl(aq) \rightarrow MCl_2(aq) + H_2(g) \qquad (Eq.\ 1)$$

We can rewrite Equation 1 as the net ionic equation

$$M(s) + 2\ H_3O^+(aq) \rightarrow$$
$$M^{2+}(aq) + 2\ H_2O(l) + H_2(g) \qquad (Eq.\ 2)$$

We see from these equations that the metal atom is oxidized when it reacts with HCl solution. The electrons lost from the metal atom are transferred to hydrogen ions, and hydrogen gas (H_2) is produced. Equation 1 also shows that the positive charge on the metal ion is counterbalanced by negatively-charged ions, called **anions**; the specific anions here are chloride ions (Cl^-). The electrically neutral combination of a cation and an anions, such as MCl_2, is called a **salt**. Many metallic salts are soluble in water.

The intensity with which different metals react with 1 M HCl solution varies. From laboratory observations of several metals in HCl solution, we can determine for each reaction whether or not H_2 is formed and, if so, with what intensity the gas is evolved. We can then determine the comparative reactivities of the metals with HCl solution. Note that some metals routinely react with components of the atmosphere, producing reaction products that coat the metal surface. These

---

Copyright © 1992 by Chemical Education Resources, Inc., P.O. Box 357, 220 S. Railroad, Palmyra, Pennsylvania 17078
No part of this laboratory program may be reproduced or transmitted in any form or by any means, electronic or mechanical, including photocopying, recording, or any information storage and retrieval system, without permission in writing from the publisher. Printed in the United States of America

products must be removed from the metal surface before a reaction of the metal with any solution can be observed.

The reaction of metals with solutions of salts of other metals is also revealing. Some metals react with the solution of a salt of another metal as indicated in Equation 3, using $PbSO_4$ as a typical metallic salt.

$$M(s) + FeSO_4(aq) \rightarrow MSO_4(aq) + Fe(s) \quad \text{(Eq. 3)}$$

We can rewrite Equation 3 as the net ionic equation

$$M(s) + Fe^{2+}(aq) \rightarrow M^{2+}(aq) + Fe(s) \quad \text{(Eq. 4)}$$

Notice electrons are transferred from the metal atom to the iron(II) ion (Fe^{2+}) in this example. The metal is oxidized, and the Fe^{2+} ion is reduced.

Because solutions of many metal ions are colorless, often we must examine the reaction mixture carefully to establish whether or not a reaction has occurred. A change in the appearance of the metal, a solution color change, or the formation of a precipitate all indicate that a reaction of this type has occurred. Note that sometimes, because metallic salt solutions are acidic, H_2 gas is produced when the metal is added. In such cases, the formation of H_2 does *not* indicate a reaction between the metal and the metallic salt. The fact that a metal does or does not react with the solution of another metallic salt can be used to establish the comparative reactivities of several metals.

In this experiment, you will examine samples of magnesium (Mg), zinc (Zn), and copper (Cu), carefully noting their appearances. Then you will add a strip of each metal to a dilute HCl solution. Comparing your initial and final observations of the metals and the solutions, you will determine whether or not a reaction has occurred, and, if so, with what intensity. Using these results, you will arrange the three metals in decreasing order of reactivity.

Next, you will add a sample of each of the metals to solutions of magnesium sulfate ($MgSO_4$), zinc sulfate ($ZnSO_4$), and copper(II) sulfate ($CuSO_4$). Based on your initial and final observations of the metals and of the solutions, you will determine whether or not any reactions occur. Using these results, you will arrange the metals in decreasing order of reactivity.

Procedure

Chemical Alert

0.5*M* copper(II) sulfate—toxic and irritant
1*M* hydrochloric acid—toxic and corrosive
0.5*M* magnesium sulfate—irritant
0.5*M* zinc sulfate—irritant

Caution: Wear departmentally approved eye protection while doing this experiment.

Note: The numbers appearing in parentheses indicate the specific lines on your Data Sheet on which the indicated data should be entered.

I. Observing the Appearance of Metals

1. Examine sample strips of Mg, Zn, and Cu. Record your observations on your Data Sheet, including such characteristics as as color, texture, and luster (1, 2, 3).

II. Reacting Metals with 1*M* HCl Solution

2. Label three test tubes "1," "2," and "3." Place the three test tubes in a test tube rack or a 150-mL beaker.

Caution: 1*M* HCl solution is corrosive, toxic, and can cause burns. Avoid contact with your eyes, skin, and clothing. Avoid inhaling vapors and ingesting the compound.

If you spill any HCl solution, immediately notify your laboratory instructor.

---

3. Use a 10-mL graduated cylinder to measure and transfer 1 mL of 1*M* HCl in each test tube.

Note: If the metal strips you are using are not shiny, sand their surfaces with abrasive paper.

---

4. Hold Test Tube 1 at a 45° angle. Carefully slide a Mg strip down the side of the test tube into the HCl solution. Observe the appearance of the solution and of the metal surface for 30 s after adding the metal. Record your observations on your Data Sheet (4).

5. Repeating the procedure in Step 4, add a Zn strip to Test Tube 2, and a Cu strip to Test Tube 3. Record your observations on your Data Sheet (5, 6).

6. Record your observations of each metal and solution after 10 minutes (7, 8, 9).

Note: In Step 7, you will decant the liquid from any solid remaining in the test tube. **Decantation** is a process for separating the liquid and the solid in a reaction mixture. First the solid is allowed to settle to the bottom of the container. Then the liquid is carefully poured from the container, without disturbing the solid. The liquid is called the **supernatant liquid** or the **supernate**.

---

7. Label a 250-mL beaker "Discarded Filtrates." Carefully decant the supernatant liquid in Test Tube 1 into your labeled beaker. Wash any remaining metal twice with 3 mL of distilled water each time. Pour the rinses into the drain, diluting with a large amount of running water.

Transfer any Mg remaining in Test Tube 1 into the container specified by your laboratory instructor and labeled, "Discarded Mg."

Rinse the test tube twice with tap water, using 5 mL each time.

8. Repeat Step 7 using Test Tubes 2 and 3. Discard any Zn in the "Discarded Zn" container and any Cu in the "Discarded Cu" container.

III. Reacting Metals with Metallic Salt Solutions

Note: In this part of the experiment, you will use 0.5M solutions of $MgSO_4$, $ZnSO_4$, and $CuSO_4$, and strips of Mg, Zn, and Cu. Your laboratory instructor will make assignments from Table 1. If you do not perform all nine assignments yourself, you may be asked to share data with students who performed the assignments you were not given.

---

9. Measure 1 mL of your first assigned metallic salt solution in a 10-mL graduated cylinder. Transfer the solution to a test tube.

Table 1   *Laboratory assignments for the reaction of metals with metallic salt solutions.*

| | 0.5M metallic salt solution | | |
metal	$MgSO_4$	$ZnSO_4$	$CuSO_4$
Mg	A	D	G
Zn	B	E	H
Cu	C	F	I

Observe the appearance of the salt solution. Record your observations on your Data Sheet, next to the appropriate Laboratory Assignment letter.

Note: If the metal strips you are using are not shiny, sand their surfaces with abrasive paper.

---

10. Tilt the test tube at a 45° angle. Carefully slide a strip of the assigned metal down the side of the test tube into the solution.

Observe the appearance of the metal surface and of the solution for 30 s after addiing the metal. Record your observations on your Data Sheet.

11. Record your observations of the metal and the solution on the Data Sheet after 10 min.

12. Decant the supernatant liquid into the "Discarded Filtrates" beaker. Rinse any solid metal remaining in the test tube twice, each time using 3 mL of distilled water. Discard the metal in the appropriately labeled containers, either "Discarded Mg," "Discarded Cu," or "Discarded Zn."

13. Thoroughly rinse the graduated cylinder and the test tube twice with 5 mL of tap water each time, and twice more with 5 mL of distilled water each time. Pour the rinses into the drain, diluting with a large amount of running water.

14. Do the other specified laboratory assignments, following the procedure in Steps 9–13, recording your observations in the appropriate spaces on your Data Sheet.

Caution: Wash your hands thoroughly with soap or detergent before leaving the laboratory.

Post-Laboratory Questions

(Use the spaces provided for the answers and additional paper if necessary.)

1. List the three metals you studied in this experiment in decreasing order of reactivity. Briefly describe the specific experimental evidence upon which you based your order.

2. (1) Briefly explain why it is necessary to clean the metal surfaces before adding the metal to the test solutions.

(2) What two pieces of evidence led you to conclude that Zn did or did not react with $0.5M$ $CuSO_4$ solution?

(3) Would your experimental results have been different if you had used larger pieces of metal? Briefly explain.

3. Aluminum (Al) is more reactive than Zn. Ingestion of aluminum ion (Al^{3+}) has been tentatively associated with some medical problems in humans.

(1) Based on these statements, briefly explain why it is recommended not to cook acidic foods in aluminum pans.

(2) Which would be less risky: to cook acidic foods in a freshly scoured aluminum pan, or to cook them in an aluminum pan that has been in a cupboard for many weeks? Briefly explain.

Data Sheet

I. Observing the Appearance of Metals

metal   observations

Mg   (1)

Zn   (2)

Cu   (3)

II. Reacting Metals with 1 *M* HCl Solution

metal   initial observation   observations after 10 min

Mg   (4)   (7)

Zn   (5)   (8)

Cu   (6)   (9)

III. Reacting Metals with Metallic Salt Solutions

with MgSO$_4$ solution

metal		initial observations	observations after 10 min
Mg	A		
Zn	B		
Cu	C		

with ZnSO$_4$ solution

metal		initial observations	observations after 10 min
Mg	D		
Zn	E		
Cu	F		

with CuSO$_4$ solution

metal		initial observations	observations after 10 min
Mg	G		
Zn	H		
Cu	I		

Pre-Laboratory Assignment

1. Reactions of metals with $1M$ HCl solution that produce H_2 gas can proceed with considerable intensity.

(1) Briefly describe the hazards you need to be aware of when you work with $1M$ HCl solution.

(2) Give two reasons why it would be unwise to almost completely fill a test tube with $1M$ HCl solution before adding a piece of metal. Briefly explain.

2. Briefly explain what is meant by the following terms as they pertain to this experiment.

(1) comparative reactivities of metals

(2) cation

(3) oxidation

(4) reduction

3. Iron (Fe) is more reactive than gold (Au).

(1) A student places a piece of Fe in 3 mL of $0.5M$ $Au(NO_3)_3$ solution in Test Tube 1. In Test Tube 2, the student places a piece of Au in 3 mL of $0.5M$ $Fe(NO_3)_2$ solution. In which test tube does the student see evidence of a reaction? Briefly explain.

(2) The student is given a piece of lead (Pb) and a $0.5M$ lead(II) nitrate solution, $Pb(NO_3)_2$. The student discovers that Pb reacts with $0.5M$ $Au(NO_3)_3$ solution, but Pb does not react with $0.5M$ $Fe(NO_3)_2$ solution. Is Pb more or less reactive than Au? Than Fe? Briefly explain.

(3) List the three metals, Pb, Au, and Fe, in order of decreasing reactivity.

Synthesizing Aspirin

prepared by **Robert L. Glogovsky**, Elmhurst College, IL

Purpose of the Experiment

Prepare acetylsalicylic acid (aspirin) and investigate two of its chemical properties.

Background Information

Hippocrates, the ancient Greek physician, knew of the curative powers of willow tree bark. Native Americans brewed willow bark tea for medicinal purposes long before Columbus' time. The component in the bark that is responsible for the bark's medical benefits, salicin, was identified in 1827, but its useful derivative, salicylic acid, was not synthesized until 1853. After that date, salicylic acid was used as an analgesic (pain reliever) and antipyretic (fever reducer), even though some users suffered troublesome side effects. In 1893, Felix Hoffman, Jr., a chemist working for Friedrich Baeyer and Company, synthesized the acetyl derivative of salicylic acid, called acetylsalicylic acid, or aspirin. Users found this compound to be less irritating to the stomach and more palatable than salicylic acid or its salts. In 1899 Baeyer began marketing envelopes and capsules filled with powdered aspirin. Aspirin tablets were introduced in 1915. Today aspirin is one of the most popular over-the-counter drugs.

Aspirin remains structurally unchanged in the human stomach, which is an acidic environment. In the intestinal tract, an alkaline environment, aspirin's acid function is neutralized, causing formation of the sodium salt, sodium acetylsalicylate. This product is absorbed into the bloodstream through the intestinal walls. It is then transported throughout the body. The biochemical action of aspirin is complex and not yet completely understood.

A primary side effect of aspirin is its tendency to irritate the stomach lining, causing the loss of about 0.5 mL of blood for each 500-mg tablet ingested. For this reason, aspirin is often buffered and combined with other ingredients to reduce stomach irritation. Other side effects include runny nose and Reye's syndrome in children.

Carboxylic acids and acid anhydrides react with alcohols to form esters in **esterification reactions**. Aspirin is the product of the esterification of acetic an-

salicylic acid   acetic anhydride

acetylsalicylic acid   (Eq. 1)

Copyright © 1994 by Chemical Education Resources, Inc., P.O. Box 357, 220 S. Railroad, Palmyra, Pennsylvania 17078
No part of this laboratory program may be reproduced or transmitted in any form or by any means, electronic or mechanical, including photocopying, recording, or any information storage and retrieval system, without permission in writing from the publisher. Printed in the United States of America

hydride with the phenolic hydroxyl group of salicylic acid, as shown in Equation 1. Because this reaction is acid catalyzed, we add small amounts of a mineral acid, such as sulfuric (H_2SO_4) or phosphoric acid (H_3PO_4), to the reaction mixture. Note that this reaction is reversible.

In the presence of water, esters can undergo a **hydrolysis reaction** that, in the case of acetylsalicylic acid, generates the carboxylic acid, acetic acid, and alcohol. In the procedure used in this experiment, the decomposition of acetylsalicylic acid by heating, with water present as vapor, leads to the formation of acetic acid, which vaporizes, causing a vinegar-like odor.

Iron(III) chloride solution ($FeCl_3$) reacts with phenols to produce colored complexes, ranging from blue to green to red-brown, depending on the substitution pattern of the phenol. Aliphatic alcohols, acids, and other functional groups do not respond to the addition of $FeCl_3$ solution. Salicylic acid has both a phenolic alcohol group and a carboxyl group. Thus, we could expect either a positive reaction with the phenolic alcohol or negative reaction depending on the way the molecule reacts with the phenolic alcohol-carboxyl group combination when $FeCl_3$ solution is added to salicylic acid.

In this experiment, you will take a measured mass of salicylic acid and add a measured volume of acetic anhydride. You will determine the yield of crystallized aspirin. Then you will test a portion of your synthesized aspirin with $FeCl_3$ solution in order to establish whether salicylic acid behaves as a phenolic alcohol or as a carboxylic acid in the reaction. You will also determine the outcome of the reaction of acetylsalicylic acid with moist air in the presence of heat. Finally, you will process the reaction mixture for disposal.

Procedure

Chemical Alert
acetic anhydride—strongly corrosive and lachrymator
95% ethanol—flammable and highly toxic
1% iron(III) chloride—corrosive
salicylic acid—toxic and irritant
concentrated sulfuric acid—highly toxic, corrosive, and oxidant

---

Caution: Wear departmentally approved eye protection while doing this experiment.

I. Synthesizing Aspirin

Note: If the balances you are using do not have a tare function on them, your laboratory instructor will describe the weighing procedure you should follow and how you should record your data.

If weighing papers are not available in your laboratory, you will be given instructions for weighing and transferring the solid in this experiment.

Unless your laboratory instructor tells you otherwise, record all masses to the nearest milligram (0.001 g).

---

1.   Prepare a boiling-water bath by half filling a 600-mL beaker with tap water. Attach a large ring support to a ring stand. Place the beaker through the ring. Adjust the ring so that the beaker is stabilized while sitting on a hot plate, as shown in Figure 1. Heat the beaker and the water to the boiling point for use in Step 4.

2.   Tare a clean piece of weighing paper. Weigh about 2.1 g of salicylic acid. Record on your Data Sheet the mass of the salicylic acid to the nearest milligram (0.001 g).

Transfer the solid into a 125-mL Erlenmeyer flask.

Caution: Acetic anhydride is strongly corrosive and a lachrymator. Use a fume hood when working with this reagent. Prevent eye, skin, and clothing contact. Avoid inhaling vapors.

Concentrated sulfuric acid is a highly toxic, strongly corrosive oxidant. The acid can cause severe burns. Prevent eye, skin, and clothing contact.

If you spill either substance on yourself, ***immediately*** rinse with a large amount of running water. If you spill any on the laboratory bench, immediately notify your laboratory instructor.

---

3.   In a ***fume hood***, use a Pasteur pipet to carefully add 4 mL of acetic anhydride to the Erlenmeyer flask containing the salicylic acid.

Using another Pasteur pipet, carefully add 5 drops of concentrated H_2SO_4 solution to the flask.

Place the pipets in a 150-mL beaker, with the tips down.

Gently swirl the flask and its contents to thoroughly mix. Some of the solid will remain undissolved.

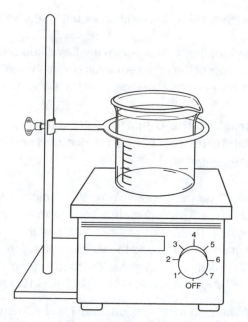

Figure 1 *Stabilizing a beaker on a hot plate*

4. Place the flask in the boiling-water bath. Clamp the flask in position as shown in Figure 2. Heat the flask and its contents in the bath for 15–20 min.

5. While the flask is heating, prepare an ice-water bath by half filling a 600-mL beaker with tap water and a few pieces of ice. Transfer 60 mL of distilled water to a 150-mL beaker, and cool the beaker and its contents in the ice-water bath.

6. Using the clamp as a handle, carefully remove the *hot* flask from the boiling-water bath. In 1–2 mL portions, slowly add a total of 10 mL of chilled water from a graduated cylinder to the flask in order to decom-

pose the unreacted acetic anhydride. Carefully swirl the flask between additions of chilled water.

7. Chill the flask in the ice-water bath for 10–15 min to crystallize the reaction product.

8. While you are cooling your flask, assemble your filtering apparatus, as shown in Figure 3. To do so, clamp a 500-mL filter flask to your other ring stand, and place a Büchner funnel in the flask.

9. Add 25 mL of chilled, distilled water to the reaction flask. Carefully break up any crystal lumps with a clean glass stirring rod.

10. Weigh a piece of filter paper and record this mass to the nearest 0.001 g on your Data Sheet.

Place the filter paper in your Büchner funnel, inserted in the filter flask. Moisten the paper with 1–2 mL of chilled, distilled water.

Attach a piece of pressure tubing to the side arm of the filter flask and to a filter trap attached to a water aspirator. Turn on the aspirator to draw the water in the funnel into the flask and to snugly seal the filter paper to the funnel. Leave the aspirator on.

11. Slowly pour the reaction mixture from the Erlenmeyer flask into the Büchner funnel, as follows. First, decant as much supernatant liquid as possible into the funnel. Use the stirring rod to guide the liquid from the flask onto the filter paper, in order to prevent splashing and product loss.

Use a rubber policeman attached to a second glass stirring rod to help transfer the solid from the flask into the funnel. Rinse the last crystals out of the flask into the funnel with a few milliliters of chilled, distilled water. Disconnect and turn off the aspirator.

12. Add 15 mL of chilled, distilled water to the funnel. Reconnect the aspirator, and turn it on. Dry the solid

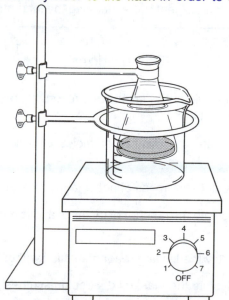

Figure 2 *Suspending the flask in the boiling-water bath*

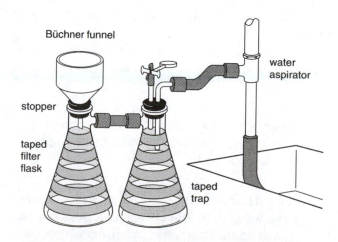

Figure 3 *A vacuum filtration apparatus*

by drawing air through the filter for 15 min. When the crystals look dry, disconnect and turn off the aspirator.

13. Weigh a clean, dry watch glass. Record this mass on your Data Sheet. Carefully remove the filter paper and crystals from the funnel and place them on the weighed watch glass. Weigh the watch glass, filter paper, and crystals. Record this mass on your Data Sheet. Retain the aspirin for the following tests.

14. Clean the pipets used in Step 3 by carefully adding distilled water to the beaker holding them. Draw some of this water into the pipets, and discharge the water back into the same beaker.

Pour the mixture in the beaker into a 600-mL beaker labeled "Discarded Reaction Mixtures and Rinses."

Pour the mixture in the filter flask from Steps 8–12 into the labeled 600-mL beaker.

Save the "Discarded Reaction Mixtures and Rinses" beaker and its contents for use in Part III.

II. Testing Your Synthesized Aspirin

Caution: Ethanol is a toxic and flammable liquid. Extinguish all open flames in the laboratory when doing Step 15. Avoid inhaling the vapor.

1% iron(III) chloride solution can cause burns. Avoid eye, skin, and clothing contact.

15. Add 1 mL of 95% ethanol (C_2H_5OH) and 1 drop of 1% iron(III) chloride solution to each of three 12×75-mm test tubes. Add a few crystals of salicylic acid to the first test tube. Add a few crystals of your reaction product from Part I to the second test tube. Use the third test tube as a control.

Using a clean glass stirring rod, stir the mixtures in each test tube. Observe and compare the contents of the three test tubes. Record all observations on your Data Sheet.

Pour the contents of the test tubes into the container provided by your laboratory instructor and labeled "Discarded Test Solutions."

Caution: Your laboratory instructor will give you suggestions for detecting the odor of any vapors generated in Step 16.

16. Place a few crystals of your reaction product from Part I in another 12×75-mm test tube. Attach a test tube clamp to the test tube. Using the clamp as a handle, gently heat the test tube and its contents over the low flame of a Bunsen burner until the crystals liquify.

Remove the test tube from the flame and carefully note the odor of the vapor emanating from the test tube. Record your observations on your Data Sheet.

III. Treating the Discarded Reaction Mixtures and Rinses Solution for Disposal

17. Add 3 drops of phenolphthalein solution to the solution in the "Discarded Reaction Mixtures and Rinses" beaker. Record on your Data Sheet the color of the solution after you add the indicator solution.

If the solution is colorless, proceed to Step 18.

If the solution is pink, proceed to Step 19.

18. To the colorless solution, add $1M$ NaOH solution dropwise until pink coloration persists throughout the solution.

Pour the neutralized mixture into the drain, followed by a large amount of running water.

Proceed to Step 20.

19. To the pink solution, add $1M$ HCl solution dropwise until the solution just turns colorless.

Pour the neutralized mixture into the drain, followed by a large amount of running water.

20. Wash, drain, and dry all glassware used in this experiment.

Caution: Wash your hands thoroughly with soap or detergent before leaving the laboratory.

Calculations

Do the following calculations and record the results on your Data Sheet.

1. Calculate the number of moles of salicylic acid used.

2. Calculate the mass of acetic anhydride used.

3. Calculate the number of moles of acetic anhydride used.

4. Identify the limiting reagent in this synthesis.

5. Calculate the mass of dry aspirin synthesized.

6. Calculate the theoretical yield of aspirin.

7. Calculate your percent yield of aspirin.

Post-Laboratory Questions

(Use the spaces provided for the answers and additional paper if necessary.)

1. (1) A student who was in a hurry to complete this experiment did not completely dry the crystalline product before doing the final weighing. How would this error affect the calculated percent yield of the experiment?

(2) The student in (1) did not bother to cool the water used to wash the crystals. How would this error affect the calculated percent yield of the experiment?

(3) The same student mixed the crystalline product with ethanolic 1% $FeCl_3$ solution and obtained a colored solution. Briefly explain what might have caused this result.

2. A desiccant, or drying agent, is often added to aspirin in order to prolong the aspirin's shelf life by delaying hydrolysis. How can you easily determine whether or not any aspirin you have at home has begun to hydrolize?

3. (1) How many grams of aspirin were lost in the washing process you used to purify your product, assuming you used 50 mL of water for washing? The solubility of aspirin in water is 0.33 g per 100 mL at room temperature.

(2) Recalculate your percent yield, taking into account the water solubility of aspirin when you calculate the theoretical yield.

4. Using structures for all species, write the complete esterification reaction begun below:

methanol (CH_3OH) + salicylic acid $\xrightarrow{H_2SO_4}$

name   section   date

Data Sheet

I. Synthesizing Aspirin

mass of salicylic acid, g   _____

mass of acetic anhydride, g (density = 1.08 g mL^{-1})   _____

number of moles of salicylic acid (GMM = 138 g mol^{-1})   _____

number of moles of acetic anhydride (GMM = 102 g mol^{-1})   _____

limiting reagent   _____

mass of filter paper, g   _____

mass of watch glass, g   _____

mass of filter paper and watch glass, g   _____

mass of aspirin crystals, filter paper, and watch glass, g   _____

actual yield of aspirin, g   _____

theoretical yield of aspirin (GMM: 180 g mol^{-1}), g   _____

percent yield, %   _____

II. Testing Your Synthesized Aspirin

observations

salicylic acid + ethanolic FeCl$_3$ solution

aspirin + ethanolic FeCl$_3$ solution

ethanolic FeCl$_3$ solution

conclusion based upon above observations

observations of heated synthesized aspirin

III. Treating the Discarded Reaction Mixtures and Rinses Solution for Disposal

initial color of "Discarded Reaction Mixtures and Rinses"
   solution after adding phenolphthalein solution   _____

name and molarity of solution used to neutralize
   "Discarded Reaction Mixtures and Rinses" solution   _____

name section date

Pre-Laboratory Assignment

1. Briefly describe the hazards you should be aware of when working with:

(1) acetic anhydride

(2) concentrated sulfuric acid

2. Briefly define the following terms as they pertain to this experiment:

(1) analgesic

(2) hydrolysis reaction

(3) esterification reaction

(4) antipyretic

3. Calculate the number of moles in:

(1) 2.1 g of salicylic acid (GMM: 138 g mol^{-1}).

(2) 4.0 mL of acetic anhydride (GMM: 102 g mol^{-1}, density: 1.08 g ml^{-1})

4. Based on the amounts of reagents specified in this experiment, calculate the theoretical yield of aspirin (GMM: 180 g mol^{-1}) in terms of:

(1) moles

(2) grams

5.   Briefly explain the purpose of the $FeCl_3$ test. What does a positive $FeCl_3$ test signify? What observation corresponds to a positive $FeCl_3$ test?

Preparing Soap and Determining Its Properties

prepared by **L.A. Whitaker**, Massachusetts Institute of Technology

---

Purpose of the Experiment

Prepare a soap from an oil and a fat. Determine some properties of soaps and detergents, and compare them.

Background Information

Soaps are commonly prepared from lipids, which are generally animal or vegetable fats or oils. Fats and oils are esters derived from glycerol and a variety of fatty acids. Because glycerol has three hydroxy groups, fats and oils are triesters and, specifically, triglycerides. In addition to the three-carbon glyceride skeleton, the other components of the triester are long-chain aliphatic carboxylic acids, commonly referred to as fatty acids. These compounds predominantly contain an even number of carbon atoms, ranging from 6 to 18. Some of the more common fatty acids are stearic, palmitic, oleic, and linoleic acids.

Figure 1 shows the structures for (a) a triglyceride; (b) glycerol; (c) two saturated fatty acids; and (d) two unsaturated fatty acids.

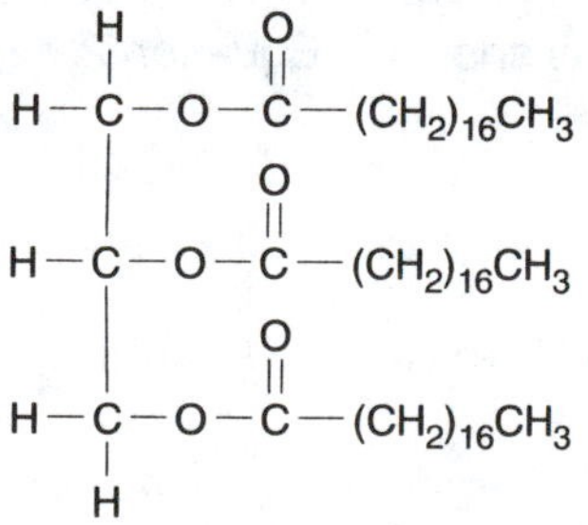

(a) tristearin, or glyceryl tristearate (a triglyceride)

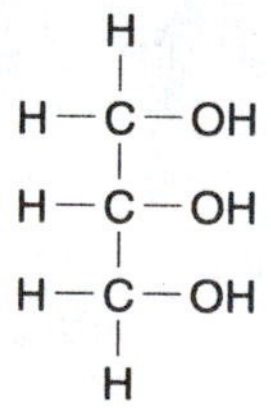

(b) glycerol

(c) two saturated acids

$CH_3(CH_2)_{14}COOH$
palmitic acid

$CH_3(CH_2)_{16}COOH$
stearic acid

$CH_3(CH_2)_7CH=CH(CH_2)_7COOH$
oleic acid

$CH_3(CH_2)_4CH=CHCH_2CH=CH(CH_2)_7COOH$
linoleic acid

(d) two unsaturated acids

Figure 1   *A triglyceride and some components of triglycerides*

Copyright © 1985 by Chemical Education Resources, Inc., P.O. Box 357, 220 S. Railroad, Palmyra, Pennsylvania 17078
No part of this laboratory program may be reproduced or transmitted in any form or by any means, electronic or mechanical, including photocopying, recording, or any information storage and retrieval system, without permission in writing from the publisher. Printed in the United States of America

I. Saturated and Unsaturated Glycerides

Animal fats such as tristearin and tripalmitin are solids at room temperature. Vegetable or plant oils such as triolein are liquids at room temperature. Why is tristearin a solid but triolein a liquid?

Tristearin is an ester of glycerol and stearic acid. All the carbon-to-carbon bonds in the hydrocarbon chain of stearic acid are single bonds. These carbons are attached to four atoms, either carbon or hydrogen. This portion of the molecule is said to be **saturated** and cannot undergo addition reactions. Stearic acid, therefore, is a saturated acid. Tristearin and glycerides containing predominantly saturated fatty acids are usually solids.

Triolein is an ester of glycerol and oleic acid. Oleic acid has one carbon-to-carbon double bond. This **unsaturated** bond may undergo addition reactions. Hence, oleic acid is said to be unsaturated. Triolein and glycerides that contain a large percentage of unsaturated fatty acids are usually oils.

How can we determine whether a glyceride is saturated or unsaturated? Let us add a reddish bromine-cyclohexane solution to a solution of oleic acid in cyclohexane. We observe that the resulting mixture is colorless. The disappearance of the red color indicates that the bromine has reacted. If the bromine adds to the carbon-to-carbon double bond in oleic acid, the reaction can be represented by Equation 1. The only product formed is 9,10-dibromooctadecanoic acid.

If the substance tested by the bromine–cyclohexane solution undergoes a substitution reaction, rather than an addition reaction, the evolution of hydrogen bromide will accompany the decoloration. Therefore, we can determine that a glyceride is un-

saturated if the bromine–cyclohexane solution is decolorized and no hydrogen bromide is detected.

An unsaturated glyceride can also be saturated by a process called catalytic hydrogenation. Hydrogen gas, under pressure and at an elevated temperature in the presence of a catalyst, adds to the unsaturated centers in the glyceride. The plant or vegetable oil becomes saturated and is converted into a solid fat. This commercial process, called **hardening**, is used to form solid food shortenings such as Crisco and Spry and foodstuffs such as margarine and peanut butter.

II. Saponification

A fat or an oil can react with either sodium or potassium hydroxide and form soap. We say that the fat, a glyceride, is saponified. The word "saponify" is derived from the Latin *sapo* (soap) and *facere* (to make). Saponification, a soap-making process, is one of the oldest reactions of applied chemistry and was reported by Pliny in the first century A.D. Before soap preparation was understood on a molecular level, a fat or oil was boiled for several hours with potassium hydroxide from terrestrial plant ash, called potash, or with sodium hydroxide from marine plant ash. After cooling, the solid soap was separated from the layer of glycerol and the water layer.

In the saponification reaction, one mole of triglyceride reacts with three moles of hydroxide ions to form one mole of glycerol and three moles of fatty acid anions. When the glyceryloleopalmitostearate reacts with three moles of aqueous sodium hydroxide, one mole each of glycerol, palmitate, stearate, and oleate are formed, as shown in Equation 2.

$$CH_3-(CH_2)_7-CH=CH-(CH_2)_7-\overset{\overset{\displaystyle O}{\|}}{C}-OH \xrightarrow[CH_2Cl_2]{Br_2} CH_3-(CH_2)_7-CHBr-CHBr-(CH_2)_7-\overset{\overset{\displaystyle O}{\|}}{C}-OH$$

9- octadecanoic acid, oleic acid   9,10-dibromooctadecanoic acid   (Eq. 1)

a mixed glyceride
glyceryl oleopalmitostearate

glycerol   palmitate   oleate   stearate   (Eq. 2)

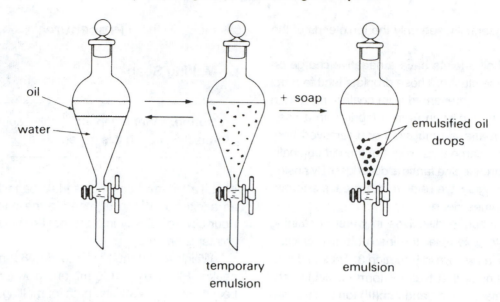

Figure 2   *Emulsion formation*

Almost all soaps are made from beef fat by melting it at a low enough temperature to avoid charring. This rendering process produces tallow. By filtering the tallow through a wire gauze, the proteinaceous material is separated from the tallow, which is then saponified. Today, Proctor and Gamble and other companies hydrolyze the beef tallow with steam, separate the glycerol from the fatty acids, and neutralize the acids with alkali to obtain soap. The use of sodium hydroxide produces a solid soap, while the use of potassium hydroxide forms a liquid soap.

III.   Cleansing Power

The cleansing power of a soap solution is associated with the ability of a small amount of soap to lower the surface tension of water. Thus, the soap solution is able to wet an object more easily. The soap solution will also emulsify the oil or grease and disperse the suspension in the aqueous medium.

Because of water's surface tension, it does not spread on a greasy surface but forms almost spherical drops. The presence of a surface-active agent such as soap lowers the surface tension of water from 72 dyne cm^{-1} to 30 dyne cm^{-1} or less. The water in such a solution can penetrate into the small holes and crevices on the surface of a soiled fabric.

Let us consider what occurs when we add a few drops of lubricating oil to water. The oil has a lower specific gravity than water and floats on the surface. When we shake the oil and water vigorously, the oil forms very small droplets, which disperse uniformly throughout the water. When we stop shaking, the oil rises to the surface, forming a separate layer again.

Now, let us add a few drops of oil to water containing a small amount of soap. When we shake the mixture vigorously, the oil droplets disperse throughout the solution, as before. However, when we allow this mixture to stand, the oil droplets do not readily coalesce to form two layers as before. Instead, we have formed an emulsion, as shown in Figure 2.

What does the soap do to the oil droplets? The answer lies in the interaction between the soap and the oil. When we add a soap such as sodium stearate ($C_{17}H_{35}COONa$) to water, the solution forms stearate ions ($C_{17}H_{35}COO^-$) and sodium ions. The charged end of the stearate ion ($-COO^-$) dissolves in water, while the hydrocarbon end ($C_{17}H_{35}$) interacts with the nonpolar oil. Thus, the hydrocarbon end of the stearate ion is attached to an oil droplet, with the ionic end projecting out from the droplet into the water, as shown in Figure 3. The oil droplet has been emulsified by the soap solution. If we look at the emulsified oil droplet in the

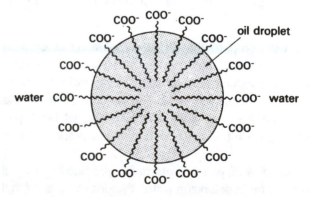

Figure 3   *An oil droplet emulsified by sodium stearate*

solution, we essentially see only the ionic ends of the soap ions.

Emulsified oil driplets have a negative charge on the surface of the cluster. These droplets tend to repel each other and are prevented from coalescing. When we remove a grease stain from a fabric with a soap solution, the grease is emulsified and removed from the fabric. At the same time, some of the dirt originally sticking to the grease and fabric is dislodged. By rinsing the fabric thoroughly, we remove the grease and dirt. The fabric becomes clean.

If sodium stearate dissolves in water containing calcium ions, for example, the insoluble salt calcium stearate forms, as shown in Equation 3. Thus, a disadvantage of soaps is that they do form insoluble salts with calcium, magnesium, and iron(III) ions, which are present in hard water. Such insoluble salts form a scum with soapy water, commonly seen as bathtub rings.

$$2\ C_{17}H_{35}COO^-(aq) + Ca^{2+}(aq) \rightarrow$$
$$(C_{17}H_{35}COO^-)_2Ca^{2+}(s) \quad \text{(Eq. 3)}$$

On the other hand, a synthetic detergent such as sodium lauryl sulfate, $CH_3(CH_2)_{11}OSO_3^-\ Na^+$, does not form insoluble calcium, magnesium, and iron(III) salts. These detergents have the same emulsifying properties as soaps and work very well as cleaning agents in hard water because they do not form a scum.

In this experiment, you will dissolve olive oil in melted fat, to which you will add a warm solution of sodium hydroxide. When the mixture becomes viscous, you will pour it into a mold to age. The glycerol will remain in the soap, imparting useful properties. During aging, carbon dioxide from the air will react with the excess sodium hydroxide in the soap, forming sodium carbonate, as shown in Equation 4.

$$2\ NaOH(s) + CO_2(g) \rightarrow Na_2CO_3(s) + H_2O(l) \quad \text{(Eq. 4)}$$

You will compare the effect of solutions of calcium chloride, magnesium chloride, and ferric chloride on a solution of your soap and on a solution of a synthetic detergent. Then you will determine the ability of the soap and the detergent to emulsify olive oil. Next, you will determine the acidity of a specially prepared soap solution.

In addition, you will determine the solubility of an oil and of a fat in deionized water, cyclohexane, and ethyl alcohol. You will also determine the degree of unsaturation of olive oil and of a fat.

Procedure

I. Making Soap

> **Caution:** Wear departmentally approved eye protection during this experiment.

Weigh a 150-mL beaker on a triple-beam balance. Add an additional 30.0 g of mass to the balance beams. Pour olive oil slowly into the beaker until the balance pointer is on zero.

Weigh a 250-mL beaker. Add 68.0 g of fat to the beaker. Place about 100 mL of tap water in a 600-mL beaker, to use as a water bath for melting the fat. Place the beaker with the fat in the water bath. Heat the water bath on a hot plate until all the fat has melted. Remove the beaker containing the melted fat from the water bath. While stirring, slowly add the olive oil to the melted fat. Set this mixture in an ice bath to cool to 40 °C.

> **Caution:** Handle solid sodium hydroxide and sodium hydroxide solutions with care. Sodium hydroxide is very caustic and hygroscopic. Avoid skin contact. If you spill any solid sodium hydroxide or solutions, call your laboratory instructor immediately.

Weigh a clean, dry 150-mL beaker. Weigh 12.5 g of sodium hydroxide in the beaker. While stirring rapidly, add 32.0 mL of deionized water to the solid. Continue stirring until the solid dissolves completely.

If you use sodium hydroxide pellets, make sure you move each pellet with the stirring rod. If a pellet sticks to the beaker, it will dissolve slowly, even with rapid stirring.

> **Caution:** Handle the hot beaker with care. The dissolution of sodium hydroxide is highly exothermic. The temperature of this solution can reach 80 °C or higher.

After all the sodium hydroxide has dissolved, place the beaker containing the solution in an ice bath and cool to 40 °C.

Allow the liquid mixture of oil and fat and the sodium hydroxide solution to reach a temperature of 40 °C. Carefully and slowly pour the sodium hydroxide solution into the liquid mixture. Continue stirring to mix these solutions thoroughly. Immediately add 40–50 drops of soap perfume oil and one or two drops of coloring to the mixture. Stir thoroughly.

Continue to stir until the mixture has the consistency of thick pea soup.

Note: Do not wait too long before pouring the mixture from the beaker to the mold. The mixture will become too viscous to be poured.

Remove the stirring rod. When the mixture drops from the rod onto the surface and traces a pattern on the surface of the mixture, pour the soap into the mold.

Do not allow the mixture to cool too rapidly, since saponification is not yet complete. Immediately cover the mold by inverting another weighing dish over it. Leave the mold covered until you leave the laboratory. As you leave the laboratory, uncover the mold.

Scrape the soap left in the beaker onto a watch glass to use in Part III of this experiment.

Use hot water and the detergent designated by your laboratory instructor to clean your greasy glassware.

Leave the soap in the mold for at least 24 hours. Remove the soap from the mold by turning the mold upside down onto several layers of paper towels. Lean the soap against the mold so that most of the surface of the soap is exposed to the atmosphere.

Let the soap age for 10 days. Remove the thin, white, powdery layer of sodium carbonate from the soap by scraping gently with a spatula.

II. Properties of Glycerides

A. Determining Solubility

Record all observations on Data Sheet 1.

Add 5 drops of an oil to 5 mL of each of the following solvents.
  (a) deionized water
  (b) cyclohexane
  (c) ethyl alcohol
Add 0.5 g of a fat to 5 mL of each of the same three solvents.

B. Determining the Degree of Unsaturation

Dissolve 10 drops of oil in 5 mL of cyclohexane.

Caution: Avoid skin contact with the bromine solution. Do not inhale the fumes. If you spill any solution, call your laboratory instructor immediately.

Add, drop by drop, a solution of bromine in cyclohexane to the oil solution until a faint coloration persists. Record on Data Sheet 1 the number of drops of bromine solution that you add to the oil solution to get the persistent coloration.

Repeat this procedure with 0.1 g of fat.

III. Comparing the Properties of Soaps and Detergents

A. Preparing the Solutions

Solution 1:   Heat 100 mL of deionized water to 80 °C. Dissolve 1 g of your un-aged soap in the hot water.

Solution 2:   Heat 100 mL of deionized water to 80 °C. Dissolve 0.5 g of the synthetic detergent in the hot water.

B. Determining the Effects of Water Hardness on Soaps and Detergents

1. *Testing your soap (Solution 1)*
Record all observations on Data Sheet 2.
Label four (4) 15 × 125-mm test tubes A, B, C, and D.
Place 5 mL of Solution 1 in each test tube.
  (a)   Add 3 drops of 10% calcium chloride solution to test tube A. Mix thoroughly.
  (b)   Add 3 drops of 10% magnesium chloride solution to test tube B. Mix thoroughly.
  (c)   Add 3 drops of 10% ferric chloride solution to test tube C. Mix thoroughly.
  (d)   Add 5 mL of tap water to test tube D. Mix thoroughly.

2. *Testing the detergent (Solution 2)*
Record all observations on Data Sheet 2.
Label four (4) 15 × 125-mm test tubes E, F, G, and H.
Place 5 mL of Solution 2 in each test tube.
  (a)   Add 3 drops of 10% calcium chloride solution to test tube E. Mix thoroughly.
  (b)   Add 3 drops of 10% magnesium chloride solution to test tube F. Mix thoroughly.
  (c)   Add 3 drops of 10% ferric chloride solution to test tube G. Mix thoroughly.
  (d)   Add 5 mL of tap water to test tube H. Mix thoroughly.

3.   Determining emulsifying action

Record all observations on Data Sheet 2.

Place 5 mL of Solution 1 in a 15 × 125-mm test tube. Add 4 drops of oil. Shake vigorously. Let stand for 10 minutes.

Repeat this procedure with 5 mL of Solution 2.

4.   Determining acidity

Record all observations on Data Sheet 2.

Place 5 mL of the soap solution (Solution 3), prepared by your laboratory instructor, in a 15 × 125-mm test tube. Add 3 drops of phenolphthalein solution to the soap solution.

Post-Laboratory Questions

(Use the spaces provided for the answers and additional paper if necessary.)

1. In preparing your soap, you dissolved 12.5 g of sodium hydroxide in 32.0 mL of deionized water. Your product was a solid cake of soap. What happened to the water you added when preparing your soap during the course of the experiment?

bromine solution to Compound B, the bromine color remains after you add 1 drop of the test solution. Which substance, Compound A or Compound B, is more likely to be an oil? Explain concisely.

4. Linseed oil has been added to paint to serve as a drying oil. Explain.

2. What happened to the glycerol formed during the saponification reaction?

5. Why does butter have a greater possibility of becoming rancid than margarine? Explain.

3. Suppose you are assigned two unknowns, Compound A and Compound B. One unknown is a fat, and the other is an oil. You dissolve a small amount of Compound A in cyclohexane. After you add 4 drops of a cyclohexane solution containing bromine, the reaction mixture retains the bromine color. When you add the

Data Sheet 1

II. Properties of Glycerides

A. Determining Solubility

solvent	oil	fat
deionized water		
cyclohexane		
ethyl alcohol		

Comments:

B. Determining the Degree of Unsaturation

glyceride	number of drops of bromine solution
oil	
fat	

Comments:

　12 • Preparing Soap; Determining Its Properties

Data Sheet 2

III. Properties of Soaps and Detergents

ionic solution	your soap (Solution 1)	detergent (Solution 2)
calcium chloride		
magnesium chloride		
ferric chloride		
tap water		

Comments:

Emulsifying action

soap (Solution 1)	detergent (Solution 2)

Comments:

Acidity

soap (Solution 3)

Comments:

Pre-Laboratory Assignment

1. Write a formula representing corn oil, a triglyceride that contains palmitic, oleic, and linoleic acids.

2. Write equations for the reaction of the glyceride written in Question 1 above with each of the following:
(a)   an aqueous solution of sodium hydroxide;

(b)   hydrogen in the presence of a catalyst and at elevated temperature and pressure;

(c)   a 5% solution of bromine in cyclohexane.

3. Write a chemical equation for the reaction of an aqueous solution of sodium stearate with each of the following reagents:
(a)   a 10% solution of magnesium chloride:

(b)   dilute hydrochloric acid.

4. Do the triglycerides in olive oil contain mainly saturated or unsaturated fatty acids? Explain concisely.

5. If a solution of Epson salts prepared by dissolving magnesium sulfate heptahydrate in water is added to a solution containing sodium lauryl sulfate, will a precipitate form? Explain clearly.

A Sequence of Chemical Reactions

prepared by **Guy B. Homman**, late of Emporia Kansas State University

---

Purpose of the Experiment

Determine the percent recovery of copper from a sequence of chemical reactions involving copper and copper species.

Background Information

One interesting study of a group of chemical systems involves starting with a substance and transforming it into another substance, then repeating this process several times, ending eventually with the starting material. The efficiency of the overall transformation can be evaluated in terms of the percent recovery of the starting material.

In this experiment, copper metal is transformed into various copper species by a series of sequential chemical reactions. The chemical transformations are shown in Equation (1).

$$Cu(s) \xrightarrow{\text{Step 1}} Cu^{2+}(aq) \xrightarrow{\text{Step 2}} Cu(OH)_2 \xrightarrow{\text{Step 3}}$$

$$CuO(s) \xrightarrow{\text{Step 4}} Cu^{2+}(aq) \xrightarrow{\text{Step 5}} Cu(s) \quad \text{(Eq. 1)}$$

Step 1 involves an **oxidation–reduction reaction**, shown in Equation (2), where copper metal is oxidized by the nitrate ion in acid solution to copper(II) ion, which is soluble in acid solution.

$$Cu(s) + 4\,H_3O^+(aq) + 2\,NO_3^-(aq) \rightarrow$$
$$Cu^{2+}(aq) + 2\,NO_2(g) + 6\,H_2O(l) \quad \text{(Eq. 2)}$$

At the same time, the nitrate ion is reduced by the copper to a brown gas, nitrogen dioxide. The resulting solution is blue because of the presence of hydrated copper(II) ion.

In Step 2, sodium hydroxide solution is slowly added to the acid solution of copper(II) ion. At first, the sodium hydroxide reacts with the hydronium ion from the nitric acid, as shown in Equation (3).

$$H_3O^+(aq) + OH^-(aq) \rightarrow 2\,H_2O(l) \quad \text{(Eq. 3)}$$

This reaction is a **neutralization reaction**. When the nitric acid is completely neutralized, the hydroxide ion will react with the copper(II) ion, as seen in Equation (4), forming an insoluble precipitate, which for convenience is represented by $Cu(OH)_2$.

$$Cu^{2+}(aq) + 2\,OH^-(aq) \rightarrow Cu(OH)_2(s) \quad \text{(Eq. 4)}$$

This reaction is a **double replacement reaction**, which essentially goes to completion because of the formation of the insoluble precipitate.

In the laboratory, the progress of these two reactions can be followed by the use of red litmus paper. When the sodium hydroxide solution is first added, the solution will have no effect on red litmus paper. Likewise, as long as there is any copper(II) ion in the solution, red litmus paper will not change color. When the copper(II) ion has completely reacted, the solution will

---

Copyright © 1983 by Chemical Education Resources, Inc., P.O. Box 357, 220 S. Railroad, Palmyra, Pennsylvania 17078
No part of this laboratory program may be reproduced or transmitted in any form or by any means, electronic or mechanical, including photocopying, recording, or any information storage and retrieval system, without permission in writing from the publisher. Printed in the United States of America

become basic because of excess hydroxide ion. At this point, a drop of the solution will turn red litmus blue, indicating that the precipitation is complete.

When the solution containing the insoluble precipitate is heated in Step 3, a **decomposition reaction** occurs, as shown in Equation (5), with the blue precipitate being changed into a more dense, black, crystalline copper(II) oxide.

$$Cu(OH)_2(s) \rightarrow CuO(s) + H_2O(l) \quad (Eq. 5)$$

Addition of sulfuric acid to the copper(II) oxide (which is soluble in acid solution) in Step 4 forms copper(II) ion. This **double replacement reaction** is shown in Equation (6).

$$CuO(s) + 2 H_3O^+(aq) \rightarrow Cu^{2+}(aq) + 3 H_2O(l) \quad (Eq. 6)$$

In Step 5, zinc metal is added to the acid solution of copper(II) ion. The zinc reduces the copper(II) ion, forming copper and zinc ion. This **oxidation–reduction reaction** is shown in Equation (7).

$$Cu^{2+}(aq) + Zn(s) \rightarrow Cu(s) + Zn^{2+}(aq) \quad (Eq. 7)$$

The completion of this reaction is noted by the disappearance of the blue color from the solution. At the same time, some of the zinc reacts with the sulfuric acid solution, as shown in Equation (8), forming zinc ion and hydrogen gas. In this reaction, zinc is oxidized by hydrogen ion.

$$Zn(s) + 2 H_3O^+(aq) \rightarrow Zn^{2+}(aq) + H_2(g) + 2 H_2O$$
$$(Eq. 8)$$

From the mass of copper used in Step 1 and the mass of copper obtained in Step 5, the percent recovery of copper from the sequential reactions can be calculated.

$$\frac{percent}{recovered} = \left(\frac{mass\ of\ Cu\ obtained\ in\ Step\ 5}{mass\ of\ Cu\ used\ in\ Step\ 1} \right)(100)$$

Procedure

> **Caution:** Wear departmentally approved eye-protection while doing this experiment.

Step 1

1. Weigh a tared watch glass to the nearest centigram and record on the Data Sheet the mass of the watch glass.

2. Weigh to the nearest centigram a 0.35 g to 0.40 g sample of copper wire or turnings on the watch glass and record on the Data Sheet the mass of the watch glass and copper. If copper turnings are used, roll the sample into a compact ball.

3. Place the copper sample in a clean 250-mL beaker.

> **Caution:** Concentrated nitric acid should be handled with care. Avoid breathing the fumes.

4. In a fume hood, carefully add from 2.5 to 3.0 mL of concentrated nitric acid from a 10-mL graduated cylinder, a drop at a time, to the copper.

5. Support the beaker and contents on an asbestos-centered wire gauze on an iron ring, mounted on a ring stand in the fume head.

6. Place a Bunsen burner under the ring, turn on the gas, and light the gas.

7. Heat the beaker and contents gently, but do not boil, until the copper is dissolved.

8. Dilute the resulting solution with 10 mL of distilled water before beginning Step 2.

Step 2

> **Caution:** Handle sodium hydroxide with care. Avoid contact with skin.

1. Add 6M sodium hydroxide to the acidic copper(II) ion solution from Step 1 a little at a time, with stirring.

2. During the addition of sodium hydroxide solution, test the copper(II) ion solution by dipping a clean glass stirring rod into the solution and then touching the drop of solution on the rod to a piece of red litmus paper that has been placed on a paper towel.

3. Stop the addition of sodium hydroxide solution when the test indicates an alkaline reaction.

4. Retain the beaker and contents for use in Step 3.

Step 3

1. Increase the volume of the solution from Step 2 to about 100 mL by adding distilled water.

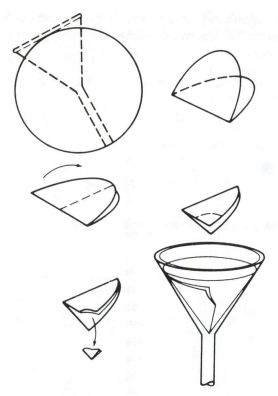

Figure 1 *Folding a piece of filter paper*

2. Place the beaker and contents on an asbestos-centered wire gauze on an iron ring, mounted on a ring stand.

3. Boil the contents of the beaker gently, with stirring, for 4 min.

4. Prepare a piece of qualitative filter paper as shown in Figure 1 and place the paper in a filtering funnel mounted in a utility clamp attached to a ring stand.

5. Add distilled water from a wash bottle to the funnel and press the edges of the filter paper against the funnel, so that the stem of the funnel fills with a column of water.

6. Filter the suspended solid and discard the filtrate, which should be colorless.

7. While the filtering is being done, heat with the flame of a Bunsen burner about 40 mL of distilled water in a 150-mL beaker, placed on an asbestos-centered wire gauze on a ring attached to a ring stand.

8. Transfer the last traces of the solid into the funnel, using a stream of distilled water from a wash bottle.

9. Wash the residue on the filter paper three times with 5-mL portions of hot distilled water. Allow each portion of water to drain through the solid before adding the next portion.

10. Retain the solid for use in Step 4.

11. Discard the filtrate according to the directions of your laboratory instructor.

Step 4

> *Caution:* Sulfuric acid should be handled with care. Avoid contact with the skin.

1. Dissolve the solid copper(II) oxide by adding about 10 mL of 3M sulfuric acid directly to the filter paper containing the residue from Step 3. Allow the acid solution to drain through the funnel into a 250-mL beaker. If the solid does not completely dissolve, pour the liquid from the beaker into the funnel again. Repeat this procedure until the solid is completely dissolved.

2. Wash the empty filter paper with cold distilled water and collect the washings in the beaker containing the acid solution.

3. Retain the acid solution for use in Step 5.

Step 5

1. Add 2 g of zinc metal to the acidic copper(II) solution. The rate of reaction will depend on surface area.

 If any blue color remains after the zinc reacts, add 1 g of zinc. Dissolve excess zinc by adding a small amount of H_2SO_4.

2. Allow the precipitated copper to settle and decant the supernatant liquid from the solid by carefully pouring the liquid so that the solid remains in the beaker. Discard the liquid, but be careful not to lose any solid. Some liquid will remain with the solid in this process.

3. Wash the solid copper three times by adding 20 mL of distilled water, stirring, and allowing the solid to settle. Each time, carefully pour out the liquid so that the solid remains in the beaker.

4. Weigh to the nearest centigram a clean, dry evaporating dish and record on the Data Sheet the mass of the dish.

5. Transfer the solid copper into the weighed evaporating dish, using a stream of distilled water from the wash bottle to aid in making a quantitative transfer.

6. Carefully pour out most of the water from the solid in the evaporating dish.

7. Place a 250-mL beaker three-quarters full of water on an asbestos-centered wire gauze on an iron ring mounted on a ring stand.

8. Place the evaporating dish on top of the beaker and heat the water to boiling to dry the copper.

9. When the copper is dry, carefully remove the hot evaporating dish with a towel.

10. With the towel, wipe the outside of the dish dry.

11. Allow the evaporating dish and contents to cool to room temperature.

12. Weigh the evaporating dish and contents to the nearest centigram and record the mass on the Data Sheet.

13. Submit your product as indicated by your laboratory instructor.

14. Discard the filtrate according to the directions of your laboratory instructor.

Calculations

Calculate the percent copper recovered.

$$\text{percent recovered} = \left(\frac{\text{mass of Cu obtained in Step 5}}{\text{mass of Cu used in Step 1}} \right)(100)$$

Post-Laboratory Questions

(Use the spaces provided for the answers and additional paper if necessary.)

1. A 0.350-g sample of copper metal was used to carry out the sequence of reactions described in this experiment.

(a) How many moles of copper metal is this?

_____
answer

(b) How many moles of hydrated copper(II) ion would be produced by dissolving the copper metal in nitric acid solution?

_____
answer

(c) How many moles of $Cu(OH)_2(s)$ would be formed?

_____
answer

(d) How many moles of $CuO(s)$ would be formed?

_____
answer

(e) How many moles of zinc metal would be required to completely reduce the copper(II) ion obtained by dissolution of $CuO(s)$ in sulfuric acid?

_____
answer

2. How does this experiment illustrate the conservation of mass?

3. The mole concept is central to a study of chemistry. Does this experiment suggest that, like mass, moles are conserved? Explain.

4. How many atoms of copper are in a 0.350-g sample of copper metal? How many atoms of copper should be present in the CuO formed in this experiment? How many atoms of copper should be recovered following the sequential reactions?

_____
answer

_____
answer

_____
answer

5. Dalton's atomic theory assumed that atoms could be neither created nor destroyed. Does the Dalton theory account for conservation of mass in the sequential reactions of this experiment? Explain, relying on the answers given to Questions 1, 2, 3, and 4.

name   section   date

Data Sheet

Step 1

mass of watch glass and copper, g   _____

mass of watch glass, g   _____

mass of copper used, g   _____

Step 5

mass of evaporating dish and copper, g   _____

mass of evaporating dish, g   _____

mass of copper obtained, g   _____

percent of copper recovered   _____

Pre-Laboratory Assignment

1. **(a)** What is the color of the solution obtained when a concentrated nitric acid solution is added to copper metal? What species is responsible for the color?

(b) If the solution formed is tested with red litmus paper, what will be observed?

(c) If sodium hydroxide is added to the solution, what color will red litmus paper show if copper(II) ion is present?

(d) What color will red litmus paper show when no copper(II) ion is present? Explain. What should be the color of the solution when the litmus paper changes color?

(e) What is the residue in the solution when sufficient sodium hydroxide solution has been added to cause a color change in the litmus paper?

(f) After heating this residue, what compound remains?

(g) If 3*M* sulfuric acid solution is added to the compound, what will be the resulting color of the solution? What species is responsible for the color?

_____
answer

(h) If sufficient zinc metal is added to the solution, what will be the resulting color of the solution? Explain.

2. A student starting with 0.351 g of copper metal carried out the steps of this experiment. If, after the last step, 0.346 g of copper metal was recovered, calculate the percent recovery of copper.

_____
answer

Stoichiometry of the Reaction of Magnesium with Hydrochloric Acid

prepared by **H.A. Neidig**, Lebanon Valley College
and **J.N. Spencer**, Franklin and Marshall College

Purpose of this Experiment

Determine the stoichiometry of the reaction of magnesium with hydrochloric acid.

Background Information

The nature of the products formed when an acid reacts with a metal will depend on the reduction potential of the metal. If the reduction potential is less than zero, the products formed will be the cation of the metal and hydrogen gas (H_2). The reaction of such a metal (M) with hydrochloric acid (HCl) is represented by Equation 1.

$$M(s) + 2\,H_3O^+(aq) \rightarrow M^{2+}(aq) + H_2(g) + 2\,H_2O(l) \quad \text{(Eq. 1)}$$

In this experiment, you will react a measured mass of magnesium (Mg) with a specified excess of standard HCl solution. The H_2 released during the reaction will displace water from a container. From the volume of displaced water, you will calculate the number of moles of H_2 formed. From the titration of the final reaction mixture with a standard sodium hydroxide (NaOH) solution, you will calculate the number of moles of unreacted HCl in the reaction mixture. You will use these data to calculate the number of moles of HCl that reacted. Then, you will find the ratio of the number of moles of HCl reacting to the number of moles of Mg reacting. From this molar ratio, you will write an equation showing the stoichiometry of the reaction.

Procedure

Chemical Alert

1 M hydrochloric acid—toxic and corrosive
magnesium—flammable
0.5 M sodium hydroxide—toxic and corrosive

Caution: Wear departmentally approved eye protection while doing this experiment.

I. Weighing the Mg

Note: In the next step, do not handle the vial directly with your fingers or hands. Wrap a strip of paper tissue or paper toweling around the vial, so that you can handle the vial by grasping the ends of the paper tissue strip, as shown in Figure 1.

---

1. Measure to the nearest tenth of a milligram (0.0001 g) the mass of a clean, dry vial. Record this mass on your Data Sheet. Place a sample of Mg turnings within the range of 0.0800 g to 0.1200 g in the vial.

Copyright © 1989 by Chemical Education Resources, Inc., P.O. Box 357, 220 S. Railroad, Palmyra, Pennsylvania 17078
No part of this laboratory program may be reproduced or transmitted in any form or by any means, electronic or mechanical, including photocopying, recording, or any information storage and retrieval system, without permission in writing from the publisher. Printed in the United States of America

Figure 1   *Handling a vile*

Measure to the nearest tenth of a milligram the mass of the turnings and vial. Record this mass on your Data Sheet.

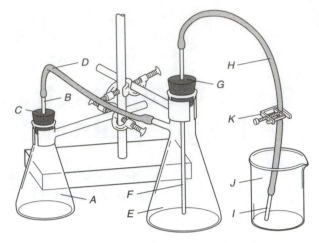

Figure 2   *Apparatus for the experiment*

II. Assembling the Apparatus

> **Note:** The letters in parentheses refer to those appearing in Figure 2.

2.   Arrange an apparatus as shown in Figure 2, using a 125-mL Erlenmeyer flask (*A*) as the reaction flask.

> **Note:** Your laboratory instructor will give you directions for inserting glass tubing into rubber stoppers.

Carefully insert a fire-polished 6-cm length of glass tubing (*B*) into a one-hole rubber stopper (*C*). Adjust the tubing (*B*) so that it extends just below the bottom of the stopper. Attach one end of a 25-cm piece of plastic tubing (*D*) to the end of the glass tubing that extends above the stopper. Insert the stopper assembly into the 125-mL Erlenmeyer flask (*A*).

Attach the other end of the plastic tubing (*D*) to the side arm of a 500-mL filter flask (*E*). Carefully insert a fire-polished 25-cm length of glass tubing (*F*) into a one-hole rubber stopper (*G*). Adjust the tubing (*F*) so that when this stopper assembly is placed in the filter flask, the bottom of the tubing (*F*) reaches almost to the bottom of the flask. Be sure the tubing does not touch the bottom of the flask. Attach one end of a 33-cm length of plastic tubing (*H*) to the end of the glass tubing that extends above the top of the stopper (*G*). Insert a fire-polished 5-cm length of glass tubing (*I*) into the other end of the plastic tubing (*H*). Insert this stopper assembly into the filter flask (*E*). Place the end of the glass tubing (*I*) in a 150-mL beaker (*J*).

> **Caution:** 1*M* hydrochloric acid is a corrosive, toxic solution that can cause skin irritation. Prevent contact with your eyes, skin, and clothing. Avoid inhaling the vapors and ingesting the compound.
>
> If any acid is spilled, clean it up following the directions of your laboratory instructor.

3.   Remove the stopper assembly from the 125-mL Erlenmeyer flask. Dispense from a buret 25.00 mL of standard HCl solution into the clean, dry 125-mL Erlenmeyer flask. Record the exact molarity and volume of the HCl solution on your Data Sheet.

> **Note:** In Step 4, handle the Erlenmeyer flask with extreme care so that the vial remains in an upright position. Avoid any direct contact between the Mg and the HCl solution, as shown in Figure 3.

4.   Carefully insert the glass vial containing the Mg into the Erlenmeyer flask containing the HCl solution. Keep the vial in an upright position so that there will be no contact between the HCl solution and the Mg.

5.   Replace the one-hole stopper (*C*) containing the glass tubing and the plastic tubing into the Erlenmeyer flask. Carefully clamp the flask in position in the assembled apparatus.

6.   Remove the stopper (*G*) from the 500-mL filter flask, and fill the flask three-quarters full with water. Add 25 mL of water to the 150-mL beaker.

7.   Remove the complete stopper assembly from the filter flask. Invert the assembly and place your finger

Figure 3 *Positioning of vial in the Erlenmeyer flask*

firmly over the end of the glass tubing (*I*) that extends into the beaker (*J*). Completely fill the glass and plastic tubing with water.

> **Note:** This filled piece of tubing will function as a siphon for transferring water back and forth from the filter flask to the 150-mL beaker.

Hold the end of the filled glass and plastic tubing (*I*) at a height lower than the water level in the filter flask. Without removing your finger from the end of the tubing, reinsert the stopper assembly into the flask. Lower the covered end of the tubing (*I*) beneath the surface of the water in the beaker, and remove your finger. Be sure that the end of the glass tubing (*I*) remains below the water level in the beaker.

Remove any air bubbles from the tubing by raising and lowering the beaker so that the water flows back and forth through the tubing between the filter flask and the beaker. Avoid transferring all of the water from the beaker, because if this happens, the siphon will not continue to function.

8. check the apparatus to make sure that it is airtight by raising the beaker so that there is a large difference between the water level in the beaker and the water level in the filter flask. If the apparatus is airtight, the water levels in the flask and in the beaker will remain unchanged. If the water levels change, find the leak in the apparatus. Recheck the apparatus after fixing the leak.

9. Attach a pinch clamp (*K*) to the plastic tubing (*H*). Close the clamp to prevent the flow of water from the tubing. Replace the 150-mL beaker containing the water with a clean, dry 150-mL beaker that has been weighed to the nearest centigram. Ask you laboratory

instructor to check your apparatus before you proceed with this experiment.

III. Reacting Mg with HCl Solution

> **Caution:** Hydrogen is a flammable gas. Extinguish all flames in the laboratory before beginning Step 10.

10. Release the clamp holding the Erlenmeyer flask and gently shake the flask so that the vial containing the Mg overturns and the HCl solution comes in contact with the Mg. Immediately release the pinch clamp (*K*) from the plastic tubing (*H*), allowing the displaced water to flow into the beaker.

11. Carefully raise the beaker so that the water level in the beaker is approximately equal to the water level in the filter flask. Continue to swirl the Erlenmeyer flask until all of the Mg has been removed from the vial. Reclamp the Erlenmeyer flask and gently shake the reaction flask from time to time until all of the Mg has reacted, which will require 10–15 min.

12. Allow the apparatus to stand for 5 min after all of the Mg has reacted and there is no further evolution of gas. Because the reaction is exothermic, the apparatus and contents should be allowed to cool to room temperature. Equalize the pressure in the apparatus by raising and lowering the beaker until the levels of the water in the beaker and in the filter flask are the same. Reattach the pinch clamp (*K*) to the plastic tubing (*H*). Close the clamp to prevent the flow of water from the tubing. Measure the mass of the 150-mL beaker and displaced water to the nearest centigram (0.01 g). Record this mass on your Data Sheet.

13. Measure and record on your Data Sheet the temperature in the laboratory.

14. Measure and record on your Data Sheet the barometric pressure in the laboratory.

15. Obtain from your laboratory instructor the density and vapor pressure of water at the temperature corresponding to that of the laboratory. Record these data on your Data Sheet.

IV. Titrating the Reaction Mixture

16. Clean a 50-mL buret, following the procedure described by your laboratory instructor.

17. Obtain about 60 mL of NaOH solution in a 150-mL beaker. Record on your Data Sheet the molarity of the NaOH to four significant digits.

Caution: 0.5M NaOH is a corrosive, toxic solution that can cause burns. Prevent contact with your eyes, skin, and clothing. Avoid ingesting the compound.

    If any NaOH solution is spilled, clean it up following the directions of your laboratory instructor.

---

18. Rinse the buret thoroughly with a 5-mL portion of the standard NaOH solution. Be sure that the rinse solution comes into contact with as much of the inner surface of the buret as possible. Drain the rinse solution through the tip of the buret and discard it, following the directions of your laboratory instructor. Repeat this procedure with two additional 5-mL portions of the NaOH solution.

19. Close the buret stopcock. Fill the buret to a level above the zero calibration mark with the NaOH solution. Allow any air bubbles trapped in the solution to rise to the surface. Briefly open the stopcock to be sure that the buret tip is filled with solution and that there are no air bubbles in the tip. If the meniscus is above the zero calibration mark, open the stopcock and drain the solution through the buret tip until the level reaches the calibrated portion of the buret. The level does not have to align exactly with the zero calibration mark as long as the initial volume is read and recorded. Touch the buret tip to a wet glass surface to remove the hanging droplet. Read the buret to the nearest 0.01 mL. Record this initial buret reading on your Data Sheet.

20. Disconnect the Erlenmeyer flask containing the reaction mixture from the assembly. Add four drops of phenolphthalein indicator solution to the flask. Place this flask under the buret and lower the buret until the tip is below the lip of the flask, as shown in Figure 4.

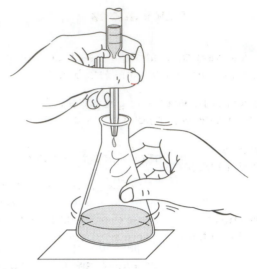

Figure 4 *Positioning the buret for titration*

21. Continuously swirl the flask with one hand and control the stopcock with your other hand. Slowly add the NaOH solution to that in the flask. The phenolphthalein indicator will produce flashes of pink color as the titration approaches the end point. As the color persists for longer periods, decrease the amount of NaOH solution being added. Continue the titration by adding NaOH solution dropwise from the buret. When the color appears faintly throughout the solution and persists for 15 s, even after swirling, the titration has reached the end point.

22. Read the buret to the nearest 0.01 mL. Record this final buret reading on your Data Sheet.

23. Do a second determination and, if time permits, a third determination.

Note: Before leaving the laboratory, obtain from your laboratory instructor the density and vapor pressure of water at the laboratory temperature.

Caution: Wash your hands thoroughly with soap or detergent before leaving the laboratory.

Calculations

Do the following calculations for each determination and record the results on your Data Sheet.

1. Calculate the volume of gas collected during the reaction from the volume of water displaced.

$$\begin{array}{l}\text{mass of} \\ \text{displaced water, g}\end{array} = (\text{mass of beaker + water})$$
$$\qquad\qquad - (\text{mass of beaker}) \quad \text{(Eq. 1)}$$

$$\begin{array}{l}\text{volume of} \\ \text{displaced water, mL}\end{array} = \frac{\text{mass of displaced water, g}}{\text{density of water, g mL}^{-1}}$$

$$= \begin{array}{l}\text{volume of gas} \\ \text{collected, mL}\end{array} \quad \text{(Eq. 2)}$$

2. Calculate the pressure of the H_2 collected.

$$P_{H_2} = P_{bar} - P_{H_2O} \quad \text{(Eq. 3)}$$

P_{bar} is the barometric pressure in the laboratory and P_{H_2O} is the vapor pressure of water at the temperature at which the volume measurement was made.

3. Calculate the number of moles of H_2 collected during the reaction, using the ideal gas equation,

$$n = \frac{PV}{RT} \quad \text{(Eq. 4)}$$

where R is the gas constant, 8.21×10^{-2} L-atm K^{-1} mol^{-1}, T is the final temperature in the reaction flask in Kelvin, P is the pressure of the gas in atmospheres, and V is the volume of the gas in liters.

4. Calculate the number of moles of HCl initially present in the reaction flask.

initial number of moles of HCl =
$\qquad$ (volume of HCl, L) (molarity of HCl, M) (Eq. 5)

5. Calculate the number of moles of Mg reacting.

$$\begin{array}{l}\text{number of} \\ \text{moles of Mg} = \\ \text{reacting}\end{array} \frac{\text{mass of Mg, g}}{\text{gram atomic mass of Mg, g mol}^{-1}}$$
$$\qquad\qquad\qquad\qquad\qquad \text{(Eq. 6)}$$

6. Calculate the volume of NaOH solution used in the titration.

$$\begin{array}{l}\text{volume of NaOH solution} \\ \text{used in the titration, mL}\end{array} =$$
$$\left(\begin{array}{l}\text{final buret} \\ \text{reading, mL}\end{array}\right) - \left(\begin{array}{l}\text{initial buret} \\ \text{reading, mL}\end{array}\right) \quad \text{(Eq. 7)}$$

7. Calculate the number of moles of NaOH solution added.

$$\begin{array}{l}\text{number of moles} \\ \text{of NaOH added}\end{array} = \left(\begin{array}{l}\text{volume of} \\ \text{NaOH solution} \\ \text{used, L}\end{array}\right)\left(\begin{array}{l}\text{molarity} \\ \text{of NaOH} \\ \text{solution, } M\end{array}\right)$$

8. Calculate the number of moles of HCl remaining in the flask at the end of the reaction.

$$\begin{array}{l}\text{number of moles} \\ \text{of HCl remaining}\end{array} = \begin{array}{l}\text{number of moles} \\ \text{of NaOH added}\end{array} \quad \text{(Eq. 8)}$$

9. Calculate the number of moles of HCl reacting.

$$\begin{array}{l}\text{number of moles} \\ \text{of HCl reacting}\end{array} =$$
$$\left(\begin{array}{l}\text{initial number} \\ \text{of moles of HCl} \\ \text{used as a reactant}\end{array}\right) - \left(\begin{array}{l}\text{number of moles of} \\ \text{HCl titrated in the} \\ \text{reaction mixture}\end{array}\right) \quad \text{(Eq. 9)}$$

10. Calculate the ratio of the number of moles of HCl reacting to the number of moles of Mg reacting.

11. Calculate the ratio of the number of moles of H_2 collected to the number of moles of Mg reacting.

12. Repeat calculations 1–11 for your second and third determinations.

13. Calculate the average ratio of the number of moles of H_2 collected to the number of moles of Mg reacting.

14. Write a stoichiometric equation for the reaction of Mg and HCl solution on the basis of the experimental data.

Post-Laboratory Questions

(Use the spaces provided for the answers and additional paper if necessary.)

1. Summarize the results of your experiment below:

number of moles of Mg(s) reacting   _____

number of moles of HCl(aq) reacting   _____

number of moles of H_2(g) generated   _____

(1)   Compare the ratio of number of moles of Mg reacting to the number of moles of HCl consumed to the ratio of the number of moles of Mg reacting to the number of moles of H_2 produced.

(2)   Explain the similarity or difference between the two molar ratios in terms of the stoichiometry of the reaction.

(3)   On the basis of possible errors in the experimental procedure, which of the experimental values is the least reliable? Briefly explain.

2. A student doing this experiment had some problems with the procedure, as described below. Explain what effect each problem would have had on the magnitude of the results calculated.

(1)   The vial containing the Mg tipped over when the student first placed it in the Erlenmeyer flask. The student was able to return it to an upright position almost immediately, and the student continued the experiment.

(2) The student neglected to open the pinch clamp on the plastic tubing between the filter flask and the beaker when the Mg came in contact with the HCl solution. After the stopper popped out of the Erlenmeyer flask, the student quickly replaced the stopper, opened the clamp, and proceeded with the experiment.

(4) The student forgot to add indicator to the titration mixture until 12.50 mL of titrant had been added. When the indicator was finally added, the solution was colorless.

(3) The student began titrating the unreacted HCl with standard NaOH solution and realized that there was a large air bubble in the buret tip. The bubble came out as the first few milliliters of titrant were added.

Data Sheet

	determination		
	1	2	3
mass of Mg and vial, g	_____	_____	_____
mass of vial, g	_____	_____	_____
mass of Mg, g	_____	_____	_____
number of moles of Mg used as reactant	_____	_____	_____
volume of HCl added to 125-mL Erlenmeyer flask, mL	_____	_____	_____
molarity of HCl, M	_____	_____	_____
number of moles of HCl used as reactant	_____	_____	_____
mass of beaker and displaced water, g	_____	_____	_____
mass of beaker, g	_____	_____	_____
mass of displaced water, g	_____	_____	_____
temperature in laboratory, °C	_____	_____	_____
barometric pressure in laboratory, torr	_____	_____	_____
volume of H_2 collected, mL	_____	_____	_____
pressure of H_2 collected, torr	_____	_____	_____
number of moles of H_2 collected	_____	_____	_____
molarity of NaOH solution, M	_____	_____	_____
volume of NaOH solution used, mL	_____	_____	_____
final buret reading, mL	_____	_____	_____
initial buret reading, mL	_____	_____	_____
volume of NaOH solution added, mL	_____	_____	_____

number of moles of unreacted HCl
   in reaction mixture   _____   _____   _____

number of moles of HCl reacting   _____   _____   _____

ratio of number of moles of HCl reacting
   to number of moles of Mg reacting   _____   _____   _____

ratio of number of moles of H_2 collected
   to number of moles of Mg reacting   _____   _____   _____

average ratio of number of moles of H_2
   collected to number of moles of Mg reacting   _____

Stoichiometric equation for the reaction of Mg and HCl solution on the basis of the experimental data:

Pre-Laboratory Assignment

1. Review Experiment 2, **The Gas Burner and Glass Working**, for a discussion of glass working. Review Experiment 4, **Volume Measurement of Liquids**, for a discussion of burets and titration techniques.

2. A student preparing to do this experiment filled the 150-mL beaker with a colorless liquid that was supposed to be water. Later, the student feared that 1.0M HCl had been used mistakenly instead of water.

(1) Explain how the student could use a tiny piece of Mg to determine whether the solution in the beaker was 1.0M HCl or water. What, specifically, would the student observe if the beaker contained HCl solution? If the beaker contained water?

(2) The student concluded that the beaker contained HCl solution, which is toxic and corrosive. Suggest a procedure the student could follow, using reagents available for this experiment, to make the solution safe enough to be poured into the drain, diluting with a large amount of running water.

(3) Describe any safety precautions the student should take in carrying out the procedure mentioned in (2).

3. A student investigated the stoichiometry of the reaction of zinc (Zn) with HCl solution and reported the following data. When 0.2158 g of Zn reacted with 10.00 mL of 1.000M HCl, 82.062 g of water was displaced. A total of 36.00 mL of $9.501 \times 10^{-2}M$ NaOH solution was required to titrate the HCl remaining in the reaction mixture at the end of the reaction. The room temperature was 27.0 °C and the barometric pressure was 777 torr. The gram atomic mass of Zn is 65.38 g mol^{-1}. The density of water is 27.0 °C is 0.9965 g mL^{-1} and its vapor pressure is 27.0 torr. $R = 8.21 \times 10^{-2}$ L-atm K^{-1} mol^{-1}. (Hint: See Calculations.)

(1) Calculate the volume of the displaced water from its mass. Note that the volume of the water will be equal to the volume of the gas collected.

_____
answer

(2) Calculate the pressure of the H$_2$ collected.

_____
answer

(3) Calculate the number of moles of H$_2$ collected, using the ideal gas equation.

_____
answer

(4)   Calculate the number of moles of HCl originally present in the reaction flask.

_____
answer

(5)   Calculate the number of moles of HCl remaining in the flask at the end of the reaction.

_____
answer

(6)   Calculate the number of moles of HCl reacting.

_____
answer

(7)   Calculate the number of moles of Zn reacting.

_____
answer

(8)   Calculate the ratio of the number of moles of HCl reacting to the number of moles of Zn reacting.

_____
answer

(9)   Calculate the ratio of the number of moles of H_2 collected to the number of moles of Zn reacting.

_____
answer

(10) Write a stoichiometric equation for the reaction of Zn and HCl solution on the basis of the experimental data.

Determining the Proportionality Constant, *R*, in the Ideal Gas Equation

by **John H. Bedenbaugh**, **Thomas S. Heard**, and **Angela O. Bedenbaugh**,
University of Southern Mississippi

Purpose of the Experiment

Determine the proportionality constant in the ideal gas equation by collecting a measured volume of oxygen produced by quantitatively decomposing hydrogen peroxide. Relate this volume to the stoichiometry of the reaction and to the gas laws.

Background Information

A gas sample can be characterized using four variables: the pressure (P) exerted by the gas; the volume (V) and temperature (T) of the gas sample; and the mass of the sample, which can be used to determine the number of moles (n) of gas in the sample. The relationships between volume and the other three variables were characterized by Boyle (V and P at constant T and n), Charles (V and T at constant P and n), and Avogadro (V and n at constant T and P). The **ideal gas equation**, Equation 1, is a statement of the interrelationship of P, V, n, and T for a single sample of an ideal gas.

$$PV = nRT \qquad \text{(Eq. 1)}$$

The symbol, R, in Equation 1 is a proportionality constant known as the **gas constant**. Equation 1 can be rearranged as shown in Equation 2.

$$R = \frac{PV}{nT} \qquad \text{(Eq. 2)}$$

We can conclude two things from Equation 2. First, if we collect a gas sample of known composition and measure its mass, temperature, volume, and pressure, we can determine the numerical value of R in the ideal gas equation. Second, the value of R will depend on the units we choose for P, V, and T.

We can prepare oxygen gas (O_2) conveniently, safely, and quantitatively by the enzyme-catalyzed decomposition of hydrogen peroxide (H_2O_2) in a dilute aqueous solution, as shown in Equation 3. The enzyme used to catalyze the reaction is catalase.

$$2\,H_2O_2(aq) \xrightarrow{\text{catalase}} 2\,H_2O(l) + O_2(g) \qquad \text{(Eq. 3)}$$

Dried baker's yeast is exceptionally rich in catalase. When the yeast comes in contact with H_2O_2 solution, there is an immediate and rapid evolution of O_2. We can decompose a measured amount of H_2O_2 in a closed system designed so that the volume of O_2 produced can be measured.

Because the volume of any gas varies directly with temperature and indirectly with pressure, we must

Copyright © 1990 by Chemical Education Resources, Inc., P.O. Box 357, 220 S. Railroad, Palmyra, Pennsylvania **17078**
No part of this laboratory program may be reproduced or transmitted in any form or by any means, electronic or mechanical, including photocopying, recording, or any information storage and retrieval system, without permission in writing from the publisher. Printed in the United States of America

measure and record the temperature and pressure at which we collect the gas. Temperature is measured with a thermometer. Atmospheric pressure is measured with a barometer, which is usually calibrated in units of inches of mercury (in. Hg). Readings taken with such a barometer are converted to torr, using Equation 4.

$$1 \text{ in. Hg} = 25.4 \text{ torr} \qquad \text{(Eq. 4)}$$

The total internal pressure, P_{total}, of the closed system is equal to the atmospheric pressure in the laboratory. The total pressure includes the pressure caused by the collected O_2 gas and by the small amount of water vapor distributed throughout the air in the system. We can use the data in Table 1 and the water temperature to determine the vapor pressure of the water. We then calculate the pressure of the gaseous O_2, using Equation 5.

$$P_{total} = P_O + P_{H_2O} \qquad \text{(Eq. 5)}$$

Table 1 *Vapor pressure of water at selected temperatures*

temperature, °C	water vapor pressure, torr
18	15.5
19	16.5
20	17.5
21	18.7
22	19.8
23	21.1
24	22.4
25	23.8
26	25.2
27	26.7
28	28.3

We can determine the number of moles of O_2 produced from the number of moles of H_2O_2 decomposed. Because the H_2O_2 solution contained both H_2O_2 and H_2O, we must first calculate the amount of H_2O_2 in the solution. The mass of H_2O_2 in the solution can be determined from the total mass of solution used and the mass percent H_2O_2 in the solution. We then use Equation 6 to determine the solution mass, from the measured volume of solution used and the reported density of the solution.

$$\begin{pmatrix} \text{mass of } H_2O_2 \\ \text{solution, g} \end{pmatrix} = \begin{pmatrix} \text{volume of } H_2O_2 \\ \text{solution, mL} \end{pmatrix} \begin{pmatrix} \text{density of } H_2O_2 \\ \text{solution, g mL}^{-1} \end{pmatrix}$$

$$\text{(Eq. 6)}$$

Then we calculate the mass of H_2O_2 in the solution, using Equation 7.

mass of H_2O_2, g =

$$\begin{pmatrix} \text{mass of} \\ \text{solution, g} \end{pmatrix} \begin{pmatrix} \dfrac{\text{g } H_2O_2}{1 \text{ g } H_2O_2 \text{ solution}} \end{pmatrix} \quad \text{(Eq. 7)}$$

Because Equation 3 expresses the molar relationship between reactants and products for this reaction, we can determine the number of moles of H_2O_2 represented by the mass of H_2O_2 calculated in Equation 7. This is done using Equation 8.

$$\begin{matrix} \text{number of} \\ \text{moles of } H_2O_2 \end{matrix} = (\text{mass } H_2O_2, \text{ g}) \begin{pmatrix} \dfrac{1 \text{ mol } H_2O_2}{34.02 \text{ g } H_2O_2} \end{pmatrix}$$

$$\text{(Eq. 8)}$$

The coefficients in Equation 3 show that the ratio of the number of moles of H_2O_2 reacting to the number of moles of O_2 produced is 2 to 1. Using this ratio, we can find the number of moles of O_2 that were produced, using Equation 9.

$$\begin{matrix} \text{number of moles} \\ \text{of } O_2 \text{ produced} \end{matrix} = \begin{pmatrix} \text{number of moles} \\ H_2O_2 \text{ reacted} \end{pmatrix} \begin{pmatrix} \dfrac{1 \text{ mol } O_2}{2 \text{ mol } H_2O_2} \end{pmatrix}$$

$$\text{(Eq. 9)}$$

With all the variables now characterized, we can substitute experimental data for them into Equation 2 and determine the value of R. The standard units in which the variables are expressed are as follows: V in liters (L), P in atmospheres (atm), and T in kelvins (K). We will follow this convention when we use the experimental data in the following calculations.

Consider the following experimental results. A student measured 5.00 mL of H_2O_2 solution into the large test tube. The laboratory instructor reported that the solution had a density of 1.01 g mL^{-1} and contained 2.36% H_2O_2. Before the yeast was placed in contact with the H_2O_2 solution, the water level in the buret was aligned with that in the leveling bulb. After alignment, the buret reading was 0.07 mL. After the reaction was complete, the final buret reading was 41.91 mL. The water temperature was 22 °C, and the barometric pressure was 30.95 in. Hg. From these data, we can determine the value of R in the ideal gas equation.

Before we find the water vapor pressure at 22 °C, we need to convert the barometric pressure to torr, using Equation 4.

$$\begin{matrix} \text{barometric} \\ \text{pressure,} \\ \text{torr} \end{matrix} = (30.95 \text{ in. Hg}) \begin{pmatrix} \dfrac{25.4 \text{ torr}}{1 \text{ in. Hg}} \end{pmatrix}$$

$$= 786 \text{ torr}$$

Using Table 1, we find that the vapor pressure of H_2O at 22 °C is 19.8 torr. We can now determine the pressure of O_2 in the system, using Equation 5.

$$\text{pressure of } O_2, \text{torr} = 786 \text{ torr} - 19.8 \text{ torr}$$
$$= 766 \text{ torr}$$

This pressure can be converted to atmospheres, using Equation 10.

$$1 \text{ atm} = 760 \text{ torr} \qquad \text{(Eq.10)}$$

$$\text{pressure of } O_2, \text{atm} = (766 \text{ torr})\left(\frac{1 \text{ atm}}{760 \text{ torr}}\right)$$

$$= 1.01 \text{ atm}$$

Under these described laboratory conditions, the volume change measured after the decomposition reaction was complete is

$$41.91 \text{ mL} - 0.07 \text{ mL} = 41.84 \text{ mL}$$

This volume can be expressed in liters, using Equation 11.

$$1 \text{ L} = 1000 \text{ mL} \qquad \text{(Eq. 11)}$$

$$\text{volume } O_2, \text{L} = (41.84 \text{ mL})\left(\frac{1 \text{ L}}{1000 \text{ mL}}\right)$$

$$= 4.184 \times 10^{-2} \text{ L}$$

Then we convert the Celsius water temperature to kelvins, using Equation 12.

$$\text{Celsius temperature} + 273 = \text{kelvins} \qquad \text{(Eq. 12)}$$

$$22 \text{ °C} + 273 = 295 \text{ K}$$

We find the number of moles of O_2 collected by substituting the experimental data into Equations 6, 7, 8, and 9.

$$\text{mass of } H_2O_2 \text{ solution, g} = (4.00 \text{ mL})(1.01 \text{ g mL}^{-1})$$
$$= 4.04 \text{ g}$$

$$\text{mass of } H_2O_2, \text{g} = (4.04 \text{ g solution})\left(\frac{3.00 \times 10^{-2} \text{ g } H_2O_2}{1 \text{ g solution}}\right)$$

$$= 0.121 \text{ g } H_2O_2$$

$$\text{number of moles of } H_2O_2 = (0.121 \text{ g } H_2O_2)\left(\frac{1 \text{ mol } H_2O_2}{34.02 \text{ g } H_2O_2}\right)$$

$$= 3.56 \times 10^{-3} \text{ mol } H_2O_2$$

$$\text{number of moles of } O_2 \text{ produced} = \left(\frac{3.56 \times 10^{-3} \text{ mol}}{H_2O_2}\right)\left(\frac{1 \text{ mol } O_2}{2 \text{ mol } H_2O_2}\right)$$

$$= 1.78 \times 10^{-3} \text{ mol } O_2$$

Finally, we substitute these data into Equation 2 and determine the value of R. The units of R will be L atm mol^{-1} K^{-1}.

$$R = \frac{(1.01 \text{ atm})(4.184 \times 10^{-2} \text{ L})}{(1.78 \times 10^{-3} \text{ mol})(295 \text{ K})}$$

$$= 8.05 \times 10^{-2} \text{ L atm mol}^{-1} \text{ K}^{-1}$$

We can compare our **experimentally determined** value for R with the **accepted** value, 8.21×10^{-2} L atm mol^{-1} K^{-1}, by calculating the percent error in our determination, using Equation 13.

$$\text{percent error} = \frac{\left(\begin{array}{c}\text{accepted}\\\text{value of } R\end{array}\right) - \left(\begin{array}{c}\text{experimental}\\\text{value of } R\end{array}\right)}{\text{accepted value of } R}(100\%)$$

$$= \frac{\left(\begin{array}{c}8.21 \times 10^{-2} \text{ L atm}\\\text{mol}^{-1} \text{ K}^{-1}\end{array}\right) - \left(\begin{array}{c}8.04 \times 10^{-2} \text{ L atm}\\\text{mol}^{-1} \text{ K}^{-1}\end{array}\right)}{8.21 \times 10^{-2} \text{ L atm mol}^{-1} \text{ K}^{-1}}(100\%)$$

$$= 2.1\% \text{ error}$$

You will follow the procedure outlined in making your own determination of the value of R in the ideal gas equation.

Procedure

Chemical Alert

3% hydrogen peroxide—irritant and mild oxidant

Caution: Wear departmentally approved eye protection while doing this experiment.

I. Assembling the Gas-Producing and Gas-Collecting Apparatus

Note: The letters in parentheses refer to those appearing in Figure 1.

---

1. Obtain the equipment shown in Figure 1.

Note: If you are not using a detached buret barrel in your equipment set up, your laboratory instructor will give you specific directions for any necessary change in the following procedure.

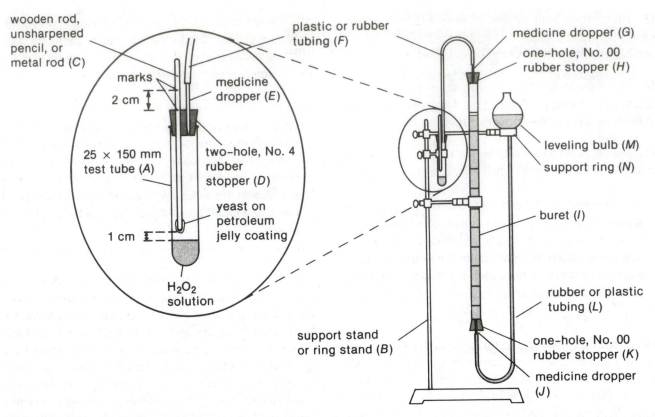

wooden rod, unsharpened pencil, or metal rod (C)

plastic or rubber tubing (F)

medicine dropper (G)

one–hole, No. 00 rubber stopper (H)

marks

2 cm

medicine dropper (E)

25 × 150 mm test tube (A)

two–hole, No. 4 rubber stopper (D)

leveling bulb (M)

support ring (N)

yeast on petroleum jelly coating

buret (I)

1 cm

H_2O_2 solution

rubber or plastic tubing (L)

support stand or ring stand (B)

one–hole, No. 00 rubber stopper (K)

medicine dropper (J)

Figure 1   *Measuring the volume of O_2 produced from the decomposition of H_2O_2 in a closed system*

2.   Clamp a clean, dry 25 × 150-mm test tube (A) to a support stand or ring stand (B).

3.   Use indelible ink or a knife to mark the wooden rod, metal rod, or unsharpened pencil (C) in two places, as directed by your laboratory instructor.

Note: Your laboratory instructor will give you directions for lubricating the rod and medicine dropper and for inserting them into the rubber stopper.

4.   Lubricate the bottom 2 cm of the rod with a small amount of petroleum jelly.

5.   Carefully insert the lubricated end of the rod (C) into one hole of a two-hole, No. 4 rubber stopper (D). Move the rod back and forth through the stopper so that the portion of the rod that will move through the stopper becomes lubricated. Be sure the rod can be easily moved back and forth.

6.   Adjust the rod so that the lower of the two marks made in Step 3 coincides with the top of the rubber stopper (D).

Caution: Be careful when inserting the lubricated medicine dropper into the stopper. Grasp the dropper as close to the point of insertion as possible and apply pressure there.

7.   Lubricate the dropper end of a medicine dropper with a small amount of petroleum jelly. Avoid putting any jelly over the dispensing end of the dropper.

8.   Carefully insert the lubricated medicine dropper (E) into the other hole of the stopper (D).

9.   Attach one end of the 24-cm length of plastic tubing (F) to the barrel of the medicine dropper (E).

10.   Insert the stopper assembly into the test tube (A).

11.   Carefully insert another lubricated medicine dropper (G) into one of the one-hole, No. 00 rubber stoppers (H). Attach the other end of the plastic tubing (F) to the barrel of the second medicine dropper (G).

12.   Insert the stopper (H) into the top of the buret barrel (I).

13.   Clamp the buret to the support stand (B).

14.   Insert a third lubricated medicine dropper (J) into the second one-hole, No. 00 rubber stopper (K).

15. Attach one end of the 50-cm length of plastic tubing (*L*) to the dropper (*J*) and place this stopper assembly into the bottom of the buret barrel (*I*).

16. Place the leveling bulb (*M*) in the ring support (*N*) attached to the support stand (*B*). Attach the other end of the tubing (*L*) to the leveling bulb.

17. Fill the buret barrel with tap water until the leveling bulb (*M*) is approximately half full of water.

> ***Note:*** In order to obtain useful data from this experiment, the gas-producing and gas-collecting system ***must be*** air tight during the reaction and while taking the final buret reading. Any leak in the system will result in a low volume of collected O_2.

18. Move the leveling bulb up and down, by adjusting the ring support (*N*), until the water levels in the bulb and buret are aligned. Inspect the tubing (*F*) connecting the test tube and the buret as well as the tubing (*L*) connecting the buret and the bulb (*M*). Be sure there are no air bubbles in the tubing nor any air or water leaks in the system.

> ***Note:*** Have your laboratory instructor approve your apparatus in writing before you begin Step 19.
>
> _____
> initials

II. Determining the Volume of O_2 Produced from the H_2O_2 Solution

> ***Note:*** Your laboratory instructor will inform you whether you are to use a measuring device other than a 5-mL pipet, such as a 4-mL pipet, a 50-mL buret, a dispensing buret, or an automatic pipetter. If so, your instructor will give you specific directions for altering the following procedure.

> ***Caution:*** 3% H_2O_2 solution is an irritant and a mild oxidant that can cause burns. Despite the fact that you are using a dilute solution, you should prevent contact between the solution and your eyes, skin, or clothing. Avoid ingesting the solution.

19. Obtain about 20 mL of H_2O_2 solution in a 125-mL Erlenmeyer flask from your laboratory instructor. Record the code number, density, and mass percent H_2O_2 of this solution on your Data Sheet.

> ***Note:*** Your laboratory instructor will give you additional directions for using a volumetric pipet.

> ***Caution:*** Never use your mouth to draw a solution into a pipet. Always use a rubber bulb to fill a pipet.

20. Rinse a clean, 5-mL volumetric pipet with about 1 mL of your H_2O_2 solution, using the following procedure. Draw the solution into the pipet using a rubber bulb. Quickly disconnect the rubber bulb and place your index finger over the top of the pipet to prevent the solution from draining out. Hold the pipet in a nearly horizontal position with the tip slightly lower than the end covered by your finger. Rotate the pipet so that the rinse solution contacts as much of the inner surface of the pipet as possible. Briefly lift your index finger during this process to allow the solution to enter the upper stem of the pipet. Drain the solution through the tip of the pipet into a beaker.

21. Repeat the rinsing procedure with two additional portions of the solution. Discard the rinse solutions into the container specified by your laboratory instructor and labeled, "*Discarded H_2O_2 Rinse Solutions.*"

22. Remove the tubing (*F*) from the medicine dropper (*E*).

23. Remove the rubber stopper (*D*) with the remaining attachments from the test tube (*A*).

> ***Note:*** After you discharge a solution from a volumetric pipet, there will be a small amount of liquid left in the tip of the pipet. Do not blow this liquid out of the pipet. This pipet is calibrated to deliver the designated volume of solution ***excluding*** the small amount remaining in the tip.

24. Pipet a 5.00-mL portion of your H_2O_2 solution into the clean, dry 25 × 150-mm test tube, as follows. As you begin to transfer the solution to the test tube, hold the pipet so that the tip of the pipet is near the bottom of the test tube. As you slowly lift your index finger and release the solution into the test tube, gently press the tip of the pipet against the wall of the test tube. Allow the solution to flow down the wall to avoid splattering. Keep the tip of the pipet above the surface of the solution. After you have delivered the solution, continue holding the tip of the pipet against the wall of the test tube for an additional 15 s. **Do not blow out the small amount of solution remaining in the pipet.**

25. Before replacing the rubber stopper (*D*) in the test tube, place a small amount of petroleum jelly on the bottom end of the rod (*C*). Use your finger to uniformly lubricate the bottom 2 to 3 cm of the rod.

26. Gently roll the lubricated end of the rod (*C*) in a small amount of baker's yeast, placed on a clean, dry watch glass. Lightly tap the rod to remove any yeast not firmly adhering to the jelly.

> **Note:** All loose yeast **must** be removed from the rod before beginning Step 27. If even a **single** particle of yeast falls prematurely into the H_2O_2 solution, the reaction will begin before you have closed the system. Do not allow the yeast to contact the solution in any way while the system is open. If this should occur, repeat Steps 24–26, using a clean, dry rod and a clean, dry test tube.

27. Carefully replace the rubber stopper (*D*) into the test tube (*A*) so that no yeast particles drop into the H_2O_2 solution.

28. Again check the system for leaks. Move the leveling bulb up and down. Inspect the tubing (*L*) connecting the buret and the bulb. Be sure there are no air bubbles in the tubing nor any water leaks in the system.

> **Note:** When the water levels in the buret and bulb are aligned, the internal pressure in the buret exactly equals the barometric pressure outside the buret in the the laboratory.

29. Reconnect the tubing (*F*) to the medicine dropper (*E*). Carefully adjust the leveling bulb so that the water level in the bulb aligns with that in the buret, which should be at or just below the 0.00 mL mark. Record the water volume in the buret to the nearest hundredth of a milliliter (0.01 mL) on your Data Sheet, as your first buret reading.

30. Gently press the rod down in the two-hole stopper until the upper mark on the rod is level with the top of the stopper. As the evolved O_2 begins to displace the water in the buret, the water level in the buret will begin to fall. Each time the water level drops by 10 mL, lower the leveling bulb so that the water levels in the bulb and buret are approximately aligned.

31. When you can no longer see any bubbling, loosen the test tube clamp and gently agitate the test tube and the solution. Then check to be sure the water level in the buret does not move any further. When the water level does not change, the reaction is complete. While pressing down on the two-hole stopper to hold it in place, carefully withdraw the rod to its original position, as indicated by the lower mark on the rod. Use a twisting motion while gently pulling.

32. Lower the leveling bulb until the water levels in it and the buret are exactly aligned. The internal and external pressures are then equal. Record the buret reading to the nearest 0.01 mL on your Data Sheet, as your second buret reading.

33. Immediately determine the temperature of the water in the leveling bulb to the nearest degree. Record this water temperature on your Data Sheet.

34. Obtain from Table 1 the water vapor pressure at the water temperature. Record this pressure on your Data Sheet.

35. Record the barometric pressure in the laboratory on your Data Sheet, including the units in which your barometer is calibrated.

36. Repeat the determination two more times, beginning with Step 19 through Step 35. After each determination, you must throughly clean and rinse the test tube, wipe the rod clean, and use a fresh sample of the same H_2O_2 solution.

37. Clean your glassware and your gas-producing and gas-collecting apparatus as directed by your laboratory instructor.

> **Caution:** Wash your hands thoroughly with soap or detergent before leaving the laboratory.

Calculations

Do the following calculations and record your results on your Data Sheet.

1. Convert the barometric pressure reading to an equivalent reading in torr, using Equation 4.

2. Determine the pressure of the collected O_2, in torr, using Equation 5.

3. Convert the pressure, in torr, of collected O_2 to atmospheres, using Equation 10.

4. Determine the volume of collected O_2 by subtracting the first buret reading from the second buret reading.

5. Convert this volume, in milliliters, to liters, using Equation 11.

6. Convert the water temperature, in Celsius, to kelvins, using Equation 12.

7. Calculate the mass of H_2O_2 solution used in the determination, using Equation 6.

8. Determine the mass of H_2O_2 in 5.00 mL of your solution, using Equation 7.

9. Calculate the number of moles of H_2O_2 reacting, using Equation 8.

10. Determine the number of moles of O_2 produced, using Equation 9.

11. Calculate the proportionality constant, R, using Equation 2.

12. If you only did one determination, calculate the percent error in your determination, using Equation 12.

13. Repeat Calculations 1–11 for each additional determination you made.

14. Calculate the mean value of R.

15. Calculate the percent error in the mean value of R, using Equation 13.

Post-Laboratory Questions

(Use the spaces provided for the answers and additional paper if necessary.)

1. Suppose that a student doing this experiment ran into some procedural and calculation difficulties. State the effect each of the following would have on the calculated value of R determined from this experiment. Briefly explain.

(1)   After the reaction was completed, the student failed to withdraw the rod to its original position before taking the second buret reading.

(2)   The student neglected to take into account the pressure of water vapor in the system when doing the calculations.

---

name   section   date

(3) The H_2O_2 solution had decomposed slightly after standing for a while, so its concentration was somewhat lower than the laboratory instructor had reported.

2. (1) Using your textbook as a reference, explain what is meant by the molar volume of a gas.

(2) What assumption did you make about the molar volume of O_2 when you performed your calculations for this experiment? Explain briefly.

(4) The student failed to read the barometric pressure at the time of the experiment. When she read the pressure the next day for use in her calculations, it was 0.50 in. Hg higher than it had been during the experiment.

(3) Briefly describe how you must revise the procedure and the calculations so this experiment could be used to determine the molar volume of O_2 at STP.

Data Sheet

II. Determining the Volume of O_2 Produced from the H_2O_2 Solution

code number of H_2O_2 solution   _____

density of H_2O_2 solution, g mL^{-1} _____

actual mass percent H_2O_2 in solution, %   _____

	determination		
	1	2	3
volume of H_2O_2 solution used, mL	_____	_____	_____
water temperature, °C	_____	_____	_____
H_2O vapor pressure at water temperature, torr	_____	_____	_____
barometric pressure, in. Hg	_____	_____	_____
barometric pressure, torr	_____	_____	_____
pressure of collected O_2 at water temperature, torr	_____	_____	_____
pressure of collected O_2 at water temperature, atm	_____	_____	_____
water temperature, K	_____	_____	_____
second buret reading, mL	_____	_____	_____
first buret reading, mL	_____	_____	_____
volume of O_2 collected at laboratory T and P, mL	_____	_____	_____
volume of O_2 collected at laboratory T and P, L	_____	_____	_____
mass of the H_2O_2 solution, g	_____	_____	_____
mass of H_2O_2 in the solution, g	_____	_____	_____
number of moles of H_2O_2 reacting	_____	_____	_____
number of moles of O_2 collected	_____	_____	_____
calculated value of R, L atm mol^{-1}K^{-1}	_____	_____	_____
mean value of R		_____	
percent error, %		_____	

Pre-Laboratory Assignment

1. Three-percent solutions of H_2O_2 are obtainable without prescription in drug stores and are considered safe for home use, if used according to instructions. Nevertheless, there are some hazards associated with such products. Describe the dangers involved if an H_2O_2 solution spills on your laboratory bench. Tell what action you should take.

(4)   Convert this pressure, in torr, to atmospheres.

_____
answer

(5)   Calculate the volume, in milliliters, of collected O_2.

_____
answer

(6)   Convert this volume, in milliliters, to liters.

2. A student performed the experiment described in this module. The 5.00-mL mass of the 2.15% percent by mass H_2O_2 solution used was 5.03 g. The water temperature was 23 °C, and the barometric pressure was 31.2 in. Hg. After the student immersed the yeast in the peroxide solution, she observed a 38.60-mL volume change in system volume.

(1)   Convert the barometric pressure to torr.

_____
answer

(7)   Convert the water temperature, in Celsius, to kelvins.

_____
answer

_____
answer

(2)   Obtain the water vapor pressure at the water temperature.

(8)   Calculate the mass of 5.00 mL of the H_2O_2 solution.

_____
answer

_____
answer

(3)   Calculate the pressure, in torr, exerted by the collected O_2 at the water temperature.

(9)   Calculate the mass of H_2O_2 in 5.00 mL of the solution.

_____
answer

_____
answer

(10) Calculate the number of moles of H_2O_2 reacting.

3. Briefly explain why it is essential that none of the yeast comes into contact with the H_2O_2 solution before the three stoppers are firmly in place to completely close the system.

———————
answer

(11) Calculate the number of moles of collected O_2.

———————
answer

(12) Determine the proportionality constant, R, in L atm mol^{-1} K^{-1}.

———————
answer

(13) Calculate the percent error for this determination.

———————
answer

Evaluation of 0 K

prepared by **William F. Kieffer**, The College of Wooster

Purpose of the Experiment

Determine the number of degrees difference between 0 °C and 0 K.

Background Information

Temperatures are measured using a thermometer. A thermometer contains a substance called **thermometric fluid**, whose volume increases with increasing temperature. The thermometric fluid expands into a calibrated capillary tube so that temperature can be read directly. Thermometers commonly used in the laboratory contain mercury as the thermometric fluid. Other substances perform satisfactorily, including alcohol in most household Fahrenheit thermometers and even liquified pentane in special thermometers for low-temperature work.

The scale of a thermometer is established by the arbitrary choice of some temperature as the zero point of the scale and some other higher temperature to be a chosen number of degrees above zero. The familiar Celsius, formerly called the Centigrade, scale sets 0 as the temperature of liquid and solid water in equilibrium and 100 as the temperature of liquid and gaseous water in equilibrium, both under 1 atm pressure. Obviously, this choice puts 0 on the Celsius scale at a much higher temperature than is 0 on the kelvin, or absolute, scale.

We can relate Celsius and kelvin temperatures by using Equation 1, where Y is a constant.

$$T \text{ (kelvin)} = t \text{ (Celsius)} + Y \qquad \text{(Eq. 1)}$$

Note that this equation requires that all three terms, T, t, and Y, be expressed in identical units. Because t is in degrees Celsius, the units of both T and Y must be equal in magnitude to Celsius degrees.

The apparatus you will assemble in this experiment constitutes a gas thermometer, in which air is the thermometric fluid. You will observe the volume of a sample of air at two temperatures. In your experiment, you can expect air to behave as an ideal gas, expanding in proportion to the kelvin temperature. Therefore, the data you obtain will allow you to relate your observed Celsius readings to the corresponding values on the kelvin scale.

The kinetic molecular theory provides us with an equation of state for an ideal gas, as shown in Equation 2,

$$V = \left(\frac{nR}{P}\right)T \qquad \text{(Eq. 2)}$$

where n is the number of moles of gas in a sample, V is its volume at pressure P, R is the gas law constant in appropriate units, and T is the kelvin temperature. Equation 2 can be read as Charles' or Gay-Lussac's law: **the volume of an ideal gas sample at constant pressure is proportional to its kelvin temperature**, if we recognize the term nR/P as the constant of proportionality.

Copyright © 1989 by Chemical Education Resources, Inc., P.O. Box 357, 220 S. Railroad, Palmyra, Pennsylvania 17078
No part of this laboratory program may be reproduced or transmitted in any form or by any means, electronic or mechanical, including photocopying, recording, or any information storage and retrieval system, without permission in writing from the publisher. Printed in the United States of America

If n and P are unchanged, we can compare two observed volumes of the sample at different temperatures, T_{high} and T_{low}.

$$\frac{V_{high}}{V_{low}} = \frac{\left(\frac{nR}{P}\right)T_{high}}{\left(\frac{nR}{P}\right)T_{low}} \qquad \text{(Eq. 3)}$$

If we represent the kelvin temperatures, T_{high} and T_{low}, in terms of the comparable Celsius readings, t_{high} and t_{low}, Equation 3 becomes

$$\frac{V_{high}}{V_{low}} = \frac{t_{high} + Y}{t_{low} + Y} \qquad \text{(Eq. 4)}$$

Note that when all of the measured quantities, V_{high}, t_{high}, V_{low}, and t_{low} are substituted in Equation 4, we can solve for the unknown term, Y.

The postulate central to the kinetic molecular theory is the concept of an ideal gas. An **ideal gas** is one in which there are no intermolecular forces among the molecules, the molecules have only point volumes, and the average kinetic energy of the molecules depends only on the kelvin temperature. Although this concept was unknown to Charles in 1787 and Gay-Lussac in 1802 when they formulated the law that bears their names, it is significant to note that they limited the law's applicability to **permanent gases**. They considered such gases to be ones they thought could not be liquified; for example, O_2, N_2, H_2, and CO. In fact, we now know that these gases can be liquified if the temperature is low enough to make their weak intermolecular forces effective. We also recognize that under low pressures and at tem-peratures far above their boiling points, **all** substances in the gaseous state approach ideal gas behavior because, under these conditions, intermolecular forces are least effective.

Since air is a mixture of the low-boiling-point gases N_2, O_2, Ar, and only trace amounts of others, we are justified in expecting air to closely follow predicted ideal-gas behavior. Water, which boils at 100 °C, certainly does not follow the Charles–Gay-Lussac law at room temperature. Therefore, the air sample used in this experiment must contain very little water vapor. However, laboratory air at room temperature, even at the extreme condition of 100% humidity, contains only about 2% water vapor. So, you can expect such a gas sample to behave satisfactorily as your thermometric fluid.

Two other properties of ideal gas molecules, no measurable volume and an average kinetic energy proportional to the kelvin temperature, allow you to use data from this experiment to evaluate 0 K by an alter-nate method. The kinetic molecular theory pre-dicts that at 0 K there will be no molecular motion. Hence the gas molecules will occupy no volume, because the molecules themselves take up no space. Consequently, if you plot your data on a volume–temperature graph and extrapolate the straight line connecting your two points to 0 on the volume axis, the resulting intercept on the temperature axis can be read as 0 K. The number of degrees between this intercept and 0 °C must be the value of Y in Equation 1. The instructions in the Calculations section of this module suggest an appropriate procedure for this graphical evaluation.

In this experiment, you will construct a gas thermometer in which the thermometric fluid is a sample of dry air, a gas. You will measure the volume of the air sample conveniently and accurately by weighing an equal volume of water. You will determine the volume of your air sample, V_{high}, at the temperature established by boiling water, t_{high}, where t_{high} is in degrees Celsius. You will then lower the temperature of the air sample to that of running tap water to give a temperature, t_{low}, in degrees Celsius. You will then determine the new volume, V_{low}, at t_{low}. By using the relationship between volume and temperature, you will find the number of degrees between 0 °C and 0 K.

Procedure

Caution: Wear departmentally approved eye protection while doing this experiment.

Assemble the apparatus shown in Figure 1. Use a dry 125-mL Erlenmeyer flask, fitted with a one-hole rubber stopper. Insert an 8-cm length of glass tubing into this rubber stopper which we will call No. 1. Position the tubing so that it almost reaches the lower end of this stopper. To the section of glass tubing projecting from stopper No. 1, attach rubber stopper No. 2 so that the bottom of No. 2 touches the top of No 1. Fasten a 10-cm length of plastic or rubber tubing to the end of the glass tubing projecting through stopper No. 2. Loosely attach a partially open Hoffman screw pinch clamp near the end of the plastic tubing. Throughout this experiment, the Erlenmeyer flask, rubber stoppers Nos. 1 and 2, glass tubing, plastic tubing, and pinch clamp will be referred to as the **flask assembly**. Weigh the flask assembly to the nearest tenth of a gram (0.1 g). Record this mass on your Data Sheet.

Place the flask assembly in a beaker containing sufficient tap water such that the water level is even

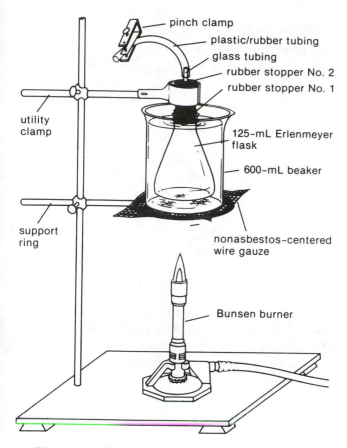

pinch clamp

plastic/rubber tubing

glass tubing

rubber stopper No. 2

rubber stopper No. 1

utility clamp

125–mL Erlenmeyer flask

600–mL beaker

support ring

nonasbestos–centered wire gauze

Bunsen burner

Figure 1   *The assembled apparatus*

with the bottom of stopper No. 1. Keep the flask in position with an extension clamp placed around stopper No. 2. Position the flask so that it is at least 1 cm above the bottom of the beaker and does not touch the side of the beaker. In this way, the flask's temperature will not exceed that of the water in the beaker.

> *Caution:* While heating and boiling water in the next step, use extreme care so that you do not scald yourself.

Heat the water in the beaker until it boils, and continue boiling for 5 min. Make sure the pinch clamp is open.

> *Note:* Do not allow the thermometer to touch the side or bottom of the beaker.

Measure the temperature of the boiling water with a mercury thermometer. Record this temperature as t_{high} on your Data Sheet.

Keep the water boiling while you firmly close the pinch clamp. Tighten the extension clamp securely

around the flask neck. Using this clamp as a handle, remove the flask from the beaker of boiling water and submerge it in a pan or trough containing cold tap water. Be sure that the flask neck and plastic tubing are kept below the surface of the water so that when the pinch clamp on the plastic tubing is opened, no air and only water can enter the flask.

Carefully open the pinch clamp on the plastic tubing so that some water will be forced into the flask by the contraction of the air sample. Allow the flask to cool for 10 min. If possible, run cold tap water over the flask to cool it and the trapped air inside it. Measure the temperature of the water in the pan or trough, using the same thermometer used previously. Record this temperature as t_{low} on your Data Sheet.

While keeping the tubing submerged, adjust the flask in the pan or trough so that the water levels inside and outside the flask are the same. Thus, the pressure of the enclosed air and the atmospheric pressure outside the flask are equal. Close the pinch clamp and remove the flask assembly from the pan or trough. Carefully remove any water in the plastic tubing above the pinch clamp by using a cotton swab. Then open the clamp so that any water in the tubing below the clamp will run into the flask. Remove the utility clamp and dry the flask thoroughly with a towel. Reweigh the flask assembly to determine the mass of water forced into it. Record all data on your Data Sheet.

Place a label on the neck of the flask. Mark on the label the level of the bottom of the stopper. Remove the rubber stoppers, the glass and plastic tubing, and the pinch clamp. Fill the flask to the brim with water. Replace the stopper assembly and push it down to the level previously marked on the label. This manipulation should fill the glass and plastic tubing; if not, add water to do so. Close the pinch clamp again and carefully remove any water in the tubing above the pinch clamp. Thoroughly dry the flask assembly. The volume of water now filling the apparatus is the same as the volume of air in the original sample. **This volume will not be the same as the nominal flask volume engraved on the neck of the flask.** Reweigh the water-filled flask assembly, measure the water temperature, and calculate the mass of the water in the flask. Record all data on your Data Sheet.

Repeat the experiment with a second clean, dry flask. A dry flask is required because it is very difficult to remove all of the water vapor from the one you used in the first determination. Your second determination should run more smoothly, due to the experience gained in the first determination. Complete the calculations with the data from your first determination while heating the second flask in the boiling water. The results of these calculations may indicate some manipulative error to be avoided during the second trial. If your experimental data are not acceptable to your laboratory instructor, try to account for the poor result by recalling some manipulative error. Take care not to repeat any such mistakes while doing your second determination.

Record all data on your Data Sheet.

If time permits, do a third determination.

> **Caution:** Wash your hands thoroughly with soap or detergent before leaving the laboratory.

Reference Table	Specific volume of water [a]
temperature, °C	volume, mL g^{-1}
4	1.0000
10	1.0003
15	1.0009
18	1.0014
20	1.0018
22	1.0022
25	1.0029
27	1.0035

[a] *Handbook of Chemistry and Physics*, 63rd ed.; Chemical Rubber Publishing Company: Cleveland, OH, 1982–83; p F–5.

Calculations

Do the following calculations for each determination and record the results on your Data Sheet.

1. Using Equation 5, calculate the volume of water forced into the flask from the values of mass of water and the appropriate values for the specific volume of water (see Reference Table).

volume, mL =
$$(mass, g)(specific\ volume,\ mL\ g^{-1}) \quad (Eq.\ 5)$$

You should recognize that unless four significant digits are justified by the magnitude and precision of your measurements, the value for the specific volume of water can be taken as 1.00 mL g^{-1} over the temperature range involved.

2. Calculate the volume of water required to fill the flask, using Equation 5.

3. The volume of water completely filling the flask is the volume of air trapped at the higher temperature, V_{high}. Calculate the value of V_{low} by subtracting the volume of water drawn into the flask from the value of V_{high}.

4. Calculate algebraically the value of Y, using Equation 4.

5. Prepare a graph of air volume versus temperature in degrees Celsius, as follows. Turn your sheet of graph paper so that the longer dimension is hori-

zontal. You will plot temperature on the longer, horizontal axis, the abscissa. You will plot volume on the shorter, vertical axis, the ordinate. Label the volume axis from 0 upward to a value large enough to include V_{high} values conveniently near the top of the scale. Locate 0 °C at a point about three-fourths of the distance along the temperature axis. Label the axes as indicated in Figure 2.

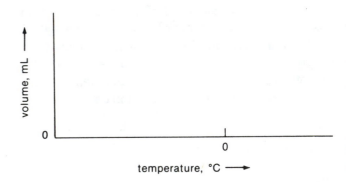

Figure 2   *Volume vs. temperature graph*

Plot the volume–temperature data from your first determination, and extrapolate the straight line connecting the two points to 0 volume. The point at which this line intersects the temperature axis will indicate 0 K in degrees Celsius. The difference between this value and 0 °C is the value of Y. The units for Y are in degrees Celsius.

Post-Laboratory Questions

(Use the spaces provided for the answers and additional paper if necessary.)

1. A student doing this experiment was not very attentive to the details of the procedure. Predict possible effects of the following situations, and state whether the situation would result in a higher, lower, or unchanged value for Y. Briefly explain each of your answers.

(1) When the student initially immersed the flask assembly in the beaker, the water level reached the bottom of the stopper, as specified in the Procedure. As the student heated the flask, the water in the beaker boiled away until, by the end of the heating period, the water level had fallen close to the middle of the flask.

(2) The student was unable to close the pinch clamp completely, prior to transferring the flask from the boiling water to the cold water bath.

(3) The student used two different mercury thermometers during the experiment, one to meas-ure the temperature of the boiling water in the beaker, and the other to measure the temperature of the water in the cold water bath.

(4) The student closed the pinch clamp without equalizing the water levels inside and outside the flask, while it was resting on its side in the cooling bath.

2. The outcome of this experiment, the value of Y, depends on an extrapolation of a line connecting two data points.

    (1) Briefly explain this statement.

    (2) What assumptions are you making about the behavior of your gas sample when you make the extrapolation?

Data Sheet

determinations

	1	2	3
temperature of boiling water, t_{high}, °C	_____	_____	_____
temperature of running water, t_{low}, °C	_____	_____	_____
mass of the flask assembly full of water, g	_____	_____	_____
mass of the flask assembly, g	_____	_____	_____
mass of water required to fill the flask assembly, g	_____	_____	_____
volume of air required to fill the flask assembly, V_{high}, mL	_____	_____	_____
mass of the flask assembly plus water forced in by cooling, g	_____	_____	_____
mass of water forced in by cooling, g	_____	_____	_____
volume of water forced in by cooling, mL	_____	_____	_____
volume of air in flask at lower temperature, V_{low}, mL	_____	_____	_____
value of Y obtained graphically, °C	_____	_____	_____
value of Y calculated algebraically, °C	_____	_____	_____
best calculated value for absolute zero, °C	_____	_____	_____

Pre-Laboratory Assignment

1. Suppose a student who was unprepared for this experiment neglected to open the pinch clamp on the flask assembly prior to heating the flask assembly in boiling water. Describe the problems this situation would create.

3. Explain why it is essential that the inside of the flask be absolutely dry before it is fitted with the stopper assembly and weighed.

2. A student performed this experiment and collected the following data:

volume of air at t_{high} (100 $^\circ$C) = 260 mL
volume of air at t_{low} (24.0 $^\circ$C) = 205 mL

(1) Plot the volume on the ordinate and the temperature on the abscissa. Draw a line connecting the two points.

(2) Extrapolate the line to the point on the x-axis where $V = 0$, and determine the value of Y.

_____
answer

(3) Calculate the percent error between the value of Y found by this graphical procedure and the accepted value of $-273\ ^\circ$C.

_____
answer

The Visible Atomic Spectrum of Hydrogen

prepared by **M. Gillette** and **S. R. Johnson**, Indiana University at Kokomo

---

Purpose of the Experiment

Prepare a calibration curve for your spectroscope, using a helium discharge tube. Determine the scale readings for the four visible lines of hydrogen, using a hydrogen discharge tube. Calculate ΔE and n_2 for each of the visible emission lines of hydrogen.

Background Information

I. The Bohr Atom

As early as the 5th century B.C. the Greek philosopher Democritus was taught that all things were composed of atoms. The term "atom" is a combination of the Greek words for "not" and "to cut". However, the physical structure and chemical behavior of atoms could not be described until critical experiments were done and sensitive measurements were made. The technology necessary to perform these studies was not available to the ancients. Therefore, their extensive use of metals, alloys, ceramics, and dyes was the result of trial and error development. It wasn't until the mid-eighteenth and nineteenth centuries that accurate measuring devices were developed and experiments were performed that resulted in a better understanding of the nature of the atom.

In 1885, J. J. Balmer observed that when hydrogen gas was excited by an electric current it emitted light. When resolved by a prism, the light produced a distinct and reproducible visible spectrum. Balmer and J. Rydberg identified the wavelengths of the spectral lines and showed that they were related by a simple mathematical formula, as shown in Equation 1,

$$\frac{1}{\lambda} = R\left(\frac{1}{n_1^2} - \frac{1}{n_2^2}\right) \qquad \text{(Eq. 1)}$$

where $n_1 = 2$, n_2 is a small whole number greater than 2, that is, 3, 4, 5, and so on, R is a constant, and λ is the observed wavelength.

In 1897, J. J. Thomson devised an experiment that demonstrated the existence of negatively-charged particles, now called **electrons**, as part of the structure of the atom. In the early 1900s, E. Rutherford showed that the electrons in the atom are balanced in number by positively-charged particles called **protons**. Rutherford and two co-workers, H. Geiger and E. Marsden, used the now-legendary gold foil experiment to show that electrons are distributed over a large volume surrounding a dense central nucleus containing the protons.

A physical picture of electron distribution, based on the classical Newtonian physics that governed science at that time, included completely random distribution of

Copyright © 1988 by Chemical Education Resources, Inc., P.O. Box 357, 220 S. Railroad, Palmyra, Pennsylvania 17078
No part of this laboratory program may be reproduced or transmitted in any form or by any means, electronic or mechanical, including photocopying, recording, or any information storage and retrieval system, without permission in writing from the publisher. Printed in the United States of America

electrons about the nucleus. It became evident to such people as Albert Einstein that what had been thought to be laws for the behavior of readily observable objects were not necessarily applicable on the atomic level. Of particular concern to Einstein was the Newtonian concept of continuous behavior: that if an object can exist in one position and then in another, it can also exist at all intermediate positions.

In 1913 Niels Bohr, in his explanation of Balmer's hydrogen spectral data, suggested that the electron of the hydrogen atom moved about the nucleus in a circular orbit. He proposed that this path represented a perfect balance between two forces: the coulombic attraction between the negatively-charged electron and the positively-charged nucleus; and the momentum of the moving electron. In his model, the electron was not drawn into the nucleus, nor could it fly off into space. Bohr further postulated that there was a series of circular orbits of increasing size in which the hydrogen electron could also be located and experience the same balance of forces. This explanation led Bohr to his picture of the hydrogen atom, shown in Figure 1.

Bohr described the hydrogen atom in its natural, low-energy **ground state** as having its single electron in the orbit closest to the nucleus. He labeled this orbit $n = 1$, as shown in Figure 1. Bohr proposed that when a collection of hydrogen atoms in their ground states were irradiated with light, the electrons could absorb only those wavelengths whose energy exactly equalled the energy difference between the stable $n = 1$ orbit and one of the other orbits pictured in Figure 1. Bohr designated these higher energy orbits as $n = 2, 3, 4,$ and so on. The process of energy absorption by electrons is known as **excitation**. Once excited, the electron has a natural tendency to fall back to its low energy ground state of $n = 1$. Bohr thought that the return to the ground state could occur in a single step or in a series of smaller steps, as shown in Figure 2. In any case, the total amount of energy given off would be exactly equivalent to the amount of energy absorbed in the excitation process.

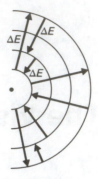

Figure 2   *Electron transitions in a hydrogen atom*

Bohr proposed that, regardless of the sequence of steps, the energy difference (ΔE) between levels would satisfy Equation 2

$$\Delta E = h\nu \qquad \text{(Eq. 2)}$$

where ν is the frequency of absorbed or emitted radiation and h is a constant, known now as Planck's constant (6.626×10^{-34} J s). Since frequency is equal to the velocity of light (c) divided by wavelength (λ), Equation 2 can be rewritten as Equation 3

$$\Delta E = \frac{hc}{\lambda} \qquad \text{(Eq. 3)}$$

to show what Bohr found: that his model of the hydrogen atom could be used to predict numerical values for the wavelengths of the hydrogen spectral lines observed by Balmer.

The Bohr theory cannot be used to predict the exact wavelengths of the spectral lines for any element other than hydrogen (and one-electron ions). This is because Equation 1 does not take into account electrostatic interactions between electrons or the shielding effect of inner electrons, which moderates the attraction between outer electrons and the nucleus. However, the general idea of specific energy states and of the $\Delta E = h\nu$ relationship is applicable to all atoms.

II. The Electromagnetic Spectrum

The **electromagnetic spectrum** can be roughly divided into regions as shown in Figure 3. Since our eyes are sensitive primarily to light with wavelengths between 390 and 770 nm, this region is therefore called the **visible region** of the spectrum. The energy equivalent of any wavelength in this region can be calculated using Equation 3. We can illustrate this for $\lambda = 770$ nm and for $\lambda = 390$ nm, as shown below.

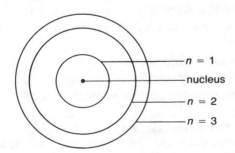

Figure 1   *Bohr's hydrogen atom*

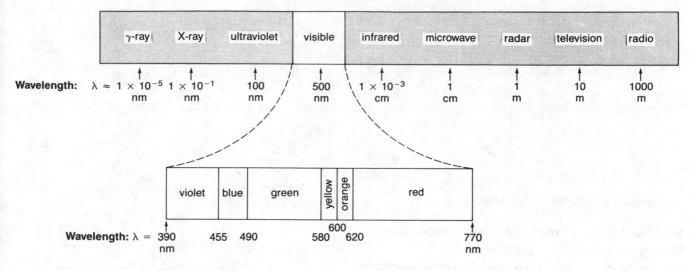

Figure 3 *Electromagnetic spectrum*

At 770 nm,

$$\Delta E = \frac{(6.626 \times 10^{-34} \text{ J s}) (2.998 \times 10^{8} \text{ m s}^{-1})}{(770 \text{ nm})\left(\dfrac{1 \text{ m}}{1 \times 10^{9} \text{ nm}}\right)}$$

$$= 2.580 \times 10^{-19} \text{ J}$$

At 390 nm,

$$\Delta E = \frac{(6.626 \times 10^{-34} \text{ J s}) (2.998 \times 10^{8} \text{ m s}^{-1})}{(390 \text{ nm})\left(\dfrac{1 \text{ m}}{1 \times 10^{9} \text{ nm}}\right)}$$

$$= 5.094 \times 10^{-19} \text{ J}$$

When excited hydrogen electrons return to the ground state, Bohr concluded, the energy emitted was in the visible region of the spectrum if the electron fell to the level $n = 2$ prior to reaching the ground state. The correlation between these transitions and Equation 1 is that n_2 represents the **highest orbit to which the electron was excited** and n_1 represents the **orbit to which the electron first falls**. The energy emitted by electrons that fall directly from an excited state to the ground state ($n_1 = 1$) is not in the energy region that can be detected visually.

In this experiment you will calculate ΔE and n_2 for each of the visible emission lines of hydrogen. A discharge tube containing hydrogen gas will be charged with enough voltage to both break the covalent bond between hydrogen atoms in the H_2 molecules in the tube and excite the electrons to higher energy levels. As the excited hydrogen electrons fall back to their ground state, they will emit energy. Some of this energy will be in the form of visible light. You will view the light through a **spectroscope**, which contains a prism. The

prism will separate the component wavelengths of the emitted light. These wavelengths will be displayed on an arbitrary scale that can be viewed through an eyepiece, as shown in Figure 4. The precise positioning of the scale varies from one spectroscope to another.

In order to determine the wavelengths of the spectral lines observed, a **calibration curve** must be determined for the particular spectroscope being used. You will prepare the curve by using a helium discharge tube. The actual wavelengths of the visible helium spectral lines are known and are shown in Figure 5. You will determine the scale readings corresponding to each helium line and draw a calibration curve for the spectroscope. A sample calibration curve is shown in Figure 6. You will use the calibration curve to determine the wavelengths of the hydrogen spectral lines. Using these wavelengths, you will calculate ΔE from Equation 3 and n_2 from Equation 1 for each hydrogen emission

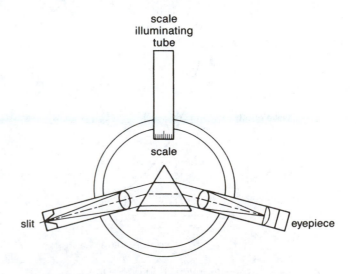

Figure 4 *Schematic of a spectroscope*

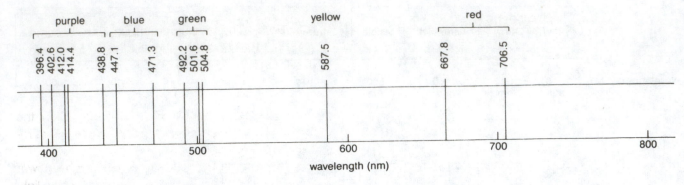

Figure 5   *Helium spectral lines*

line. The following example illustrates these calculations for a wavelength of 486.3 nm.

The amount of energy representing an emission of a wavelength 486.3 nm can be determined using Equation 3.

$$\Delta E = \frac{hc}{\lambda} = \frac{(6.626 \times 10^{-34} \text{ J s}) (2.998 \times 10^{8} \text{ m s}^{-1})}{(486.3 \text{ nm})\left(\dfrac{1 \text{ m}}{1 \times 10^{9} \text{ nm}}\right)}$$

$$= 4.085 \times 10^{-19} \text{ J}$$

We can determine the level from which the excited electron fell (n_2) by using Equation 1 and recalling that $R = 1.097 \times 10^{-2}$ nm^{-1} and $n_1 = 2$, because the wavelength of the emission is in the visible region of the spectrum.

$$\frac{1}{486.3 \text{ nm}} = 1.097 \times 10^{-2} \text{ nm}^{-1} \left(\frac{1}{2^2} - \frac{1}{n_2^2}\right)$$

Carrying out the division on the left side of the equation, we obtain

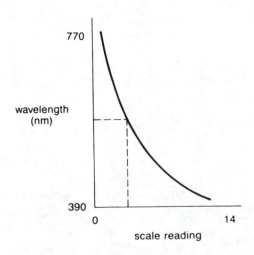

Figure 6   *Sample calibration curve for a spectroscope*

$$2.056 \times 10^{-3} \text{ nm}^{-1} =$$

$$1.097 \times 10^{-2} \text{ nm}^{-1} \left(\frac{1}{2^2} - \frac{1}{n_2^2}\right)$$

Dividing both sides by R, we can simplify the expression.

$$\frac{2.056 \times 10^{-3} \text{ nm}^{-1}}{1.097 \times 10^{-2} \text{ nm}^{-1}} = \left(\frac{1}{4} - \frac{1}{n_2^2}\right)$$

Solving both fractions, we can rewrite the equation as

$$0.1874 = 0.2500 - \frac{1}{n_2^2}$$

which we can rearrange and simplify to

$$\frac{1}{n_2^2} = 0.0626$$

We can then solve the expression for n_2.

$$n_2^2 = 15.97$$

$$n_2 = (15.97)^{1/2} = 3.996, \text{ approximately 4}$$

Therefore, the electron fell from $n = 4$ to $n = 2$. We can conclude that the energy difference between the levels of $n = 4$ and $n = 2$ is 4.085×10^{-19} J. We can depict diagrammatically the electron transition and associated energy, as shown in Figure 7.

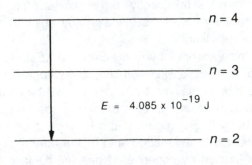

Figure 7   *Electron transition and energy change*

Procedure

> **Caution:** Wear departmentally approved eye protection while doing this experiment.

> **Note:** Your laboratory instructor will give you directions for using your power supply and spectroscope.

Place a helium discharge tube in the holders of the discharge tube power supply.

> **Caution:** Do not touch the discharge tube or the holder while the power supply is on. A severe electrical shock will result if you touch the apparatus.

Turn the power supply knob to **ON** and move the spectroscope "slit" arm toward the discharge tube until the tube and the slit nearly touch.

Adjust the light source so that it completely illuminates the end of the scale arm of the spectroscope. Adjust the eyepiece on the third arm of the spectroscope until the scale and the helium emission lines are in exact focus. The spectrum lines should be sharply defined. Ask your laboratory instructor for help if you have difficulty adjusting the spectroscope.

Record the scale reading of each of the helium emission lines in the appropriate spaces on your Data Sheet. The positioning of the lines on your Data Sheet is similar to the actual spacing of the lines you will view in your spectroscope.

> **Note:** It may be difficult to find all of the purple lines, but this can usually be done with slight adjustments in focus and a little patience. Move your eyes and not your head when making scale readings.

Your laboratory instructor should check your data for the helium emission lines before you proceed further.

> **Caution:** The discharge tubes can become very hot if the power supply is left on for a long time. Avoid touching them.

Turn the power supply knob to **OFF** and allow the spectrum tube to cool for at least one minute before removing it from the holder.

Insert a hydrogen discharge tube in the power supply and turn it on. Check the adjustment of the light source, slit, and eyepiece. Record the scale readings for the four visible emission lines of hydrogen on your Data Sheet.

Observe and qualitatively describe the spectra of neon, argon, and any other elements or compounds for which you have discharge tubes. Check with your laboratory instructor for specific instructions. The descriptions should be complete enough to allow comparisons of the spectra. A simple drawing of the spectrum of each may also prove useful. The use of colored pencils is often helpful.

Calculations

1.   Prepare a calibration curve for your spectroscope by plotting your scale readings against the known wavelengths of the helium lines. Use Figure 6 as a reference. Draw a **smooth curve** through the points, **using a French curve.**

2.   Using your calibration curve, determine the wavelength of each of the four hydrogen lines. Record these wavelengths on your Data Sheet.

3.   Using Equation 3, determine the amount of energy (ΔE) associated with each of the four wavelengths in the hydrogen spectrum. Record these energies on your Data Sheet.

4.   Using Equation 1, determine the level from which the electron fell (to $n_1 = 2$) to produce the emission. Record this level on your Data Sheet.

5.   As a means of correlating these data, indicate the electron transition and associated energy for each of the four wavelengths on the diagram on your Data Sheet. Use the diagram at the end of the Background Information as a model. With care, you should be able to include the information for all four lines on the same diagram.

Post-Laboratory Questions

(Use the spaces provided for the answers and additional paper if necessary.)

1. Three of the known wavelengths of the emission spectrum of helium are 402.6 nm, 447.1 nm, and 587.5 nm.

(1) Using Equation 1, calculate n_2 for these three lines.

λ = 402.6 nm, n_2 = _____

λ = 447.1 nm, n_2 = _____

λ = 587.5 nm, n_2 = _____

(2) Draw a diagram depicting the electron transitions that occurred to produce these emissions.

(3) Explain why your results in (2) indicate that the electron behavior of atoms of elements with more than one electron is not well described by Equation 1, if Bohr's picture of the atom is correct.

2. The ionization energy of hydrogen is reported to be 1.31×10^6 J mol^{-1} of hydrogen.

(1) Using your text or other reference, explain what is meant by ionization energy.

(2) How does ionization differ from excitation with respect to electron behavior?

(3) Compute the ionization energy for a single atom of hydrogen.

_____
answer

(4) Determine the wavelength that would characterize the amount of energy computed in (3) above.

_____
answer

(5) What type of radiation, such as ultraviolet, visible, or infrared, would be required to cause ionization of the hydrogen atom? Briefly explain.

_____   _____   _____
name   section   date

Data Sheet

	helium			hydrogen	
spectral line color	wavelength, nm	spectrum	scale reading	scale reading	wavelength (from the calibration curve), nm
	396.4				
	402.6				
purple	412.0 414.4				
	438.8				
blue	447.1 471.3				
green	492.2 501.6 504.8				
yellow	587.5				
red	667.8 706.5				

name   section   date

Pre-Laboratory Assignment

1. Read an authoritative source for a discussion of graphing techniques.

2. Why should you not touch the holder or discharge tube while the power supply to the gas tube is on?

3. The element sodium, after having been excited electrically, emits an amount of energy equal to 3.37×10^{-19} J atom^{-1}.

   (1) Calculate the wavelength of this light.

                                  _____
                                     answer

   (2) In what region of the electromagnetic spectrum would the emission fall? Briefly explain.

   (3) If the light is colored, what color is it?

4. A student performing the experiment described in this module collected the following data for helium and hydrogen.

helium lines

wavelength, nm	scale reading	wavelength, nm	scale reading
396.4	12.3	492.2	6.9
402.6	11.2	501.6	6.6
412.0	10.7	504.8	6.5
414.4	10.6	587.5	4.6
438.8	9.1	667.8	3.5
447.1	8.7	706.5	3.1
471.3	7.7		

hydrogen lines
scale readings

10.7
9.3
7.1
3.6

   (1) Plot a calibration curve relating the scale readings to the known values for the wavelengths of emission of helium.

   (2) Using the calibration curve prepared in (1), determine the wavelength of the hydrogen line at scale reading 10.7.

                         _____
                          answer

   (3) What color is the light emitted at the wavelength determined in (2)?

   (4) Calculate the energy (ΔE) corresponding to the wavelength determined in (2).

                         _____
                          answer

   (5) Determine the level (n_2) from which the electron fell to $n = 2$ to produce the emission discussed in (2).

                         _____
                          answer

Determining Aluminum(III) Concentration in Natural Water

prepared by **Kenneth E. Borst**, Rhode Island College, Providence, RI
and **Raymond McNulty**, Texas Instruments, Attleboro, MA

Purpose of the Experiment

Determine the aluminum(III) ion concentration in samples of natural and treated water using a prepared standard absorption curve at 535 nm for the aluminum(III)–Eriochrome Cyanine R complex.

Background Information

During recent years, increasing concern has been expressed as to the effect of aluminum(III) ion concentration in natural and treated waters. Aluminum(III) has been identified as one of the products of acid rain that is harmful to fish and trees. Acid rain includes all types of precipitation that are significantly more acidic than natural rain.

Brook trout show a specific toxic response to aluminum(III) levels greater than or equal to 0.2 mg L^{-1} in the pH range 4.4–5.9. Growth reduction of the fish occurs at aluminum(III) levels of 0.1–0.3 mg L^{-1}. Lakes of about neutral pH show aluminum(III) levels 10–50 times lower, that is, about 4×10^{-3} mg L^{-1}. Even such relatively low aluminum(III) levels cause young salmon to die from gill malformations, and damage to tree root systems is observed.

The Environmental Protection Agency has not set a maximum contaminant level for dissolved aluminum. However, dissolved aluminum has been associated with renal failure in humans, and the aluminum(III) concentration in kidney dialysis water must be below 1×10^{-2} mg L^{-1}. Thus, a recommendation has been made that aluminum(III) levels in drinking water be maintained below 1×10^{-1} mg L^{-1} (0.1 ppm).

Precipitation with pH below 5.6 is considered to be acid rain. Atmospheric carbon dioxide is responsible for the slight natural acidity in precipitation of pH 5.6. This acidity is the result of the following reactions.

Dissolution: $CO_2(g) \rightleftarrows CO_2(aq)$ (Eq. 1)

Hydrolysis: $CO_2(aq) + H_2O(l) \rightleftarrows$
$$H_2CO_3(aq)$$ (Eq. 2)

Dissociation: $H_2CO_3(aq) + H_2O(l) \rightleftarrows$
$$HCO_3^{-1}(aq) + H_3O^+(aq)$$ (Eq. 3)

Acid rain is considered to result form atmospheric pollutants, largely derived from fossil fuels. Some of these fuels contain sulfur and sometimes nitrogen compounds as impurities. Combustion of such fuels produces oxides of sulfur and nitrogen. These oxides react with water in the atmosphere to form sulfuric acid and nitric acid. These substances are carried by the revailing winds and later fall as acid rain. As the acid rain enters surface water, the pH of the water is lowered. The resulting acidic water can leach aluminum(III) from the minerals in the soil.

Copyright © 1986 by Chemical Education Resources, Inc., P.O. Box 357, 220 S. Railroad, Palmyra, Pennsylvania 17078
No part of this laboratory program may be reproduced or transmitted in any form or by any means, electronic or mechanical, including photocopying, recording, or any information storage and retrieval system, without permission in writing from the publisher. Printed in the United States of America

Aluminum is the third most abundant element in the Earth's outer crust, rocks, and soils. In soils, aluminum is found in some clay minerals such as kaolin, muscovite, and white mica silicates. The typical formula for the kaolin group of silicates is $Al_2Si_2O_5(OH)_4$, whereas that of white mica is $KAl_2(AlSi_3O_{10})(OH)_2(F)_2$.

Aluminum(III) is rarely found in concentrations greater than 0.5 mg L^{-1} in fresh water. However, leaching caused by acid rain introduces additional aluminum(III) to the surface water. The following **unbalanced** reaction shows the effect of acid rain, H_3O^+, on feldspar, $KAlSi_3O_8$. As a result of this reaction, aluminum(III) is leached into surface water and becomes part of our water system.

$$KAlSi_3O_8(s) + H_3O^+(aq) \rightleftarrows$$
$$Al_2Si_2O_5(OH)_4(s) + H_4SiO_4(aq)$$
$$+ K^+(aq) + Al^{3+}(aq) \quad \text{(Eq. 4)}$$

There is a direct relationship between the concentration of aluminum(III) and the alkalinity of natural waters. Increased aluminum(III) concentrations are found in areas of the United States where the pH of the natural water is about 5. The EPA classifies these natural waters as "critical" or "acidified." In some parts of the country, the natural waters are alkaline or well buffered. We would not expect to find significant concentrations of aluminum(III) in such waters.

When natural water enters a water-purification plant, aluminum(III) is often added to the waters. Part of the treatment of natural water involves the addition of coagulates to remove undesirable color from the water. Aluminum sulfate and polyaluminum chlorides are often used as coagulates. Thus, any aluminum(III) in treated water could have been introduced in several different ways.

In this experiment, a spectrophotometric method is used to determine the concentration of aluminum(III) in several different water samples. For this procedure, the aluminum(III) must be quantitatively converted to a colored complex. A suitable agent for complexing aluminum(III) is Eriochrome Cyanine R (ECR). A solution containing the aluminum(III)–ECR complex ion is red and has a maximum absorbance at 535 nm. The composition of the complex is affected by pH, with the greatest stability at a pH of approximately 6.2. The equation for formation of the complex is:

$$Al^{3+}(aq) + 2\ ECR^{3-}(aq) \rightleftarrows Al(ECR)_2^{3-}(aq) \quad \text{(Eq. 5)}$$
$$\text{(orange-red)} \qquad\qquad \text{(red)}$$

In addition to ECR, several other reagents are added to the sample being analyzed for aluminum(III). Very dilute sulfuric acid and an acetate buffer are added to maintain optimum pH. Ascorbic acid is added to complex dissolved iron and manganese, which would otherwise interfere with the aluminum(III) determination. The ascorbic acid solution must be freshly prepared, as it oxidizes if left standing.

A blank solution containing all reagents, but no aluminum(III), is prepared first for zeroing the spectrophotometer. Then, a series of standard solutions is prepared with known concentrations of aluminum(III). The percent transmittance of these soutions is measured and converted to absorbance, from which a Beer's Law curve is prepared, which is then used as a standard aluminum(III) curve. From this standard curve, the concentration of aluminum(III) in several water samples is determined.

A measured quantity of a water sample is prepared by adding the reagents to form quantitatively the aluminum(III)–ECR complex. The measured absorbance of the sample is used to calculate the concentrations of aluminum(III) in the sample.

A different blank must be used for natural and treated water samples. The compositions of the blank solutions are different in order to prevent the trace components that might be present in the water sample from interfering with the analysis. The water sample itself is used to prepare the blank. The preparation is identical to that for the previous solutions, except that a small amount of EDTA is added. EDTA is a complexing agent for metal ions. EDTA complexes have very high formation constants, which assures the removal of any possible interfering metal ions. For the synthetic water samples, the same blank can be used as the one used with the aluminum(III) standards.

This investigation will involve determining the concentration of aluminum(III) in several different types of water samples. One sample will be from natural water, such as a lake, pond, stream, or river. A second sample will be from treated water that has been processed at a purification plant for human consumption. The third sample will be a synthetic one prepared by your laboratory instructor, containing a fixed concentration of aluminum(III) that is unknown to you.

You will complex the aluminum(III) in each sample, measure the percent transmittance of the solution, and calculate its absorbance. From your standard aluminum(III) curve, you will determine the concentration of aluminum(III) in each of the samples.

Procedure

I. Sampling a Water Source

> **Note:** Samples of natural water should be collected no sooner than 24 hours prior to doing the analysis. For better and more consistent results, refridgerate the sample or place it on ice until the determination is done.

A. Selecting the Stream Site

For stream sampling, choose a location that will provide representative samples. Collect samples as near to the center of the stream as practical. Sample only flowing water; avoid stagnant or still water. If you plan to monitor the stream again, mark the original sampling site with a stake and sign.

Lake or pond sampling is more complex. A mid-lake sample is preferable to a shoreline sample. For mid-lake monitoring, select three permanent objects on shore and use them to triangulate the sampling site.

B. Collecting the Sample

Use a container specified by your laboratory instructor. Collect the sample below the surface, but avoid disturbing the bottom sediment. Sample at a depth of an arm's length in a lake or pond.

Collect a sample by lowering the bottle upside down into the water to the desired depth. Turn the bottle right side up and allow it to fill completely. Cap the filled bottle while it is underwater. Be sure no air is trapped above the sample in the bottle.

Label the bottle with your name, the date, and the sample location.

II. Obtaining Data to Prepare the Standard Aluminum Curve

> **Caution:** Wear departmentally approved eye protection during the experiment.

A. Preparing the Spectrophotometer for Use

Turn on the instrument and let it warm up for 10–15 min. Set the wavelength at 535 nm. With no cuvette in the instrument, use the on-off-zero knob to obtain a steady reading of 0% transmittance.

B. Preparing the Standard Aluminum(III) Solutions

> **Note:** You will prepare six solutions, each containing a different concentration of aluminum(III) ion. You will prepare these solutions from a stock standard solution.

1. Preparing Blank Solution #1: Label a clean 100.00-mL volumetric flask "Blank Solution #1."

> **Caution:** Handle dilute sulfuric acid with care.

Use a graduated cylinder to measure each of the following and then add them to the flask.

20.0 mL of buffer solution
2.0 mL of ascorbic acid solution
2.0 mL of $2.0 \times 10^{-2} N(1.0 \times 10^{-1} M)$ H_2SO_4

Pipet 10.00 mL of ECR solution into the flask. Mix thoroughly. Add distilled water to the calibration mark. Mix the solution thoroughly again.

2. Determining the percent transmittance of Blank Solution #1: Fill a clean, dry cuvette with Blank Solution #1. Use this blank to set the instrument to 100% transmittance at 535 nm. Record on Data Sheet 1 the percent transmittance of Blank Solution #1.

3. Preparing standard aluminum(III) solutions:

> **Note:** The aluminum(III) complex is not stable after 20 min. Determine the percent transmittance of the following solutions within 10–20 min after preparation.

Record on Data Sheet 1 the aluminum(III) concentration in mg L^{-1} of the aluminum(III) stock solution.

Prepare Standard Solutions #1, 2, 3, 4, and 5, each in a clean 100.00-mL volumetric flask. Follow the procedure used in preparing Blank Solution #1, but add aluminum(III) stock solution as follows: 1.00 mL for Standard Solution #1, 2.00 mL for #2, 4.00 mL for #3, 6.00 mL for #4, and 8.00 mL for #5.

4. Determining the percent transmittance of the standard aluminum(III) solutions: Measure the percent transmittance of each solution within 10–20 min after preparation. Record on Data Sheet 1 the percent transmittance of each solution.

> *Note:* Your laboratory instructor will inform you whether your investigation is to involve samples prepared from a stock solution, treated water, natural water, or some combination of all three.

III. Determining the Concentration of Aluminum(III) in a Synthetic Water Sample

Record on Data Sheet 2 the code number of the synthetic water sample assigned to you.

A. Preparing the Solution of the Synthetic Water Sample

Label a clean 50.00-mL volumetric flask "Synthetic Water Solution". Pipet 25.00 mL of the synthetic water sample into the flask.

Use a graduated cylinder to measure each of the following and add them to the sample in the flask:

10.00 mL of buffer solution
1.0 mL of ascorbic acid solution
1.0 mL of $2.0 \times 10^{-2}N(1.0 \times 10^{-1}M)$ H_2SO_4

Pipet 5.00 mL of ECR solution into the flask. Mix the solution thoroughly. Add distilled water to the calibration mark. Mix thoroughly again.

Be sure to make your determination within 10–20 min after preparing the synthetic water solution.

B. Determining the Percent Transmittance of the Synthetic Water Solution

At 535 nm, set the instrument at 100% transmittance with Blank Solution #1. Fill a clean, dry cuvette with your synthetic water solution and place the cuvette in the instrument. Determine the percent transmittance at 535 nm. Record the transmittance on Data Sheet 2.

Do a second and third determination if time permits.

IV. Determining the Concentration of Aluminum(III) in a Natural Water Sample

Record on Data Sheet 3 the code number of the natural water sample assigned to you.

A. Preparing the Natural Water Sample for Analysis

1. Estimating the volume of H_2SO_4 to be used: Measure 25.0 mL of the natural water sample in a graduated cylinder. Transfer this portion of your sample to a porcelain evaporating dish or 250-mL Erlenmeyer flask. Add 2 drops of methyl orange indicator. Titrate with $2.0 \times 10^{-2}N(1.0 \times 10^{-1}M)$ H_2SO_4 to a faint pink color. Record the volume of titrant used on Data Sheet 3.

2. Preparing Blank Solution #2: Label a clean 50.00-mL volumetric flask "Blank Solution #2". Pipet 25.00 mL of the natural water sample into this flask.

> *Note:* Prepare the following solutions by using the same volume of acid as was used in the titration in Step 1 of Part IV*A*, plus 1.0 mL.

Use a graduated cylinder to measure each of the following, and add each to the sample in the flask:

the volume of $2.0 \times 10^{-2}N(1.0 \times 10^{-1}M)$
   H_2SO_4 to be used
1.0 mL of ascorbic acid solution
10.0 mL of buffer solution

Pipet 5.00 mL of ECR solution into the flask. Measure 1.0 mL of $1 \times 10^{-1}M$ EDTA solution in a graduated cylinder, and add this solution to the flask. Mix thoroughly. Add distilled water to the calibration point. Mix thoroughly again.

Be sure to make your determination within 10–20 min after preparing your natural water solution.

3. Preparing the solution of the natural water sample: Follow the procedure used to prepare Blank Solution #2, but *do not* add the EDTA solution. Label this flask "Natural Water Solution."

B. Determining the Percent Transmittance of the Natural Water Solution

At 535 nm, set the instrument at 100% transmittance with Blank Solution #2. Place a cuvette containing the natural water sample in the instrument. Measure the percent transmittance. Record the transmission on Data Sheet 3.

Do a second and third determination if time permits.

V. Determining the Concentration of Aluminum(III) in a Treated Water Sample

Record on Data Sheet 4 the code number of the treated water sample assigned to you.

A. Preparing the Treated Water Sample for Analysis

1. Estimating the volume of H_2SO_4 to be used:

Note: Some water from water treatment plants will have a pH of 8.0 or above. In such cases, the following procedure must be adjusted to avoid exceeding a total solution volume of 50.00 mL. Your laboratory instructor will describe the changes that should be made.

Measure 25.0 mL of the treated water sample in a graduated cylinder. Transfer this portion of the sample to a porcelain evaporating dish or 250-mL Erlenmeyer flask. Add 2 drops of methyl orange indicator. Titrate with $2.0 \times 10^{-2} N(1.0 \times 10^{-1} M)$ H_2SO_4 to a faint pink color. Record on Data Sheet 4 the volume of titrant used.

2. Preparing Blank Solution #3: Label a clean 50.00-mL volumetric flask "Blank Solution #3." Pipet 25.00 mL of the treated water sample into this flask.

Note: Prepare the following solutions by using the same volume of acid as was used in the titration in Step 1, Part VA, plus 1.0 mL.

Use a graduated cylinder to measure each of the following, and add each to the sample in the flask:

the volume of $2.0 \times 10^{-2} N(1.0 \times 10^{-1} M)$
  H_2SO_4 to be used
1.0 mL of ascorbic acid solution
10.0 mL of buffer solution

Pipet 5.00 mL of ECR solution into the flask. Measure 1.0 mL of $1 \times 10^{-1} M$ EDTA solution in a graduated cylinder, and add this solution to the flask. Mix thoroughly. Add distilled water to the calibration mark. Mix thoroughly again.

3. Preparing the solution of the treated water sample: Follow the procedure used to prepare Blank Solution #3, but *do not* add the EDTA solution. Label this flask "Treated Water Solution."

Be sure to make your determination within 10–20 min after preparing your treated water solution.

B. Determining the Percent Transmittance of the Treated Water Solution

At 535 nm, set the instrument at 100% transmittance with Blank Solution #3. Place a cuvette containing the treated water solution in the instrument. Measure the percent transmittance. Record the transmission on Data Sheet 4.

Do a second and third determination if time permits.

Calculations

II. Preparing the Standard Aluminum(III) Curve

(Note: Record all data calculated in this section on Data Sheet 1.)

1. Calculate the concentration of the standard solutions.

$$\frac{\left(\begin{array}{c}mg\ L^{-1}\\ of\ stock\ solution\end{array}\right)\left(\begin{array}{c}volume\ of\\ sample,\ L\end{array}\right)}{(1.00 \times 10^{-1}\ L)} = mg\ L^{-1} \quad \text{(Eq. 6)}$$

2. Convert the percent transmittance (%T) of the stock solution to absorbance (A).

$$A = 2 - \log \%T \quad \text{(Eq. 7)}$$

3. Prepare a standard aluminum(III) curve to be used for determining aluminum(III) concentrations. Plot the absorbance (A) of the standard aluminum(III) solutions against the concentration of aluminum(III) in mg L^{-1}. Use the ordinate (the y-axis) for the absorbance, with scale units of 0.000 to 0.900. Use the abscissa (the x-axis) for the aluminum(III) concentration, with scale units of 0.000 to 0.400.

III. Determining the Concentration of Aluminum(III) in a Synthetic Water Sample

(Note: Record all data calculated in this section on Data Sheet 2.)

1. Convert your three determinations of the percent transmittance (%T) of the synthetic water solution to absorbances (A), using Equation 7.

2. Use the standard aluminum(III) curve to obtain the concentration of aluminum(III) in the solution for each of the three determinations.

3. Calculate the concentration of aluminum(III) in the sample for each of the three determinations by using Equation 8, where the dilution factor is 2.

(mg L^{-1} from standard curve)(2) =
　　　　　　　　mg L^{-1} in sample　(Eq. 8)

4. Calculate the average concentration of aluminum(III) in the sample.

IV. Determining the Concentration of Aluminum(III) in a Natural Water Sample

(Note: Record all data calculated in this section on Data Sheet 3.)

1. Convert your three determinations of the percent transmittance of the natural water solution to absorbances, using Equation 7.

2. Use the standard aluminum(III) curve to obtain the concentration of aluminum(III) in the solution for each of the three determinations.

3. Calculate the concentration of aluminum(III) in the sample for each of the three determinations by using Equation 8.

4. Calculate the average concentration of aluminum(III) in the sample.

V. Determining the Concentration of Aluminum(III) in Treated Water

(Note: Record all data calculated in this section on Data Sheet 4.)

> *Note:* If your treated water sample has been fluoridated, your laboratory instructor will give you a correction factor for these calculations.

1. Convert your three determinations of the percent transmittance of the treated water solution to absorbances, using Equation 7.

2. Use the standard aluminum(III) curve to obtain the concentration of aluminum(III) in the solution for each of the three determinations.

3. Calculate the concentration of aluminum(III) in the sample for each of the three determinations by using Equation 8.

4. Calculate the average concentration of aluminum(III) in the sample.

Post-Laboratory Questions

(Use the spaces provided for the answers and additional paper if necessary.)

1. EDTA is a complexing reagent for many metal ions. In Parts IV*A* and V*A* of the Procedure, why is EDTA added to Blank Solution #2 and Blank Solution #3, but not to the solutions of the natural or treated water samples?

2. The main source of error in this experiment is volumetric measurements, either with a graduated cylinder, pipet, or microburet. Assume, in Part II*B* of the Procedure, that you dispensed 5.00 mL of the stock aluminum(III) solution *instead* of 6.00 mL when preparing Standard Solution #4. Assume the standard aluminum(III) curve you originally drew is still correct, even though the aluminum(III) concentration of Solution #4 has now changed. Use this curve for the following calculations.

(a) Calculate the absorbance (A) you would expect from this solution.

_____
answer

(b) Calculate the relative percent error in the absorbance as a result of adding only 5.00 mL of stock aluminum(III) solution to Solution #4 instead of 6.00 mL.

_____
answer

_____   _____   _____
name   section   date

3. A student did not add EDTA when preparing Blank Solution #2 in Part IV*A* of the Procedure, part of determining the aluminum(III) concentration of a natural water sample. Assume the student did not make any other mistakes in the determination. Estimate the absorbance of the student's Blank Solution #2. Briefly explain your answer.

4. When the student repeated the determination of the aluminum(III) concentration in a natural water sample, Blank Solution #1 (prepared in PartII*B* for obtaining data to prepare the standard aluminum(III) curve) was used by mistake, instead of Blank Solution #2. Would you expect the student to have a large error in this determination because of using Blank Solution #1? Briefly explain your answer.

Data Sheet 1

II. Preparing the Standard Aluminum Curve

[Al^{3+}] in the aluminum(III) stock solution, mg L^{-1} _____

Blank Solution #1: % transmittance _____   absorbance _____

standard solution number	aluminum stock solution, mL	mg L^{-1}	percent transmittance	absorbance
1	_____	_____	_____	_____
2	_____	_____	_____	_____
3	_____	_____	_____	_____
4	_____	_____	_____	_____
5	_____	_____	_____	_____

Data Sheet 2

III. Determining the Concentration of Al(III) in a Synthetic Water Sample

Name, number, or code letter of synthetic water sample _____

	determination 1	determination 2	determination 3
percent transmittance (%T)	_____	_____	_____
absorbance (A)	_____	_____	_____
[Al^{3+}] from the standard curve, mg L^{-1}	_____	_____	_____
[Al^{3+}] in the sample, mg L^{-1}	_____	_____	_____
average [Al^{3+}] in the sample	_____	_____	_____

Data Sheet 3

IV. Determining the Concentration of Al(III) in a Natural Water Sample

Name, number, or code letter of natural water sample _____

Volume of $2.0 \times 10^{-2} N$ H_2SO_4 required to reach the endpoint of a 25.0 mL sample _____

	determination		
	1	2	3
percent transmittance (%T)	_____	_____	_____
absorbance (A)	_____	_____	_____
[Al^{3+}] from the standard curve, mg L^{-1}	_____	_____	_____
[Al^{3+}] in the sample, mg L^{-1}	_____	_____	_____
average [Al^{3+}] in the sample	_____	_____	_____

Data Sheet 4

V. Determining the Concentration of Al(III) in a Treated Water Sample

Name, number, or code letter of treated water sample _____

Volume of $2.0 \times 10^{-2} N$ H_2SO_4 required to reach the endpoint of a 25.0 mL sample _____

	determination		
	1	2	3
percent transmittance (%T)	_____	_____	_____
absorbance (A)	_____	_____	_____
[Al^{3+}] from the standard curve, mg L^{-1}	_____	_____	_____
[Al^{3+}] in the sample, mg L^{-1}	_____	_____	_____
average [Al^{3+}] in the sample	_____	_____	_____

Pre-Laboratory Assignment

A student was given a stock aluminum(III) solution with a known concentration of 5.000 mg L^{-1}. The student prepared the six solutions as described in Part IIB of the Procedure and determined the percent transmittance of each solution at 535 nm.

blank solution #	volume of aluminum(III) stock solution	$[Al^{3+}]$, mg L^{-1}	%T	A
1	0.00	_____	100.0	_____
2	1.00	_____	81.0	_____
3	2.00	_____	61.0	_____
4	4.00	_____	35.2	_____
5	6.00	_____	20.1	_____
6	8.00	_____	13.2	_____

1. For each of the six solutions, calculate the concentration of aluminum(III) in mg L^{-1} and the absorbance. Enter these data in the table above.

3. An aluminum(III) solution prepared by a laboratory instructor had a percent transmittance of 46.8. Determine the concentration of aluminum(III) in this solution.

_____
answer

2. Use the data in the table to construct a standard aluminum(III) curve on the graph paper on the back of this page.

4. A solution of a natural water sample prepared as described in Part IVA of the Procedure had a percent transmittance of 14.5. Determine the concentration of aluminum(III) in this solution.

_____
answer

Spectrophotometric Determination of the Formula of a Complex Ion

prepared by **Emily P. Dudek**, Brandeis University

Purpose of the Experiment

The ratio of copper(II) ion to ethylenediamine in a complex ion, $Cu(H_2NCH_2CH_2NH_2)_n^{2+}$, will be investigated.

Background Information

For the complex ion $Cu(H_2NCH_2CH_2NH_2)_n^{2+}$, the most likely values of n are 1, 2, or 3. The Cu(II) ion in this complex is bound to the nitrogen atoms of an ethylenediamine molecule. Each Cu(II)–nitrogen interaction is viewed as a coordinate bond in which the nitrogen donates a pair of electrons toward the Cu(II) ion. In aqueous solutions of Cu(II) ions, the oxygen atoms of the water molecules form coordinate bonds with the Cu(II) ion. The usual number of bonds formed by the Cu(II) ion is six, giving rise to an elongated octahedral structure. This geometry may be visualized by placing the Cu(II) ion at the origin of an x, y, z-coordinate system and drawing four equal bonds along the x- and y-axes and two longer bonds along the z-axis. Shown in Figure 1 are structural formulas for the aqueous Cu(II) and $Cu(H_2NCH_2CH_2NH_2)_n^{2+}$ ions .

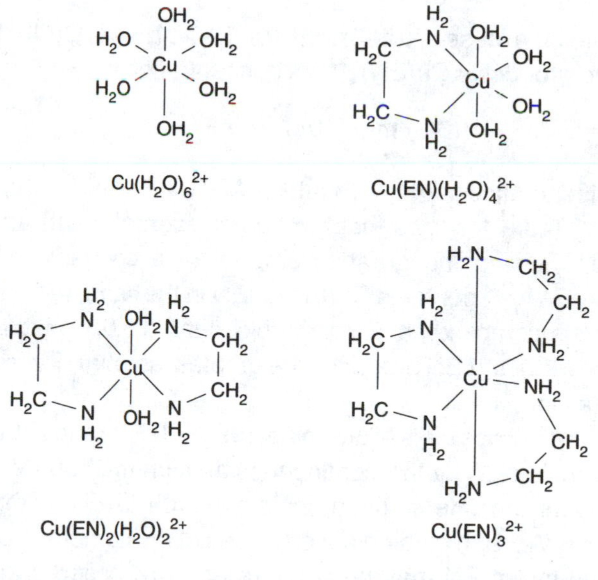

The method of continuous variation, also called Job's method, is used to establish the ratio of Cu(II) ion to ethylenediamine (abbreviated EN). Aqueous solutions of the two components, each with the same initial concentration, are combined in various proportions, but the total volume of the resulting solution is held fixed. Hence, the sum of moles of Cu(II) ion and moles

Figure 1

of EN mixed together is constant. If the absorbances of the solutions containing Cu(II) ion and ethylenediamine mixtures are proportional to the concentrations of the

---

Copyright © 1977 by Chemical Education Resources, Inc., P.O. Box 357, 220 S. Railroad, Palmyra, Pennsylvania 17078
No part of this laboratory program may be reproduced or transmitted in any form or by any means, electronic or mechanical, including photocopying, recording, or any information storage and retrieval system, without permission in writing from the publisher. Printed in the United States of America

$Cu(EN)_n^{2+}$ ion produced, then the solution having the largest absorbance is the solution prepared by mixing Cu(II) ion and EN in the same ratio as in the complex ion.

The rationale of the continuous variation method is illustrated by Table 1, which lists the concentrations of possible $Cu(EN)_n^{2+}$ complex ions formed by mixing various amounts of 10M Cu(II) and 10M EN solutions. The total volume of the mixtures is fixed at 10 mL. To understand how the projected concentrations of complex ions are computed, consider the mixture of 4 mL of a Cu(II) ion solution and 6 mL of an EN solution. If the reaction between Cu(II) and EN is complete and if the product is $Cu(EN)^{2+}$, the concentration of $Cu(EN)^{2+}$ is limited by the volume of Cu(II) solution. The Cu(II) concentration is equal to

$$\frac{(4 \text{ mL}) \,(10M)}{(10 \text{ mL of solution})} = 4M$$

An excess of $(2 \text{ mL})(10M)/10 \text{ mL} = 2M$ of unbound EN remains. If the product is $Cu(EN)_2^{2+}$, its concentration is limited by the volume of EN solution and is equal to

$$[(6/2) \text{ mL}] \,[10M/10 \text{ mL} = 3M$$

with an excess of $[(4-3) \text{ mL}][10M]/10 \text{ mL} = 1M$ Cu(II). If the product is $Cu(EN)_3^{2+}$, its concentration is

$$[(6/3) \text{ mL}] \,[10M]/10 \text{ mL} = 2M$$

with an excess of $[(4-2) \text{ mL}][10M]/10 \text{ mL} = 2M$ Cu(II).

Table 1 shows that the combination of Cu(II) and EN yielding the greatest amount of a complex ion, $Cu(EN)_n^{2+}$, contains Cu(II) and EN in the same ratio as in the complex ion. For example, if the 3.3:6.7 mixture of Cu(II) and EN produces the greatest amount of complex ion, the ion is $Cu_{3.3}(EN)_{6.7}^{2+}$, or $Cu(EN)_2^{2+}$.

It is possible to determine the Cu(II):EN ratio in the complex ion by the continuous variation method even though no one solution contains Cu(II) to EN in that ratio. A graph plotting absorbances of the complex versus mL of EN may be constructed, a so-called Job's plot, and the Cu(II) to EN ratio giving the greatest absorbance may be obtained by extrapolation. For example, Figure 2 pictures a Job's plot for a 1:2 complex ion, such as $Cu(EN)_2^{2+}$, at two different wavelengths, λ_1 and λ_2. If the 1:2 complex ion is the only absorbing species in the solution, the intersection of the two linear portions of the Job's plot corresponds to the Cu(II):EN ratio in the complex ion irrespective of the wavelength chosen.

If the Cu(II) or EN species, or both, also exhibit some absorbance in the region of the spectrum where

Table 1 *Concentrations of possible $Cu(EN)_n^{2+}$ complex ions*

volume of 10M solution, mL		concentrations in moles per liter		
Cu(II)	EN	$Cu(EN)^{2+}$	$Cu(EN)_2^{2+}$	$Cu(EN)_3^{2+}$
10	0	0	0	0
8	2	2	1	0.67
6	4	4	2	1.3
5	5	5	2.5	1.7
4	6	4	3	2
3.3	6.7	3.3	3.3	2.2
2.5	7.5	2.5	2.5	2.5
2	8	2	2	2
1	9	1	1	1
0	10	0	0	0

the complex ion absorbs strongly, the observed absorbance must be corrected for this additional factor. The resulting corrected absorbance is proportional to the concentration of the complex ion. Suppose, for example, the observed absorbance is 0.55 for the 4:6 mixture, while at the same wavelength a 10M Cu(II) solution shows an absorbance of 0.05. The Cu(II) in the mixture could exhibit an absorbance of at most $[(4/10)(0.05) = 0.02]$, and the corrected absorbance would equal $[0.55 - 0.02 = 0.53]$. This is an over-correction, since some or all of the Cu(II) in the mixture may be bound to EN. Unless the formula of the $Cu(EN)_n^{2+}$ ion is established, however, the amount of unbound Cu(II) ion in solution is not known.

When more than one $Cu(EN)_n^{2+}$ ion is formed, depending upon the relative amounts of Cu(II) and EN in solution, the relationship between absorbance of solution and concentration of individual $Cu(EN)_n^{2+}$ ions is complicated unless a wavelength can be found for each of the complex ions such that only the complex in question absorbs appreciably. For example, Figure 3

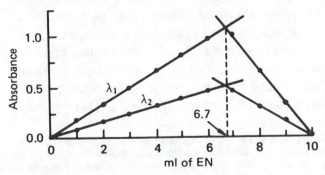

Figure 2 *Job's plot for a 1:2 complex*

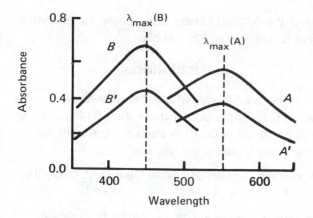

Figure 3 *Absorption spectra of two solutions of species A and two solutions of species B*

might represent the spectra of $A = Cu(EN)^{2+}$ and $B = Cu(EN)_3^{2+}$. A spectrum is a graph of absorbance versus wavelength, and λ_{max} is the wavelength at which the absorbance of a solution is a maximum. The corresponding Job's plot is given in Figure 4.

Note that in Figure 4, the linear portion of the Job's plot for the 1:1 complex occurs only from 0 to 5 mL of EN, while the linear portion for the 1:3 complex occurs only from 7.5 to 10 mL of EN. In the range of 5 to 7.5 mL of EN, where some of each of complex is in solution, the Job's plots are not linear, and linear extrapolations cannot be made.

In this experiment, you will establish the formula for the complex ion $Cu(H_2NCH_2CH_2NH_2)_n^{2+}$ by varying the relative amount of Cu(II) ion to ethylenediamine in aqueous solution, while keeping the total number of moles of the two substances constant. The measured absorbance of each prepared solution is proportional to the concentration of the $Cu(H_2NCH_2CH_2NH_2)_n^{2+}$ ion. Thus, the solution exhibiting the greatest absorbance is the solution containing copper(II) ion and ethylenediamine in the same ratio as in the complex ion.

Procedure

> **Caution:** Wear departmentally approved eye protection while doing this experiment.

Turn the spectrophotometer on and allow it to warm up.

Obtain about 90 mL of 0.020M CuSO$_4$ in a 100-mL beaker labeled Cu, and 90 mL of 0.020M EN in a 100-mL beaker labeled EN.

Prepare eight mixtures of the two solutions according to the volumes given in Table 2. The eight mixtures may be prepared in eight 18 × 150-mm test tubes. Label each test tube according to the mL of EN it will contain and place the tubes upright in a large beaker. Measure out the EN solution using a 10-mL graduated cylinder, and use an eyedropper to add the last few drops of EN solution to the cylinder. Transfer the EN solution to the test tubes as completely as possible. Do this for all eight test tubes. Then rinse the graduated cylinder and eyedropper with distilled water and use them to measure out the prescribed volumes of Cu(II) solution. When the Cu(II) solution has been added to the EN solution in a test tube (for example, when 8 mL of Cu(II) has been added to 2 mL of EN in the test tube labeled 2), cork the tube and shake it well.

Obtain two cuvettes for the spectrophotometer. Fill one about half-full with water and the other half-full with 0.020M Cu(II) solution.

Table 2 *Preparation of eight Cu(II)–EN solutions*

mL of 0.020M EN	mL of 0.020M Cu(II)
2.0	8.0
4.0	6.0
5.0	5.0
6.0	4.0
6.7	3.3
7.5	2.5
8.0	2.0
9.0	1.0

Set the wavelength control of the spectrophotometer to 475 nm, and adjust the zero control until the meter reads zero transmittance, $\%T = 0$. Insert the cuvette containing water into the sample holder, and adjust the light control to $\%T = 100$. Replace the cuvette containing water with the cuvette containing the Cu(II) solution. Record the absorbance of this solution on the Data Sheet. If the spectrophotometer has

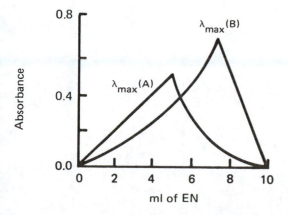

Figure 4 *Job's plots for 1:1 and 1:3 complexes*

only a %T meter, record %T and convert it to absorbance A by using the equation $A = 2 - \log(\%T)$.

Remove the cuvette containing the Cu(II) solution. Set the wavelength to 500 nm, and repeat the %$T = 0$ setting with no cuvette in the sample holder and the %$T = 100$ setting with the cuvette containing water in the sample holder. Record the absorbance of the Cu(II) solution at 500 nm. Proceed to the other wavelength settings listed on the Data Sheet. Remember to reset the %$T = 0$ and the %$T = 100$ readings for each new wavelength setting.

Empty the cuvette containing the Cu(II) solution. Rinse it first with distilled water and then with the Cu(II)–EN mixture containing 5 mL of EN. Repeat the above procedure for this solution, recording all absorbances on the Data Sheet.

In the same way, record absorbances versus wavelength data for the solution mixtures containing 6.7 and 7.5 mL of EN. For the remaining five mixtures

and for the 0.020M EN solution, record the absorbances at 500, 550, and 625 nm only.

Calculations

1. At the wavelength where the absorbance of the 0.020M $CuSO_4$ solution is more than 0.020, subtract the absorbance of Cu(II) in each mixture from the observed absorbance of the mixture,

$$\text{corrected } A = \text{observed } A - \frac{[A \text{ of Cu(II)}]\,[\text{mL of Cu(II)}]}{10 \text{ mL}}$$

and enter the corrected values on the Data Sheet.

2. On one graph, plot the spectra of the solutions containing 5.0, 6.7, and 7.5 mL of EN.

3. On one graph, construct Job's plots for wavelengths 500, 550, and 625 nm. Clearly note the wavelength corresponding to each curve of absorbance versus mL of EN.

Post-Laboratory Questions

(Use the space provided for the answer and additional paper if necessary.)

1. Do the experimental results indicate that more than one $Cu(EN)_n^{2+}$ ion is formed in solution depending upon the relative amounts of Cu(II) and EN? Explain.

3. A colored solution transmits light of that color and absorbs light of other colors. Thus a green solution transmits green light and absorbs red and blue light. The wavelength range corresponding to each color of light is approximately: red, 700–620 nm; orange, 620–590 nm; yellow, 590–570 nm; green, 570–500 nm; blue, 500–450 nm; violet, 450–400 nm. Account for the color of each of the three solutions for which an absorbance vs wavelength plot was made, with respect to the λ_{max} value it displays.

2. Give the formula and corresponding wavelength maximum for each $Cu(EN)_n^{2+}$ ion detected. Also note its color in solution.

_____   _____   _____
name   section   date

name _____   section _____   date _____

Data Sheet

          (a)   *observed absorbances*   (b)   *corrected absorbances*

wave-length (nm)	0	2.0	4.0	5.0	6.0	6.7	7.5	8.0	9.0	10
mL of 0.020M EN										
475(a)	___			___		___	___			
500(a)	___	___	___	___	___	___	___	___	___	___
525(a)	___			___		___	___			
550(a)	___	___	___	___		___	___	___	___	___
(b)	___	___	___	___	___	___	___	___	___	___
575(a)	___			___		___	___			
(b)	___			___		___	___			
600(a)	___			___		___	___			
(b)	___			___		___	___			
625(a)	___	___	___	___	___	___	___	___	___	___
(b)	___	___	___	___	___	___	___	___	___	___

• Formula of a Complex Ion

name _______________________ section __________ date __________

Pre-Laboratory Assignment

1. (a) To determine the formula of a complex ion AB_n, various amounts of a 0.50*M* solution of A were combined with a 0.50*M* solution of B. The total volume of each mixture was fixed at 10 mL. On the accompanying graph paper, draw a Job's plot of the following data and from the graph find the formula of the AB_n ion.

absorbance	*mL of B*
0.18	1.0
0.27	2.0
0.35	3.0
0.44	4.0
0.52	5.0
0.61	6.0
0.60	7.0
0.40	8.0
0.20	9.0

____
answer

(b) If the reaction between A and B is complete, what is the concentration of the AB_n ion where

(i) mL of B = 2.0

__________*M*
answer

(ii) mL of B = 9.0

__________*M*
answer

(c) Are the absorbances proportional to the concentrations of the AB_n ion where mL of B = 2.0 and 9.0? If the absorbance is not proportional to the concentration of AB_n, give a reason why it is not.

A Colorimetric Determination of Aspirin in Commercial Preparations

prepared by **Robert P. Pinnell**, Claremont McKenna, Scripps, and Pitzer Colleges

---

Purpose of the Experiment

Determine colorimetrically the amount of acetylsalicylic acid in commercially available pharmaceutical preparations.

Background Information

I. Considering Beer's Law

The characteristics of colored solutions have been of interest to chemists for a long time. Of particular interest has been the fact that colored solutions, when irradiated with white light, will selectively absorb incident light of some wavelengths but not of others. We can determine the particular wavelength or group of wavelengths absorbed by systematically exposing the solution to monochromatic light of different wavelengths and recording the responses. If light of a particular wavelength is not absorbed, the intensity of the beam directed at the solution (I_o) will match the intensity of the beam transmitted by the solution (I_t). If some of the light is absorbed, the intensity of the beam transmitted by the solution will be less than that of the incoming beam. The ratio of I_t and I_o can be used to indicate the percent of incoming light that is absorbed by the solution, as shown in Equation 1. The wavelength at which the **percent transmittance (%T)** is lowest is the wavelength to which the solution is most sensitive. This wavelength, which is used for analysis, is the **analytical wavelength**.

$$\%T = \left(\frac{I_t}{I_o}\right)(100\%) \qquad \text{(Eq. 1)}$$

Once we have determined the analytical wavelength for a particular solution, we can study the three variables that influence the specific response of the solution. These variables are the **concentration (c)** of the absorbing substance in the solution, the **pathlength (b)** of the light through the solution, and the sensitivity of the absorbing species to the energy of the analytical wavelength. When concentration is expressed in molarity (mol L^{-1}), and the pathlength is measured in centimeters (cm), the sensitivity factor is known as the **molar absorptivity (ε)** of the particular absorbing species. Molar absorptivity is a proportionality constant of a particular absorbing species with units of L mol^{-1} cm^{-1}. Its value depends on the analytical wavelength used for the analysis. The product of these three variables is **absorbance (A)**, as shown in Equation 2.

$$A = \varepsilon bc \qquad \text{(Eq. 2)}$$

This relationship is known as **Beer's law**. Thus, we can define absorbance in terms of I_o and I_t, as given in Equation 3.

Copyright © 1989 by Chemical Education Resources, Inc., P.O. Box 357, 220 S. Railroad, Palmyra, Pennsylvania 17078

No part of this laboratory program may be reproduced or transmitted in any form or by any means, electronic or mechanical, including photocopying, recording, or any information storage and retrieval system, without permission in writing from the publisher. Printed in the United States of America

$$A = \log\left(\frac{I_o}{I_t}\right) \qquad \text{(Eq. 3)}$$

A **spectrophotometer** is an instument used to study the response of solutions to light. It has two scales, one calibrated to display percent transmittances and the other calibrated to display absorbances. The absorbance scale is logarithmic with values ranging from zero to infinity. The percent transmittance scale is linear with values ranging from zero to 100. It is easier to obtain an accurate reading from the linear scale, because the distance between units is constant. Readings on the percent transmittance scale can be converted to equivalent absorbance, using Equation 4.

$$A = 2.000 - \log (\%T) \qquad \text{(Eq. 4)}$$

We can see from Equation 2 that the absorbance of a solution is directly proportional to the concentration of the absorbing substance in solution. Often several components of a solution absorb energy of the wavelength being used for the analysis. To compensate for this interference we prepare a **reference solution**, or **blank** that contains all the components of the solution, except the species being determined. The spectrophotometer is set so that 100% of the light of the chosen wavelength is transmitted by the reference solution. Thus, any absorbance by a solution will be due to the presence of the substance being studied.

If sample containers, called **cuvettes**, of uniform size are used, the absorbance of a series of solutions will, according to Beer's law, be proportional to the concentration of the absorbing substance in the solutions. The linear relationship between concentration and absorbance shown in Figure 1 is typical of many chemical systems that are said to follow Beer's law. This plot of absorbance versus concentration is referred to as a **Beer's law plot**.

Note, however, such systems cannot be shown to follow Beer's law throughout the range of all possible concentrations. This is true for a variety of reasons, one being the limitations of the spectrophotometer. The sensitivity of most spectrophotometers is greatest between 10 and 90 %T. Using only percent transmittances falling in the most sensitive meter range, and Equation 4, we can correctly determine equivalent absorbances.

For instance, for %T = 10%,

$$A = 2.000 - \log 10 = 2.000 - 1.000 = 1.00$$

and for %T = 90%,

$$A = 2.000 - \log 90 = 2.000 - 1.954 = 0.046$$

Of course, the effective concentration range for a particular absorbing substance will depend on the magnitude of the molar absorptivity of the substance at the analytical wavelength. If we know, or can determine, this value, then we can determine a concentration range over which we would expect to obtain a linear relationship between concentration and absorbance.

A variety of chemical factors can cause deviations from the expected linear relationship between concentration and absorbance. One common problem occurs when the absorbing substances are particularly sensitive to hydrogen ion concentration. If the pH of a series of solutions of one substance is kept constant, but the concentration of the absorbing substance is changed, then the equilibrium-favoring formation of the absorbing substance may shift unexpectedly. The chromate–dichromate equilibrium shown in Equation 5 is a well-known example of a chemical system that responds in this way.

$$2\ CrO_4{}^{2-}(aq) + 2\ H_3O^+(aq) \rightleftarrows$$
$$Cr_2O_7{}^{2-}(aq) + 3\ H_2O(l) \qquad \text{(Eq. 5)}$$

The relationship between absorbance and concentration is used extensively in quantitative analysis. In one application, a series of solutions of known concentrations of a substance is prepared. The percent transmittance of each is read at the analytical wavelength. After converting the percent transmittances to equivalent absorbances, a Beer's law plot can be made. If the absorbance of a solution of unknown concentration of the same substance is then measured, the concentration can be read directly from the plot. Figure 1 shows how this would be done for a solution with a calculated absorbance of 0.420.

The instrument most widely used for this type of analysis is a spectrophotometer. Figure 2 shows the arrangement of the optics in a spectrophotometer such as the Spectronic 20. The light generated by the tungsten lamp passes through the entrance slit and is

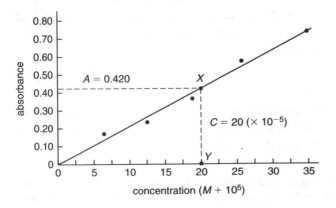

Figure 1 *A typical Beer's law plot*

reflected by the diffraction grating. The grating acts like a prism, separating the white light into monochromatic beams of various wavelengths. A desired wavelength is selected by rotating a cam that allows the light to pass through the sample and finally strike the measuring photoelectric cell or phototube. Here the light energy is converted to an electric signal.

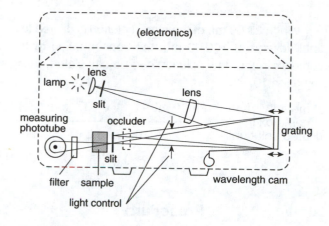

Figure 2 *Schematic of the Spectronic 20 optical system*

II. Determining the Concentration of Acetylsalicylic Acid

Acetylsalicylic acid (GMM: 180.2 g mol^{-1}), the active ingredient in aspirin tablets, is hydrolyzed rapidly and quantitatively in basic solution to produce the salicylate dianion. We will use ASA as an abbreviation for acetylsalicylic acid. If the hydrolysis is carried out with sodium hydroxide (NaOH) as the base, salicylate ion is formed as shown in Equation 5. If the reaction mixture is acidified and then mixed with a solution containing the iron(III) ion, an intensely purple complex of tetraaquo-

salicylatoiron(III) ion is formed. Equations 6 and 7 show the formation of this complex.

The reaction in Equation 7 proceeds to completion. Since each molecule of ASA forms one complex ion, the concentration of the complex formed is the same as the original concentration of ASA in the sample.

A solution of this complex exhibits the absorption spectrum shown in Figure 3. Because the complex has a high molar absorptivity, small amounts of the colored complex can be detected in solution. The composition of this complex is sensitive to pH, and the solutions must be maintained in the pH range 0.5–2.0 to avoid formation of the di- and trisalicylate complexes of iron(III).

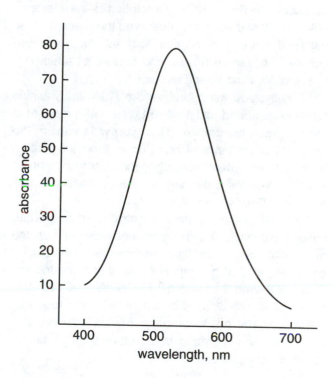

Figure 3 *Absorption spectrum of a solution of iron(III) ion and salicylate anion*

$$+ \ 3 \ OH^- \longrightarrow \qquad\qquad + \ 2 \ H_2O \ + \ CH_3COO^- \qquad (Eq. \ 5)$$

$$+ \ H_3O^+ \longrightarrow \qquad\qquad + \ H_2O \qquad\qquad (Eq. \ 6)$$

$$\xrightarrow{[Fe(H_2O)_6]^{3+}} \qquad\qquad + \ 2 \ H_2O \qquad\qquad (Eq. \ 7)$$

III.　Analyzing Commercial Aspirin Tablets

In this experiment, you will determine the amount of ASA in a commercial aspirin tablet. You will first prepare a stock solution of known concentration of sodium salicylate. Using measured amounts of this stock solution and an excess of iron(III) chloride in hydrochloric acid and potassium chloride solution ($FeCl_3$–KCl–HCl solution), you will make a series of solutions of the purple salicylate complex. You will measure the percent transmittance of each solution at 530 nm (the analytical wavelength of the complex). To make certain that only the complex is responsible for any light absorbance by the solution, you will set the spectrophotometer to 100 %T, using the $FeCl_3$–KCl–HCl solution as the reference solution.　Once you have measured the percent transmittance of each of the standard solutions, you can convert the transmittances into equivalent absorbances and draw the Beer's law plot.

　　Finally, you will hydrolyze the ASA in an aspirin tablet in base and add acid and the iron(III) reagent to form the purple complex. Then you will measure the percent transmittance of the solution. From the percent transmittance, you will calculate the absorbance of the solution. You will determine the concentration of ASA from your Beer's law plot.

　　You will calculate the concentration of the iron(III) salicylate complex in the original stock solution from the amount of ASA used in the determination. Consider, for example, a 0.200-g sample of ASA that is treated first with NaOH and then with the $FeCl_3$–KCl–HCl solution and diluted with distilled or deionized water to a total volume of 100 mL to prepare a stock solution. The concentration of complex in the stock solution can be determined as follows.

$$\text{number of mol of ASA} = (\text{mass of ASA, g})\left(\frac{1\ \text{mol ASA}}{180.2\ \text{g ASA}}\right)$$

$$= (0.200\ \text{g ASA})\left(\frac{1\ \text{mol ASA}}{180.2\ \text{g ASA}}\right)$$

$$= 1.11 \times 10^{-3}\ \text{mol} \qquad \text{(Eq. 8)}$$

The molarity of the complex in the stock solution is calculated next.

$$\text{molarity of complex in stock solution} = \frac{\text{number of mol of ASA}}{\text{total of stock solution, L}}$$

$$= \frac{1.11 \times 10^{-3}\ \text{mol ASA}}{0.100\ \text{L}}$$

$$= 1.11 \times 10^{-2}\ \text{mol L}^{-1} \qquad \text{(Eq. 9)}$$

Measured amounts of this stock solution are then diluted to prepare a set of standard solutions. The molar concentration of complex in any of the standard solutions can be determined using Equation 10.

$$\text{molarity of complex in standard solution} = \frac{\left(\substack{\text{volume of stock} \\ \text{solution added, mL}}\right)\left(\substack{M\ \text{of stock} \\ \text{solution}}\right)}{\substack{\text{total volume of standard} \\ \text{solution, mL}}}$$

$$\text{(Eq. 10)}$$

　　When 25.00 mL of the stock solution is diluted to a total volume of 100.00 mL, the concentration of the complex in the new solution can be calculated, using Equation 10.

$$\text{molarity of complex in diluted solution} = \frac{(25.00\ \text{mL})(1.11 \times 10^{-2}\ \text{mol L}^{-1})}{100.00\ \text{mL}}$$

$$= 2.78 \times 10^{-3}\ \text{mol L}^{-1}$$

Procedure

> ### Chemical Alert
> $2 \times 10^{-2}\,M$ iron(III) chloride solution—toxic and corrosive
> $1\,M$ sodium hydroxide—toxic and corrosive

> **Caution:**　Wear departmentally approved eye protection while doing this experiment.

I.　Operating a Spectrophotometer

> **Note:**　If the instrument you are using for this experiment is not a Spectronic 20, your laboratory instructor will give you directions for using the specific instrument available.

Turning on the power to the instrument.　Rotate the left-hand knob, the **amplifier control**, clockwise. On some models, a red LED will light at this point. Allow about 15 min for the instrument to warm up before recording any measurements.

Cleaning cuvettes.　While waiting for the instrument to warm up, rinse two cuvettes with distilled or deionized water. Then, rinse one of the cuvettes three times with the solution to be measured. Fill this cuvette three-quarters full with the solution to be measured. Fill the other cuvette three-quarters full with distilled water or the solution you are using as the reference solution. Do not handle the lower portion of the cuvettes, because smudges or droplets of solution will affect the passage of the light beam through the cuvette. Wipe off the outside of the cuvettes with an absorbent tissue before inserting the cuvette in the clamp holder.

Setting the wavelength. Turn the **wavelength control knob**, on the top of the instrument, until the appropriate wavelength setting appears on the scale visible at the left of the knob. Do not change the setting during the experiment unless specifically instructed to do so.

Setting 0% transmittance. Make the following adjustment with no cuvette in the sample holder; under this condition, no light strikes the phototube. About 15 min after turning the instrument on, adjust the left-hand knob, the amplifier control, so that the needle on the meter points to zero on the percent transmittance scale.

Setting 100% transmittance. Turn the right-hand knob, the **light control**, counterclockwise almost to its limit before inserting a cuvette into the sample holder. Insert the cuvette containing the reference solution into the sample holder. Match exactly the index line on the cuvette with the index line on the holder. Close the top of the holder tightly. Turn the right-hand knob clockwise until the needle points to 100 on the percent transmission scale. Immediately remove the cuvette to avoid fatiguing the phototube, and proceed to the sample measurement.

Check 0 and 100% transmittance. After the cuvette is removed from the sample holder, an occluder automatically drops into the light beam path. The needle should then point to zero. Each time the wavelength is changed, and during any extensive series of measurements at the same wavelength, the 0 and 100% transmittance settings should be checked. If necessary, reset these two settings using the procedures described above.

II. Preparing a Beer's Law Plot

Weigh to the nearest milligram (0.001 g) on a piece of tared weighing paper or a tared weighing dish approximately 0.4 g of reagent grade ASA. Transfer the sample to a 125-mL Erlenmeyer flask. Record the exact mass of ASA and weighing paper on your Data Sheet 1.

> *Caution:* Sodium hydroxide solutions are corrosive and can cause severe burns. Prevent eye, skin, and clothing contact.

Measure 10 mL of 1M NaOH in a clean, dry, graduated cylinder. Add the NaOH to the ASA in the 125-mL Erlenmeyer flask. Heat the mixture to boiling on a hot plate to hydrolyze the ASA. Care should be exercised to avoid splattering and loss of contents. Rinse the inside walls of the flask with small amounts of distilled or deionized water to ensure the complete reaction of all of the ASA.

Quantitatively transfer the resulting solution of sodium salicylate to a 250-mL volumetric flask through a glass funnel. Thoroughly rinse the flask and funnel with distilled water so that the rinse water flows into the volumetric flask. Add distilled water to the solution in the flask until the bottom of the meniscus is at the base of the flask neck. Stopper the flask. While firmly holding the stopper with your forefinger, invert the flask 10 times to thoroughly mix the solution.

After allowing the trapped air bubbles to rise to the surface, add distilled water until the bottom of the meniscus coincides with the calibration mark on the flask neck. Stopper the flask. While firmly holding the stopper with your forefinger, invert the flask 10 times to thoroughly mix the solution. Label this flask Solution 1.

> *Caution:* Iron(III) chloride solutions containing potassium chloride and hydrochloric acid are corrosive. Prevent eye, skin, and clothing contact.

> *Caution:* Never use your mouth to draw a solution into a pipet. Always use a rubber bulb to fill a pipet.

Rinse a clean, 5-mL pipet with about 1 mL of Solution 1. Draw the solution into the pipet using a rubber bulb. Quickly disconnect the rubber bulb and place your index finger over the top opening to prevent the water from draining out of it. Hold the pipet in a nearly horizontal position. Rotate the pipet so that the rinse solution contacts as much of the inner surface of the pipet as possible. Remove your finger briefly during this process to allow the solution to enter the upper stem of the pipet. Drain the solution through the tip of the pipet into a beaker.

Note: After you discharge a solution from a pipet, there will be a small amount of liquid left in the tip of the pipet. Do not blow this liquid out of the pipet. The pipet is calibrated to deliver 10.00 mL of solution *excluding* the small amount remaining in the tip of the pipet.

Repeat the rinsing procedure with two additional portions of the solution. Discard the rinse solution following the directions of your laboratory instructor.

Pipet a 5.00-mL portion of Solution 1 into a clean 100-mL volumetric flask. As you release the solution into the flask, hold the tip of the pipet against the side of the flask. Allow the solution to flow down the side of the flask to avoid splattering. After you deliver the solution from the pipet, continue to hold the tip of the pipet against the side of the flask for an additional 15 s.

Add $2.0 \times 10^{-2} M$ FeCl$_3$–KCl–HCl solution (pH = 1.6) until the bottom of the meniscus is at the base of the volumetric flask neck. Stopper the flask. While firmly holding the stopper with your forefinger, invert the flask 10 times to thoroughly mix the solution.

After allowing the trapped air bubbles to rise to the surface, continue adding FeCl$_3$–KCl–HCl solution until the bottom of the meniscus coincides with the calibration mark on the flask neck. Stopper the flask. While firmly holding the stopper with your fore-finger, invert the flask 10 times to thoroughly mix the solution. Label this flask Solution A.

If the same 100-mL volumetric flask is to be used for the preparation of standard Solutions A through E, transfer each solution to a clean, labeled, 125-mL flask. Thoroughly rinse the 100-mL volumetric flask with small portions of distilled water.

In a similar fashion, prepare solutions labeled B, C, D, and E by diluting 4.00-, 3.00-, 2.00-, and 1.00-mL portions of the sodium salicylate stock solution (Solution 1) with the FeCl$_3$–KCl–HCl solution. Organize the flasks to ensure that portions are removed from the sodium salicylate stock solution (Solution 1) and not one of the flasks labeled A through E.

Fill a cuvette three-quarters full with Solution A. Be careful to avoid leaving fingerprints on the cuvette. Carefully wipe the outside of the cuvette with an absorbent tissue. Check to make certain that you have removed all fingerprints.

Set the instrument at 100% transmittance with a second cuvette three-quarters full of the reference solution (the FeCl$_3$–KCl–HCl solution). Then, immediately insert the cuvette containing Solution A into the sample holder. Match exactly the index line on the cuvette with the line on the holder. Close the top of the holder tightly. Immediately read the percent transmittance to three significant digits, and record it on Data Sheet 1.

Remove the cuvette from the sample holder. Rezero the instrument, if the zero setting has changed. Check 100% transmittance before each sample measurement.

Following this procedure, determine the percent transmittance of Solutions B, C, D, and E. Record these transmittances on Data Sheet 1.

Discard all solutions following the directions of your laboratory instructor.

III. Analyzing Commercial Aspirin Tablets

Record your unknown number on Data Sheet 2.

Measure the mass of a 125-mL Erlenmeyer flask to the nearest milligram (0.001 g). Record this mass on Data Sheet 2. Add one commercial tablet to the flask, reweigh the flask, and record the mass on Data Sheet 2. Weigh two more samples (one tablet per flask) in a similar manner. Be sure to label the flasks Sample 1, 2, and 3, respectively.

To each flask, add 10 mL of a 1M NaOH solution and heat to boiling, taking care to avoid loss of solution. Carry out the following operations on each sample consecutively.

Cool the solution by running tap water over the outside of the flask. Quantitatively transfer the sample to a 250-mL volumetric flask. Dilute to the calibration mark on the flask neck and mix thoroughly as described earlier. Starch fillers may give the solution a milky appearance that will disappear upon acidification. Some buffering agents such as aluminum hydroxide will not dissolve in this concentration of base. In such cases allow the solid to settle to the bottom of the flask.

Note: If there is a precipitate present, use your pipet to remove solution from the top portion of the liquid only so that you will not draw any precipitate into your pipet.

Using a clean volumetric pipet, transfer a 5-mL portion into a 100-mL volumetric flask, and dilute to the mark with $2 \times 10^{-2} M$ FeCl$_3$–KCl–HCl solution and mix thoroughly as described earlier.

Should cloudiness persist in the mixture, transfer the acid solution to a clean, dry 125-mL Erlenmeye flask, without diluting, and warm slightly on a hot plate until the solution is clear.

Set your instrument to 100% transmittance with your reference solution. Immediately determine the percent transmittance of each solution at 530 nm and record the readings on Data Sheet 2. Zero the instrument and check 100% transmittance with the reference solution after measuring the percent transmittance of each unknown solution.

Discard your reaction mixture following the directions of your laboratory instructor.

Do a second determination, and if time permits, a third one.

> *Caution:* Wash your hands thoroughly with soap or detergent before leaving the laboratory.

Calculations

Do the following calculations and record the results on Data Sheets 1 and 2.

II. Preparing a Beer's Law Plot

1. Calculate the number of moles of ASA used, using Equation 8.

2. Calculate the molarity of the complex in the 250-mL stock solution (Solution 1), using Equation 9.

3. Calculate the molarity of the complex for each standard solution (A, B, C, D, and E) prepared by dilution of the stock solution, using Equation 10.

4. Convert the percent transmittances to equivalent absorbances for each standard solution, using Equation 4.

5. Plot the calculated molar concentration of the iron(III)–salicylate complex on the abscissa (x-axis) versus the calculated absorbance at 530 nm of each solution on the ordinate (y-axis). Draw the best straight line through these points. The extrapolated line should pass through the origin.

6. Find the slope of your Beer's law plot.

7. Determine the molar absorptivity of the iron(III)–salicylate complex from the slope of your Beer's law plot, using Equation 11.

$$\varepsilon = \frac{A}{bc} \qquad \text{(Eq. 11)}$$

Do the following calculations for each determination and record the results on Data Sheet 2.

III. Analyzing Commercial Aspirin Tablets

8. Convert the percent transmittances to equivalent absorbances, using Equation 4.

9. From the calculated absorbances and the Beer's law plot you prepared in Part I for known concentrations of ASA, determine the concentration of ASA.

10. Calculate the mass of ASA in each tablet, using Equation 12.

$$\begin{array}{c}\text{mass ASA} \\ \text{in tablet}\end{array} = \left(\dfrac{\begin{array}{c}\text{concentration} \\ \text{ASA, mol}\end{array}}{\text{L}}\right)\left(\dfrac{180.2\text{ g ASA}}{1\text{ mol ASA}}\right) \times$$

$$(100\text{ mL})\left(\dfrac{1\text{ L}}{1000\text{ mL}}\right)\left(\dfrac{250\text{ mL}}{5\text{ mL}}\right) \qquad \text{(Eq. 12)}$$

11. Find the percent ASA in each tablet, using Equation 13.

$$\begin{array}{c}\text{percent ASA} \\ \text{in tablet}\end{array} = \left(\dfrac{\begin{array}{c}\text{mass ASA} \\ \text{in tablet, g}\end{array}}{\begin{array}{c}\text{mass of} \\ \text{tablet, g}\end{array}}\right)(100\%) \qquad \text{(Eq. 13)}$$

12. Calculate the mean percent ASA in your unknown, using Equation 14.

$$\begin{array}{c}\text{mean} \\ \text{percent} = \\ \text{ASA}\end{array} \dfrac{\left(\begin{array}{c}\text{\% ASA} \\ \text{sample 1}\end{array}\right)+\left(\begin{array}{c}\text{\% ASA} \\ \text{sample 2}\end{array}\right)+\left(\begin{array}{c}\text{\% ASA} \\ \text{sample 3}\end{array}\right)}{3}$$

$$\text{(Eq. 14)}$$

Post-Laboratory Questions

(Use the spaces provided for the answers and additional paper if necessary.)

1. A student was in a hurry to carry out this determination and took some shortcuts in doing the experiment. Briefly explain whether each of the following changes would lead to high or low results, or no change.

(1) In preparing the ASA solution for the Beer's law plot, the student added 10 mL of 1 M NaOH solution but omitted heating the reaction mixture.

(2) Because all the 250-mL volumetric flasks were dirty, the student diluted the stock solution of sodium salicylate in a clean 100-mL volumetric flask.

(3) When analyzing the commercial aspirin tablet, the solution was cloudy after it had been acidified. However, the student went ahead and analyzed it.

2. Explain why the $FeCl_3$–KCl–HCl solution was used as a reference solution. Suggest a procedure you could follow to determine whether it was necessary to use this solution as a reference or whether distilled water would have been just as satisfactory.

3. It is common to see the strength of aspirin tablets listed in terms of the number of grains of ASA per tablet. Using the conversion factor of 15.4 grains per gram ASA, compute the mass, in grains, of the tablets you analyzed. Compare this with the advertised value and determine the percent error.

Data Sheet 1

II. Preparing Standard Solutions and a Beer's Law Plot

mass of ASA + flask, g   _____

mass of flask, g   _____

    mass of ASA, g   _____

number of moles of ASA   _____

concentration of complex in solution, M _____

solution	concentration, M	% transmittance	absorbance, A
A	_____	_____	_____
B	_____	_____	_____
C	_____	_____	_____
D	_____	_____	_____
E	_____	_____	_____

slope of Beer's law plot   _____

molar absorptivity for iron(III)–salicylate complex,
   M^{-1} cm^{-1} at a wavelength of 530 nm   _____

Data Sheet 2

III. Analyzing Commercial Aspirin Tablets

unknown number_____

	determination 1	2	3
mass of tablet + flask, g	_____	_____	_____
mass of flask, g	_____	_____	_____
mass of tablet, g	_____	_____	_____
percent transmittance of solution, %T	_____	_____	_____
absorbance of solution, A	_____	_____	_____
concentration of aspirin in solution, M	_____	_____	_____
mass of ASA in tablet, g	_____	_____	_____
percent ASA in tablet, %	_____	_____	_____

    mean percent   _____

Pre-Laboratory Assignment

1. Read Experiment 6 for a discussion of spectro-photometry.

2. Explain why it is important to exercise extreme care when heating the ASA sample with $1 M$ NaOH solution.

3. The determination of glucose concentration in blood serum is often based on the formation of a blue-green complex of glucose and *o*-toluidine in glacial acetic acid. This reaction is shown at the bottom of this page.

The analytical wavelength for the complex is 635 nm. A set of standard solutions was prepared by taking a known volume of a stock glucose solution, containing 1.000 g glucose dissolved in 100 mL of distilled water, and diluting to 100 mL with distilled water. These solutions and the blood samples were treated with the *o*-toluidine reagent, and the percent transmittance of each sample was read against a ref-erence solution. The data obtained are

stock glucose, mL	concentration of glucose complex, M	%T	A
5.00	_____	76.91	____
10.00	_____	59.16	____
20.00	_____	35.08	____
25.00	_____	27.10	____
30.00	_____	20.65	____
unk. solution	_____	46.17	____

(1) Calculate the molar concentration of glucose in the stock solution.

(2) For each standard solution and the unknown, convert the percent transmittance to absorbance, using Equation 4. Enter these absorbances in the table.

(3) Calculate the molar concentration of glucose complex present in each solution. Enter the concentrations in the table.

(4) Prepare a Beer's law plot, using the data obtained in (2) and (3).

(5) Using your Beer's law plot, determine the molar concentration of glucose in the unknown sample.

$$HOCH_2(CHOH)_4CHO + \quad H_2N-\underset{CH_3}{\bigcirc} \rightleftharpoons HOCH_2(CHOH)_4CH=N-\underset{CH_3}{\bigcirc} + H_2O$$

glucose   *o*–toluidine   blue–green complex

Separating and Identifying Food Dyes by Paper Chromatography

prepared by **Peter G. Markow**, Saint Joseph College, CT

Purpose of the Experiment

Determine the retention factor of seven pure food dyes in three different solvent systems. Determine the most effective of the three solvents for separating all seven dyes. Separate and identify the dyes in unknown mixtures and selected commercial products using paper chromatography.

Background Information

I. A Brief History of Food, Drug, and Cosmetic Dyes

Because the appearance of food has a major effect on its sales, dyes have been commonly added to food products for many years. Questions regarding the safety of these dyes persist, despite federal legislation of food dyes that dates back to 1906, when the Federal Food and Drug Act was passed. In 1938, the Federal Food, Drug, and Cosmetic Act became law, further restricting the use of dyes. More recent research has indicated that some of the dyes approved for use in 1938 may be toxic or carcinogenic. As recently as 1976, red 2, one of the dyes commonly used to color maraschino cherries, was removed from the listing of permissible coloring agents.

The Food and Drug Administration (FDA) currently allows the following seven certified F, D, and C dyes to be added to food products: red 3, red 40, blue 1, blue 2, yellow 5, yellow 6, and green 3. Blue 2 and green 3 are not used as frequently as are the other five certified food dyes. Yellow 5 and 6 must be specifically identified on food labels because they are known to cause hyperactivity in some children. In addition, violet 1 is used for stamping meat, citrus red 2 is used to color orange skins, and orange B is used to color sausage skins.

A large quantity of F, D, and C dyes is consumed each year. In 1982, for example, 793,143 kg of red 40 were consumed in the United States alone, the equivalent of 34.1 mg of dye per person.

II. Chemistry of Food Dyes

The structures of the seven certified food dyes are shown in Figure 1. Note the structural similarities. All but red 3 are sodium salts of sulfonic acids, containing $-SO_3Na$ groups, and are therefore water soluble. Red 3 is a sodium salt of a carboxylic acid and is also water soluble. All seven dyes have long, conjugated structures, alternating carbon-to-carbon single and double bonds. Extensive conjugated systems such as these often result in strongly colored molecules, hence, strongly colored solutions of the molecules.

III. Chromatography

Chromatography is a technique developed during the last thirty years that has revolutionized modern analytical chemistry. The word **chromatography**, origi-

---

Copyright © 1989 by Chemical Education Resources, Inc., P.O. Box 357, 220 S. Railroad, Palmyra, Pennsylvania 17078
No part of this laboratory program may be reproduced or transmitted in any form or by any means, electronic or mechanical, including photocopying, recording, or any information storage and retrieval system, without permission in writing from the publisher. Printed in the United States of America

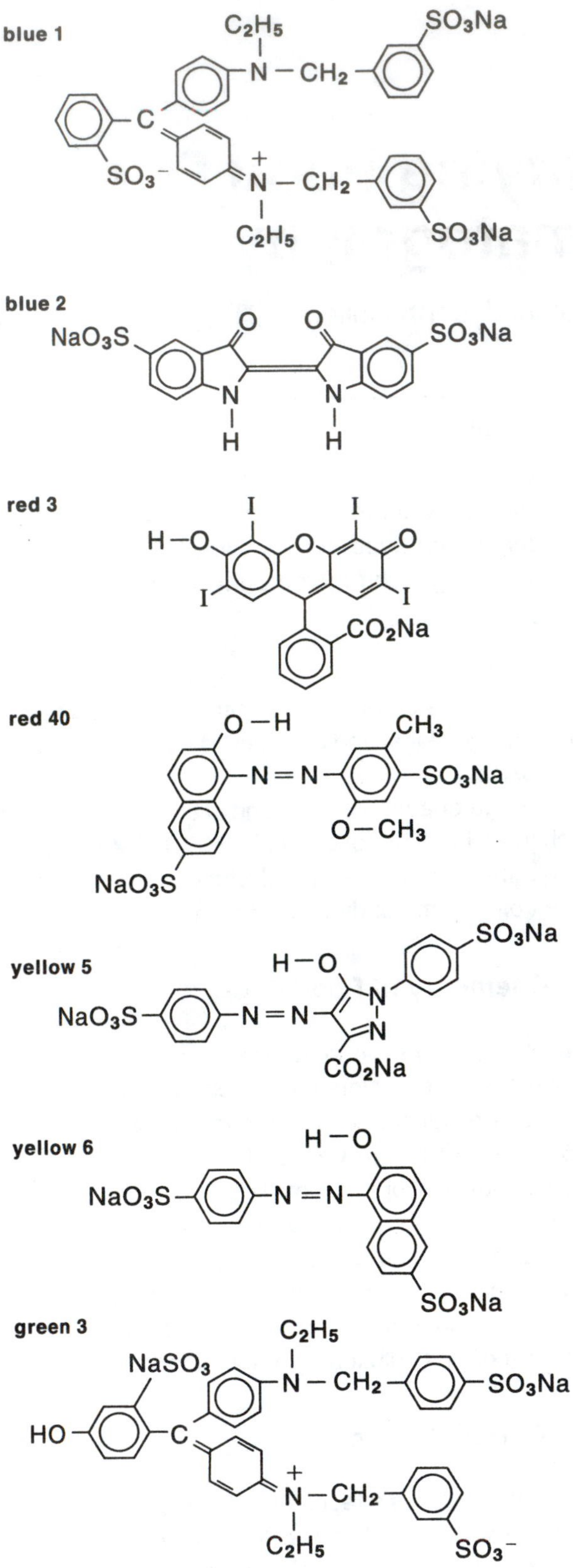

blue 1

blue 2

red 3

red 40

yellow 5

yellow 6

green 3

Figure 1 *Structures of the seven certified F, D, and C food dyes*

nating from the Greek word *chromatos*, "color," literally means "the separation of colors."

Many types of chromatography are now routinely used in laboratories around the world to separate and identify components in mixtures. The analysis of blood and urine samples for drugs and the analysis of drinking and ground water for hazardous chemicals are two examples of the versatility of chromatographic separations.

All chromatography techniques involve a stationary phase and a mobile phase. The **stationary phase** can be either a liquid or a solid. The mixture to be separated is usually placed on (or in) the stationary phase. The **mobile phase** can be either a liquid or a gas. The mobile phase moves along the stationary phase, carrying some or all of the mixture with it, resulting in the separation of the components in the mixture. In **liquid chromatography**, separation is based on the varying attractions of the components in the mixture to the mobile phase. The variation in attraction is due to different intermolecular interactions.

Paper chromatography is the simplest form of chromatography. While paper chromatography is not used in analyses for drugs or hazardous chemicals, it is extremely useful in the separation and identification of food dyes. In paper chromatography, a sample of the mixture to be separated is placed on a piece of chromatography paper, which acts as the stationary phase. One edge of the paper is placed in a solvent, such as water, alcohol, or a mixture of both, which acts as the mobile phase. Many different solvent systems are used, depending on the components to be separated.

The chromatography paper acts like a wick and draws the solvent up the paper by capillary action. Water molecules are permanently bound to the cellulose fibers of paper. The wick effect actually occurs because the solvent is attracted to the water molecules bound to the paper. The water molecules bound to the paper and the paper itself form the stationary phase.

When a sample is applied, or **spotted**, on an area of the paper near the bottom, and the bottom edge of the paper is placed in a solvent, the solvent is drawn up the paper. When the leading edge of the mobile phase, called the **solvent front**, reaches the sample, the sample components are preferentially attracted to either the stationary or mobile phase, depending on the polarities of the sample and the two phases. Recall that like solvents dissolve like solutes. However, the attraction is seldom an all-or-nothing situation. Most compounds, whether they are ionic or molecular in nature, are somewhat attracted to both phases. An equilibrium is established for the component between the two phases, as shown in Equation 1.

component–mobile phase $\rightleftarrows$
component–stationary phase   (Eq. 1)

As the solvent front moves up the paper, fresh solvent passes the spotted sample, and new equilibria are established. At the same time, any of the components that have dissolved in the mobile phase encounter fresh stationary phase, and new equilibria are established. Thus, the components of a mixture move up the paper at different rates and separate, producing a **chromatogram**.

The overall effect of all these equilibria is that the movement of the components depends on the nature of their relative attractions for the mobile and stationary phases. We characterize this movement in terms of a **retention factor** or **retardation factor** (R_f), defined in Equation 2.

$$R_f = \frac{\text{distance traveled by component, cm}}{\text{distance travelled by solvent front, cm}} \quad \text{(Eq. 2)}$$

R_f values can be as high as 1.0, if the substance moves with the solvent front, and as low as 0.0, if the substance does not move at all. The values are reproducible for a particular component–solvent system, if the experimental conditions are closely controlled. One important variable is the composition of the solvent. If one of the solvent components is volatile, then the percent composition of the solvent may change due to evaporation during the analysis. You can prevent this by keeping the container in which you are developing the chromatogram closed, so that the air in the container remains saturated with solvent vapor.

A sample containing two or more components can be separated, or **resolved**, if we choose a solvent system for which the sample components have distinctly different R_f values. You can make this choice in advance by determining the R_f values of the individual components in a variety of solvent systems and finding the best solvent system for separating your group of components.

Figure 2 illustrates the preparation of a chromatogram. Spots of the sample to be resolved are placed on the origin line of the chromatography paper. The bottom edge of the paper is placed in the solvent, and the solvent moves up the paper. After separation, the components have moved up the paper. The distance a component has moved is determined by measuring the distance from the origin line to the center of the component spot.

In Figure 2, the left-hand spot is the sample being resolved and the right-handed spot is Compound A, which we think is present in the mixture. The distance the solvent front moved is 4.0 cm, and the distance the middle component of the mixture moved is 2.0 cm. The R_f of the middle component is 2.0 cm/4.0 cm = 0.50. The distance Compound A moved is 2.0 cm, so its R_f is also 0.50. On the basis of the R_f values, we can conclude that Compound A is probably one of the components of the mixture.

Frequently, we can obtain other information from chromatographs to support the findings based on R_f calculations. For instance, when you are resolving food dyes, you can make color comparisons on the chromatograms. When you resolve green food coloring, you will observe two spots, one yellow and one blue. These two colors, in conjunction with the R_f values of the components, will help you to identify the dyes in the green food coloring.

In this experiment, you will determine the R_f values of seven F, D, and C food dyes in three different solvent systems. The solvent systems are distilled or deionized water; rubbing alcohol, which is a mixture of

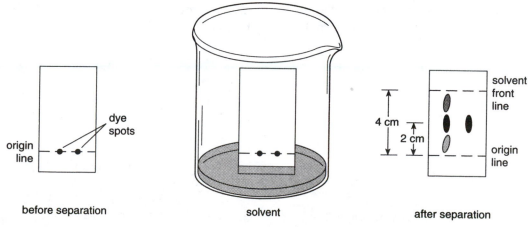

before separation   solvent   after separation

Figure 2 *Preparing a chromatogram*

isopropyl alcohol and water; and a distilled water solution containing 0.10% sodium chloride, commonly called "table salt." By comparing the R_f values, you will identify the solvent system that most effectively separates these dyes. You will also identify the solvent that gives the fastest separation of the dyes.

Finally, you will identify which of the seven dyes are present in several unknown mixtures and in various commercial products, such as food products and felt-tip pens.

Procedure

I. Choosing a Solvent

> *Caution:* Wear departmentally approved eye protection while doing this experiment.

1. Obtain three clean 250-mL beakers, three clean Petri dish covers, and a clean 10-mL graduated cylinder. Label the beakers "water," "rubbing alcohol," and "table salt."

2. Measure 7 mL of distilled or deionized water into your first beaker labeled "water," 7 mL of rubbing alcohol into the second, and 7 mL of 0.10% table salt solution into the third. Cover each beaker with a Petri dish cover, and set the beakers aside.

> *Note:* Wash your hands thoroughly with soap or detergent before touching the chromatography paper. Handle the chromatography paper by its edges. Avoid touching the front or back of the paper.

3. Obtain a 7.5 × 13.5-cm piece of chromatography paper. Using a pencil, draw the origin line 1 cm from

the bottom edge of the paper (see Figure 3). Beginning 1.5 cm from the left-hand edge, make marks on the origin line at 1-cm intervals. Label your marks in sequence: B1, B2, Y5, Y6, R3, R40, G3 (for the individual dyes); and B1 and R40, B1 and Y5, R3 and Y6, and B2 and Y5 (for the combinations of two dyes). For the individual dyes, B1 indicates "blue 1," while for the combinations of dyes B1 and Y5 indicates "blue 1 and yellow 5." Prepare two additional pieces of chromatography paper in the same way.

4. In the upper right-hand corner of the paper, label one piece of paper "Chromatogram 1, water," one "Chromatograph 2, rubbing alcohol", and the third "Chromatograph 3, 0.10% salt water."

> *Caution:* Dyes will stain clothes. Avoid contact with skin and clothing.

5. Place one drop each of the seven pure dye solutions on a glass plate or on three glass microscope slides. Space the drops carefully so they do not mix. Mark the plate or slide to identify each dye.

6. Use a clean, wooden toothpick to spot each dye onto each of your three pieces of chromatography paper. To spot each dye, place one end of the toothpick in the drop of dye, and allow the toothpick to soak up some of the dye. Then, keeping the toothpick vertical, touch the toothpick to the mark on the origin line on your paper.

Before respotting, allow the spot from each individual application to dry, in order to prevent the spot from becoming too large. Spot the dyes on their appropriate marks.

For the four combinations of two dyes, spot the first dye, and then spot the second dye directly on top of the first. Let each piece of chromatography paper air dry for a few minutes. Save the toothpicks and dyes for Part II.

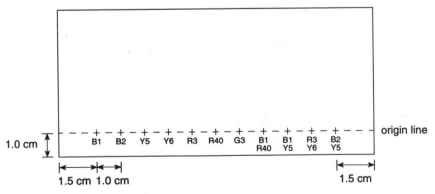

Figure 3 *Marking chromatography paper for Part I*

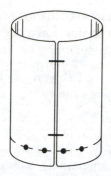

Figure 4 *Preparing spotted chromatography paper for development*

7. Carefully roll each piece of chromatography paper on its shorter axis to form a cylinder, with the dye spots on the outside and the origin line at the bottom, as shown in Figure 4. Staple the ends of the paper together, with the edges just touching, but not overlapping.

8. Remove the Petri dish covers from the beakers containing the solvents. Carefully place each labeled paper cylinder in the appropriately labeled beaker, making sure that the dye spots are at the bottom of the cylinders. Replace the covers. Be sure that the cylinders do not touch the walls of the beaker and that the dye spots are above the surface of the solvents, as shown in Figure 2. Record the starting times of your three chromatograms on Data Sheet 1.

9. Remove each cylinder from its beaker when the solvent front is about 1.5 cm from the top of the paper. On Data Sheet 1, record the ending time as you re-move each cylinder. Carefully unroll the cylinder to minimize tearing the paper. As the paper is drying, the solvent will move up the paper for a few minutes. Draw a line with your pencil marking the solvent front after the solvent has stopped moving, but *before the chromatogram is completely dry*. Allow the paper to continue air drying.

10. Discard each solvent according to the directions of your laboratory instructor. Wash and dry your beakers for use in Part II.

11. Do Calculations 1–4 to determine the best solvent system (mobile phase) for separating the dyes in Part I. Enter these results on Data Sheet 1.

II. Identifying F, D, and C Food Dyes in Unknown Dye Mixtures

12. Measure 7 mL of the best solvent system and transfer it to one of your clean 250-mL beakers. Place a Petri dish cover on the beaker, and set the beaker aside.

13. On a fourth piece of 7.5 × 13.5-cm chromatography paper, draw the origin line 1 cm from the bottom edge of the paper. Beginning 1.5 cm from the left-hand edge, make marks on the origin line at 1-cm intervals. Label your marks in sequence: B1, B2, Un#, Y5, Y6, Un#, R3, R40, Un#, and G3, where Un# represents the code number of an unknown food dye mixture (see Figure 5). Label the paper in the upper right-hand corner "Chromatogram 4."

14. Obtain three unknown dye mixtures from your laboratory instructor, and record their code numbers on Data Sheet 2.

15. Carefully spot each of the seven pure dyes on your chromatography paper, as in Step 6 of Part I.

16. Place one drop each of the three unknown dye mixtures on a glass plate or microscope slide. Space the drops carefully so they do not mix. Mark the plate or slide to identify the drop of each mixture.

17. Label the marks on the origin line of your prepared chromatography paper with the code numbers of your unknowns. Allow the spot from each individual application to dry before respotting, in order to

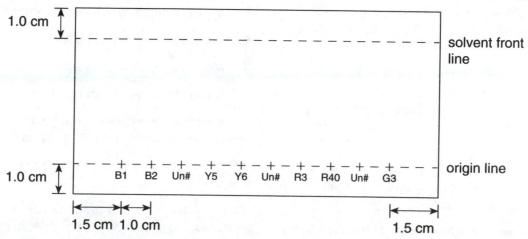

Figure 5 *Marking chromatography paper for Part II*

prevent the spot from becoming too large. Spot each of these mixtures four or five times.

18. Repeat Steps 7–10 of Part I.

III.   Identifying F, D, and C Dyes in Commercial Products

19. Measure 7 additional mL of the best solvent for separating the dyes into another clean 250-mL beaker. Place a Petri dish cover on the beaker, and set the beaker aside.

20. Obtain seven food product solutions and three felt-tip pens from your laboratory instructor. Record the code numbers of the solutions and pens on Data Sheet 3.

21. On a fifth piece of 7.5 × 13.5-cm chromatography paper, draw the origin line 1 cm from the bottom edge of the paper. Beginning 1.5 cm from the left-hand edge, make marks on the origin line at 1-cm intervals. Label each mark with the appropriate code number of the food products and pens you have selected. Label the paper in the upper right-hand corner "Chromatogram 5."

> **Note:** The dyes in some food products are much less concentrated than the dye solutions used in Parts I and II. Spot each food sample solution several times, until the intensity of the color is close to that of the original spots in Part I. Allow the spot from each individual application to dry before respotting, in order to prevent the spot from becoming too large.
>
> In contrast, the dyes in the felt-tip pens are strongly concentrated. Spot each one only once. Do not allow the pen to contact the chromatography paper for longer than one second. Note that some dyes used in these pens cannot be separated and identified using the procedures in this experiment.

22. Place one drop of each of the food product solutions on a glass plate or microscope slide. Space the drops carefully so that they do not mix. Label the plate or slides to identify the drops.

23. Use a clean wooden toothpick to spot each of the solutions on the chromatography paper at the appropriately designated marks.

24. Quickly touch the tip of each of the pens to the paper at the appropriately designated marks.

25. Repeat Steps 7–10 of Part I, using the solvent system you selected that you found best for separating the seven pure dyes.

26. Discard the used solvent system following the directions of your laboratory instructor.

> **Caution:** Wash your hands thoroughly with soap or detergent before leaving the laboratory.

Calculations

I.   Choosing a Solvent

Do the following calculations for Chromatograms 1, 2, and 3 and record your results on Data Sheet 1.

1. Measure the distance the solvent moved, in cm.

2. Draw an ellipse around the edge of each spot. Find the center of the ellipse. Measure the distance from the origin line to the center of the ellipse. Record this distance as the distance each spot moved.

3. Calculate the R_f for each F, D, and C dye, using Equation 2.

4. Determine which solvent system is the best for separating the seven F, D, and C dyes and known combinations of the dyes.

II.   Identifying F, D, and C Dyes in Unknown Dye Mixtures

Do the following calculations for Chromatogram 4 and record your results on Data Sheet 2.

5. Measure the distance the solvent moved.

6. Record the color of each spot.

7. Measure the distance each spot moved.

8. Calculate the R_f for each F, D, and C dye.

9. Find the mean R_f for each dye using the R_f val-ues you determined in Parts I and II.

10. Calculate the R_f for each component of each unknown.

11. Use the color and the mean R_f for the known dyes to identify the component dyes in each of your unknown mixtures.

III. Identifying F, D, and C Dyes in Commercial Products

Do the following calculations for Chromatogram 5 and record your results on Data Sheet 3.

12. Measure the distance the solvent moved.

13. Record the color of each spot.

14. Measure the distance each spot moved.

15. Calculate the R_f for each component of each product.

16. Using the color and mean R_f for the seven known dyes from Part II, identify the component dyes in each product.

Post-Laboratory Questions

(Use the spaces provided for the answers and additional paper if necessary.)

1. Refer to the chromatograms you developed when answering the following questions.

(1) Are there any obvious differences in the sizes of the spots after development in water as compared to those after development in rubbing alcohol? Briefly explain your answer.

(2) What correlation can you make between the size of the developed spots and the time required for development of the chromatogram?

2. A student did this experiment in a hurry and did not read or follow the directions carefully. Comment on the following problems that occurred because of the student's haste.

(1) Because there were no clean Petri dish covers available, the student left the beakers containing the developing solvents uncovered throughout the experiment.

(2) Because no pencils were in sight, the student used a felt-tip pen to mark the origin and solvent front lines on the chromatography paper.

(3) While spotting the chromatograms, the student spilled some red 3 and blue 1 dye solutions on a friend's laboratory coat. The student grabbed the 0.10% salt water solution bottle and tried to use this solution to wash the spots out. Describe the likely success of using this solvent system in removing these two dyes.

(4) The student was chatting with friends while the chromatogram was developing. By the time the student checked the chromtogram, the solvent front had reached the top of the paper.

3. Explain why it is necessary that the original spots on the chromatography paper be above the surface of the solvent when the paper is placed in the beaker for developing.

_____
name   section   date

Data Sheet 1

I. Choosing a Solvent

	Chromatogram 1	Chromatogram 2	Chromatogram 3
		solvent	
	water	rubbing alcohol	0.10% salt water
starting time	_____	_____	_____
ending time	_____	_____	_____
elapsed time, min	_____	_____	_____
distance solvent moved, cm	_____	_____	_____

F, D, and C dyes	water distance spot moved, cm	Rf	rubbing alcohol distance spot moved, cm	Rf	0.10% salt water distance spot moved, cm	Rf
blue 1	_____	_____	_____	_____	_____	_____
blue 2	_____	_____	_____	_____	_____	_____
yellow 5	_____	_____	_____	_____	_____	_____
yellow 6	_____	_____	_____	_____	_____	_____
red 3	_____	_____	_____	_____	_____	_____
red 40	_____	_____	_____	_____	_____	_____
green 3	_____	_____	_____	_____	_____	_____
blue 1 & red 40	_____	_____	_____	_____	_____	_____
blue 1 & yellow 5	_____	_____	_____	_____	_____	_____
red 3 & yellow 6	_____	_____	_____	_____	_____	_____
blue 2 & yellow 5	_____	_____	_____	_____	_____	_____

best solvent system for separating the seven dyes_____

Briefly explain your reasons for selecting this solvent system.

Data Sheet 2

II. Identifying F, D, and C Dyes in Unknown Mixtures

solvent system _____

code numbers of unknowns _____

distance solvent moved, cm _____

F, D, and C dye	distance spot moved, cm	Part II R_f	Part I R_f	mean R_f
blue 1	_____	_____	_____	_____
blue 2	_____	_____	_____	_____
yellow 5	_____	_____	_____	_____
yellow 6	_____	_____	_____	_____
red 3	_____	_____	_____	_____
red 40	_____	_____	_____	_____
green 3	_____	_____	_____	_____

	unknown mixtures		
	Un #	Un #	Un #
component spot 1			
color	_____	_____	_____
distance spot moved, cm	_____	_____	_____
R_f	_____	_____	_____
component spot 2			
color	_____	_____	_____
distance spot moved, cm	_____	_____	_____
R_f	_____	_____	_____
component spot 3			
color	_____	_____	_____
distance spot moved, cm	_____	_____	_____
R_f	_____	_____	_____
component spot 4			
color	_____	_____	_____
distance spot moved, cm	_____	_____	_____
R_f	_____	_____	_____

component dyes present in

   unknown #___: _____

   unknown #___: _____

   unknown #___: _____

Data Sheet 3

III.   Identifying F, D, and C Dyes in Commercial Products

solvent system _____

code numbers of unknowns _____

_____

distance solvent moved, cm _____

code number	component spot 1 distance spot moved,			component spot 2 distance spot moved,			component spot 3 distance spot moved,		
	color	cm	R_f	color	cm	R_f	color	cm	R_f
_____	_____	_____	_____	_____	_____	_____	_____	_____	_____
_____	_____	_____	_____	_____	_____	_____	_____	_____	_____
_____	_____	_____	_____	_____	_____	_____	_____	_____	_____
_____	_____	_____	_____	_____	_____	_____	_____	_____	_____
_____	_____	_____	_____	_____	_____	_____	_____	_____	_____
_____	_____	_____	_____	_____	_____	_____	_____	_____	_____
_____	_____	_____	_____	_____	_____	_____	_____	_____	_____
_____	_____	_____	_____	_____	_____	_____	_____	_____	_____
_____	_____	_____	_____	_____	_____	_____	_____	_____	_____
_____	_____	_____	_____	_____	_____	_____	_____	_____	_____

code number or name of product	component dyes present
_____	_____
_____	_____
_____	_____
_____	_____
_____	_____
_____	_____
_____	_____
_____	_____
_____	_____
_____	_____

Pre-Laboratory Assignment

1. Is it necessary to take any safety precautions while doing this experiment? Briefly explain.

4. A student used the procedure described in this module to study the behavior of blue 1 (B1), yellow 5 (Y5), red 3 (R3), red 40 (R40), and green 3 (G3) in two solvent systems, *n*-propanol–water (2:1), and ethanol–water (1:4). The student's measurements from each chromatogram are given below.

	n-propanol–water distance traveled	R_f	*ethanol–water* distance traveled	R_f
solvent front	4.6 cm	_____	4.8 cm	_____
B1	2.9 cm	_____	4.7 cm	_____
Y5	1.3 cm	_____	3.5 cm	_____
R3	4.0 cm	_____	0.8 cm	_____
R40	2.9 cm	_____	2.5 cm	_____
G3	3.1 cm	_____	4.7 cm	_____

2. List three desirable characteristics of an F, D, and C dye.

(1) Calculate R_f values for each of the five food dyes. Enter the R_f values in the table above.

(2) Which solvent system would be most effective for separating R3 from G3? Briefly explain.

3. Define the following terms.
   (1) mobile phase

(2) solvent front

(3) Would *n*-propanol–water be an effective solvent system for separating B1 from R40? Briefly explain.

(3) retention factor

(4) The student was given an unknown that appeared green in color and was asked to determine its composition. Describe a procedure for doing so.

6. Briefly explain why R_f values can be no greater than 1.0 and no smaller than 0.0.

5. Why is the beaker used while developing the chromatogram kept covered during this experiment?

Paper Chromatography of Selected Transition-Metal Cations

prepared by **Frank Rioux**, Saint John's University, MN,
and **Judith C. Foster**, Bowdoin College

Purpose of the Experiment

Separate four transition-metal cations by paper chromatography and identify them by spot tests.

Background Information

Chemists are frequently asked to determine the composition of multicomponent mixtures. In such cases, it is often desirable or necessary to physically separate and isolate the components, so they can be identified qualitatively and measured quantitatively. One useful separation technique is **paper chromatography**. Chromatographic paper is composed of polar cellulose fibers that readily absorb water from the atmosphere. In paper chromatography, the polar cellulose and absorbed water form the **stationary phase.**

When the bottom edge of a piece of chromatographic paper is placed in a beaker containing a solvent, the paper acts like a wick, slowly drawing the solvent up the paper by capillary action. This moving solvent is the **mobile phase**. The solvent is referred to as the **developing solvent.** Solvents used for paper chromatography usually have distinctly different polarities and chemical properties from the stationary phase.

When a substance is applied, or **spotted**, on an area of the paper near the bottom, and the bottom edge of the paper is placed in a developing solvent, the solvent is drawn up the paper. When the leading edge of the mobile phase, called the **solvent front**, reaches the substance, the substance is preferentially attracted to either the stationary or mobile phase, depending on the substance's polarity. Recall that like solvents dissolve like solutes. However, the attraction is seldom an all-or-nothing situation. Most substances, whether they are ionic or molecular in nature, are somewhat attracted to both phases. An equilibrium is established for the substance between the two phases, as shown in Equation 1.

substance-polar phase $\rightleftarrows$

substance-nonpolar phase   (Eq. 1)

As the solvent front moves up the paper, fresh developing solvent passes the spotted substance, and new equilibria are established. At the same time, any of the substance that has dissolved in the mobile phase encounters fresh stationary phase, and new equilibria are established. The overall effect of all these equilibria is that the movement of a substance depends on the nature of its relative attractions for the mobile and stationary phases. We characterize this movement in terms of a **retention factor** or **retardation factor** (R_f), defined in Equation 2.

$$R_f = \frac{\text{distance traveled by substance, cm}}{\text{distance traveled by solvent front, cm}} \quad \text{(Eq. 2)}$$

R_f values can be as high as 1.0, if the substance moves with the solvent front, and as low as 0.0, if the substance does not move at all. The values are

---

Copyright © 1989 by Chemical Education Resources, Inc., P.O. Box 357, 220 S. Railroad, Palmyra, Pennsylvania 17078
No part of this laboratory program may be reproduced or transmitted in any form or by any means, electronic or mechanical, including photocopying, recording, or any information storage and retrieval system, without permission in writing from the publisher. Printed in the United States of America

reproducible for a particular ion-and-solvent system, if the experimental conditions are closely controlled.

One important variable is the composition of the developing solvent. If one of the solvent components is volatile, then it is possible that evaporation will change the percent composition of the solvent as you develop the chromatogram. You can avoid this situation by keeping the container in which you are developing the chromatogram closed, so that the air in the container remains saturated with solvent vapor.

A sample containing two or more components can be separated, or **resolved**, by choosing a solvent system for which the sample components have distinctly different R_f values. You can make this choice in advance by spotting the individual cations that may be present in the sample on filter or chromatographic paper and determining their R_f values in a variety of solvent systems.

The affinity of a substance for the mobile phase will depend on that substance's chemical nature in solution. In this experiment, you will be working with transition-metal ions. The charge on these cations is either 2^+ or 3^+. When the cations dissolve in an aqueous HCl system, they become solvated by H_2O and Cl^- to form variously charged complexes. Using the symbol "M" to represent the transition-metal cations, we can symbolize the variety of complexes that can form:

$$[M(II)(H_2O)_x]^{2+} \text{ to } [M(II)Cl_x]^{2+-x}$$
and
$$[M(III)(H_2O)_x]^{3+} \text{ to } [M(III)Cl_x]^{3+-x}$$

Transition metals that form highly charged complexes tend to move more slowly than those that form complexes with little or no charge. Also, the equilibria governing their distribution between the two phases favor the stationary phase. Those metals that form neutral or slightly charged cations are more attracted to the mobile phase and, therefore, move more rapidly.

Once you have prepared the chromatogram, you must be able to locate and identify the metal cations on the paper. When the migrating complexes have color, you can locate them visually. Another approach involves adding a test reagent solution that forms a characteristic color with specific cations. We refer to the latter approach as a **spot test.**

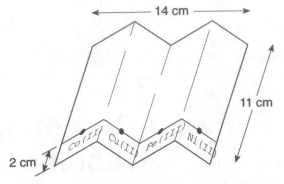

Figure 1   *Chromatographic paper folded accordion style and spotted with four cations*

In this experiment, you will be studying the paper chromatographic behavior of four cations, iron(III) (Fe^{3+}), cobalt(II) (Co^{2+}), nickel(II) (Ni^{2+}), and copper(II) (Cu^{2+}), in an acetone-hydrochloric acid system. You will also de-termine suitable spot tests for each cation. Finally, you will separate and identify the components of an unknown mixture of two or more of these four cations.

You can simultaneously determine the R_f values for each of the cations in the acetone–HCl solvent mixture on one piece of chromatographic paper if the paper is wide enough to allow each cation to be spotted separately, as shown in Figure 1.

While you are preparing this chromatogram, you will experiment with various spot tests to determine the test best suited for each cation. The test reagents you will be using are potassium ferrocyanide, $K_4Fe(CN)_6$, dimethylglyoxime (DMG), and potassium thiocyanate (KSCN). You will perform the spot tests by spotting three pieces of filter paper with each of the metal cations and exposing the spots to the various test reagent solutions, while carefully recording your observations. From these results, you will select the most unique and sensitive spot test for each of the four cations.

After you have determined the R_f values for each cation and identified suitable spot tests for each cation, you will receive two unknowns and a standard solution containing a mixture of all four cations. You will prepare chromatograms with these solutions and calculate the R_f of the cations that are separated. From the data you collect, you will identify which cations are present in your unknown solutions.

Procedure

Chemical Alert
acetone—flammable
concentrated ammonia—toxic and corrosive
0.1 *M* cobalt(II) nitrate—oxidant and irritant
0.1 *M* copper(II) nitrate—toxic and irritant
6 *M* hydrochloric acid—toxic and corrosive
0.1 *M* iron(III) nitrate—toxic and corrosive
0.1 *M* nickel(II) nitrate—suspected carcinogen
and oxidant
0.1 *M* nitric acid—toxic, corrosive, and oxidant
0.1 *M* potassium ferrocyanide—irritant
0.1 *M* potassium thiocyanate—irritant

Caution: Wear departmentally approved eye protection while doing this experiment.

Caution: Chromatographic paper is easily contaminated. Be sure your hands are clean and dry before handling it. Hold it only by what will be the two upper corners, the two corners on one of the 14-cm edges. When marking or spotting the paper, place it on a clean piece of paper to avoid contamination from the bench top.

Caution: Acetone is flammable and should not be used near an open flame.

6 *M* hydrochloric acid is a corrosive, toxic substance that can cause burns. Prevent contact with your eyes, skin, and clothing. Avoid inhaling vapors and ingesting the compound.

1.   Prepare the developing solvent by adding 10 mL of 6 *M* HCl and 35 mL of acetone to a 600-mL beaker. Immediately cover the beaker with plastic wrap held in place by a rubber band. Allow 10 min for the air in the beaker to become saturated with solvent.

2.   Obtain an 11 × 14-cm piece of chromatographic paper. Fold it into quarters, accordion style, with the folds running parallel to the 11-cm edge. With a ***pencil*** and ruler, draw a line parallel to the lower 14-cm edge, the edge that you did ***not*** handle. The line should be 2 cm from the edge (See Figure 1).

3.   Use a toothpick to practice spotting on a piece of filter paper before using the chromatographic paper. Aim for spots about 4 mm in diameter. A larger spot

may lead to decreased separation and smearing of the resolved cations.

4.   Use a separate toothpick to apply a small amount of each of the four standard solutions of Co(II), Cu(II), Fe(III), and Ni(II) cations on the pencil line.

Note: Avoid touching the chromatographic paper with your hand while labeling it in the next step.

Below the line you drew, label the paper in pencil "Chromatogram 1" and label each section of the paper with the cation present. After you have spotted and labeled the paper, dry it gently over a hot plate or by air drying.

5.   Carefully place the paper in the 600-mL beaker, with the pencil line at the bottom. Immediately re-cover the beaker with the plastic wrap. Be careful not to agitate the beaker, as waves in the developing solvent may spread the spot and decrease the separation of the cations. You must keep the beaker covered to prevent the evaporation of acetone, which would alter the composition of the developing solvent, thus affecting the development of the chro-matogram and the separation of the cations. Allow the solvent to rise to within 1 cm of the top edge of the paper before removing the paper.

6.   While you are developing Chromatogram 1, per-form the spot tests listed below. Record all observations on Data Sheet 1.

   (1)   Place one 4-mm spot of each of the cation solutions on a piece of filter paper. Use a pencil to label each spot, and dry carefully over a hot plate or by air drying. Treat each of the spots with 0.1 *M* K₄Fe(CN)₆ solution. Record your observations.

   (2)   Place 4-mm spots of the cation solutions on a second piece of filter paper, label, and dry. Treat each of the spots with 1 *M* KSCN solution in acetone. Record your observations.

Caution: Concentrated ammonia is a corrosive, toxic substance. Prevent contact with your eyes, skin, and clothing. Avoid inhaling vapors.

   (3)   Repeat the above procedure using a 1% di-methylglyoxime solution as a test reagent. Place the spotted filter paper beside a small beaker containing a small amount of concentrated ammonia sitting in a covered 600-mL beaker in the fume hood. The ammonia

fumes in this covered beaker make the paper basic. Record your observations.

Based on your observations in (1)–(3), determine which test reagent is best for each cation on Chromatogram 1. List your selections on Data Sheet 1. Verify your choices with your laboratory instructor.

7. By this time, the solvent front should be close to the top edge of the chromatographic paper (see Step 5). If it is, remove the paper, mark the final position of the solvent front with a pencil, and recover the beaker with the plastic wrap. Gently dry the chro-matographic paper, over a hot plate or by air drying.

8. Obtain from your laboratory instructor two unknown solutions, each containing one, two, three, or four of the cations. Record the number of your unknowns on Data Sheet 2.

9. Obtain another piece of chromatographic paper, and fold it in fourths as in Step 2. Draw the line 2-cm from the bottom edge of the paper, and spot the first and third sections with a 4-mm spot of the standard solution containing all four cations. Spot the other two sections with your two unknowns, one per section.

Label with a pencil this piece of paper "Chromatogram 2." Label Columns 1 and 3 "standard solutions," Column 2 "unknown 1," and Column 4 "unknown 2." Dry the paper and place it in the 600-mL beaker of developing solvent. Recover the beaker with the plastic wrap. Allow the solvent front to move to within 1 cm of the top edge of the paper. While you are developing Chromatogram 2, use the test reagents you selected in Step 6 to spot test Chromatogram 1, as described in Step 10.

10. When applying test reagent solutions to Chromatograph 1, you must use capillary tubing in order to place just a fine streak of reagent solution on the paper. Dip the clean, dry capillary tubing in the appropriate test reagent solution. Capillary action will draw up a sufficient amount of reagent. Start at the line marking the farthest advance of the solvent front, and draw the tubing rapidly down the center of the column to the starting line. Draw a line around the edge of the spot in each column with a pencil. Repeat for each cation.

Measure the distance traveled by the solvent and that traveled by each cation, as follows. Begin each measurement at the line 2-cm from the edge, and measure to the midpoint of each cation spot. Calculate the R_f value for each cation, using Equation 2. Record all measurements and R_f values on Data Sheet 1.

11. When the solvent front in Chromatogram 2 has moved to within 1 cm of the top edge of the paper, remove the paper from the beaker and recover, mark the solvent front, and dry as in Step 7.

Based on your observations in Step 10, streak each part of Column 1, containing the known solution of four cations, with the test reagent solution appropriate for the cation you expect to find there. You should spot Column 3, also containing the four known cations, as confirmation. You might find it necessary to rearrange the order of reagents when spotting these columns, due to the close proximity of some cations.

12. Measure the distance the solvent traveled and that traveled by each cation in Column 1, as you did in Step 10. Record these measurements and the test reagent used with each cation, on Data Sheet 2. Calculate the R_f value for each cation in the standard solution, and record these values on Data Sheet 2. Repeat these measurements for the cations in Column 3, calculate the R_f values for the cations, and record the measurements and values on Data Sheet 2.

13. Write on Data Sheet 2 a comparision of the R_f values calculated from the distance the solvent and the cations traveled in Column 1 compared to that traveled by the cations in Column 3.

14. Streak Columns 2 and 4 on Chromatogram 2, containing your unknowns. Indicate in the table on Data Sheet 2 which cations are present in each of your unknowns. Measure the distance each of the cations traveled. Record the test reagents you used and the distances the cations traveled, on Data Sheet 2. Calculate R_f values for the cations and record them on Data Sheet 2.

15. Compare, on Data Sheet 2, the R_f values calculated for the cations in the standard solution from Chromatogram 2 with those you calculated for the cations from Chromatogram 1.

16. Discard your developing solvent and unknown solutions following the directions of your laboratory instructor.

17. Retain your chromatograms and attach them to your Data Sheets or laboratory report to be submitted to your laboratory instructor.

> *Caution:* Wash your hands thoroughly with soap or detergent before leaving the laboratory.

Post-Laboratory Questions

1. An alternate setup for this experiment calls for the chromatographic paper to be rolled to form a cylinder, with the edges stapled together at the top and bottom, and inserted upright into the developing solvent. A student developed Chromatogram 1 using this setup and tested the spot near one of the staples by streaking that column of the paper with $K_4Fe(CN)_6$ solution. The student then observed a dark blue coloration running from the staple all the way to the solvent front.

(1) What caused the blue color?

(2) Why was the color found from the staple to the solvent front instead of in one small spot?

2. List experimental factors that might make it difficult for you to identify the cations in this experiment.

3. Predict the effect on cation identification and distance traveled by cations, if the chromatograms in this experiment were developed using a taller beaker and a taller piece of chromatographic paper.

4. Predict what will happen to the position of the cations in the column if the solvent front is allowed to move to the top edge of the paper, and the paper is removed from the beaker 10 min after this occurs. Briefly explain.

name   section   date

Data Sheet 1

Spot Tests

cation	$K_4Fe(CN)_6$	KSCN/acetone	DMG/NH_3
Co(II)	_____	_____	_____
Cu(II)	_____	_____	_____
Fe(III)	_____	_____	_____
Ni(II)	_____	_____	_____

Selecting the Test Reagent

Co(II)	_____	Fe(III)	_____
Cu(II)	_____	Ni(II)	_____

verification of selected test reagents by your laboratory instructor _____

Determining R_f Values for Chromatograph 1

distance traveled by solvent front, mm _____

cation	test reagent	distance to center of spot, mm	R_f value
Co(II)	_____	_____	_____
Cu(II)	_____	_____	_____
Fe(III)	_____	_____	_____
Ni(II)	_____	_____	_____

Data Sheet 2

Separating and Identifying the Cations in Standard Solution (Chromatograph 2)

distance traveled by solvent front, mm _____

column 1

cation	test reagent	distance to center of spot, mm	R_f value
Co(II)	_____	_____	_____
Cu(II)	_____	_____	_____
Fe(III)	_____	_____	_____
Ni(II)	_____	_____	_____

column 3

cation	test reagent	distance to center of spot, mm	R_f value
Co(II)	_____	_____	_____
Cu(II)	_____	_____	_____
Fe(III)	_____	_____	_____
Ni(II)	_____	_____	_____

Compare the R_f values for the four cations in the standard solution calculated from Column 1 with those calculated from Column 3 on Chromatogram 2.

V.   Identifying the Cations in the Unknowns (Chromatogram 2)

column 2

unknown number _____

cation	present	test reagent	distance to center of spot, mm	R_f
Co(II)	_____	_____	_____	_____
Cu(II)	_____	_____	_____	_____
Fe(III)	_____	_____	_____	_____
Ni(II)	_____	_____	_____	_____

column 4

unknown number _____

cation	present	test reagent	distance to center of spot, mm	R_f
Co(II)	_____	_____	_____	_____
Cu(II)	_____	_____	_____	_____
Fe(III)	_____	_____	_____	_____
Ni(II)	_____	_____	_____	_____

Compare the R_f values for the four cations from Chromatogram 1 with those values obtained from Chromatogram 2.

Pre-Laboratory Assignment

1. A student doing this experiment was careless and spilled about 2 mL of the 10 mL of $6M$ HCl being added to 35 mL of acetone to prepare the developing solvent.

(1)   Explain why the student had to clean up the outside of the beaker and the bench top.

(2)   How would the smaller amount of HCl used in the developing solvent have affected the results of the experiment?

2. (1)   A student ran a chromatogram for the cation, Mn^{2+}, following the method described in this experiment. The solvent front traveled 96 mm, and the distance to the center of the Mn^{2+} spot was 80 mm. Based on these data, what is the R_f value for Mn^{2+}?

(2)   Consider a chromatogram of an unknown, in which the solvent front has traveled 84 mm. If the unknown were Mn^{2+}, where would you expect to find the center of the cation spot?

_____
answer

3. Explain why initial spots with a diameter of about 4 mm are preferable to larger ones.

4. Define the terms "spotted" and "resolved."

_____
answer

5.   Which of the following equations shows the best spot test for Mn^{2+}? Briefly explain.

$$Mn^{2+}(aq) + 2\ OH^-(aq) \rightarrow Mn(OH)_2(s, \text{pale pink})$$

$$Mn^{2+}(aq) + C_2O_4{}^{2-}(aq) \rightarrow MnC_2O_4(s, \text{pale pink})$$

$$2\ Mn^{2+}(aq) + [Fe(CN)_6]^{4-}(aq) \rightarrow Mn_2Fe(CN)_6(s, \text{pale green})$$

$$2\ Mn^{2+}(aq) + 5\ BiO_3{}^-(aq) + 14\ H_3O^+(aq) \rightarrow 2\ MnO_4{}^-(aq, \text{deep purple}) + 5\ Bi^{3+}(aq) + 21\ H_2O(l)$$

Determining the Solubility of an Unknown Salt at Various Temperatures

prepared by **Enno Wolthuis**, Calvin College

Purpose of the Experiment

Determine the solubility of an unknown inorganic salt in water at various temperatures. Draw a solubility–temperature curve for the salt.

Background Information

An experienced cook knows the difference in flavor that results from the presence of salt in water used for boiling vegetables. A good cook is also careful not to make the cooking water too salty. In this case, "too salty" means that too much salt is dissolved in the water, not that undissolved salt remains suspended in the cooking water.

A **salt** is any compound formed from cations and anions, except water. The most common example is sodium chloride ($NaCl$), also called table salt or just "salt." As chemists, we are interested in knowing how much salt will dissolve in a specific volume of water. We also want to know the effect of water temperature on the amount of salt that will dissolve.

We can prepare an aqueous solution of $NaCl$ of increasing concentration by adding larger and larger masses of $NaCl$ to a constant volume of water. However, this process of increasing the concentration of an $NaCl$ solution cannot proceed indefinitely. There is a point at which no additional $NaCl$ will dissolve in the solution. We refer to a solution that contains the maximum mass of dissolved substance, or **solute**, that can be dissolved at a particular temperature as a **saturated**

solution. We call the mass of solute dissolved in 100 mL of a saturated solution the **solubility** of the solute in that **solvent**, or dissolving agent, at the specified temperature. For example, the solubility of the solute $NaCl$ in the solvent water is 36.0 g of $NaCl$ per 100 mL of water at 20 $^{\circ}$C. This means that we can dissolve no more than 36.0 g of $NaCl$ in 100 mL of water at 20 $^{\circ}$C. Sodium chloride solutions containing less than 36.0 g of $NaCl$ per 100 mL of water at 20 $^{\circ}$C are called **unsaturated** solutions.

Saturated Solutions

To prepare a saturated solution, we mix a measured volume of solvent with a greater mass of solute than can dissolve in the volume of solvent at the experimental temperature. The solute will dissolve until the solution is saturated, leaving the remaining salt undissolved.

The undissolved salt plays an active role in the saturated solution. Salt particles continually enter the solution, and dissolved solute ions continually recombine to form solid salt. The dissolution rate of solid into solution is exactly offset by the precipitation rate of solid from the solution. Hence the mass of dissolved solute

---

Copyright © 1993 by Chemical Education Resources, Inc., P.O. Box 357, 220 S. Railroad, Palmyra, Pennsylvania 17078
No part of this laboratory program may be reproduced or transmitted in any form or by any means, electronic or mechanical, including photocopying, recording, or any information storage and retrieval system, without permission in writing from the publisher. Printed in the United States of America

and the solution composition remain constant when the solution is saturated. We call the temperature at which we prepare the saturated solution the **saturation temperature**.

The solubility of a solute in a solvent depends on the nature of the solute, the nature of the solvent, the pressure, especially when the solute is in the gas phase, and the temperature. In this experiment, you will investigate the effect of temperature on the solubility of a salt in water.

The Effect of Temperature on the Solubility of a Salt

The amount of heat released or absorbed when a salt dissolves determines the effect of temperature on the solubility of that salt. Whether heat is released or absorbed depends on both the energy required to disrupt the crystal structure of the ionic solid and the energy liberated when the cations and anions interact with the solvent. If more energy is required to remove ions from the crystal lattice than is released by solvent interactions with the dissociated ions, heat is absorbed by the system during the dissolution process. Conversely, if less energy is required to remove ions from the lattice than is released when ions interact with the solvent, heat released by the system during the dissolution process.

For most salts, heat is absorbed when they dissolve in water. Therefore, the solubility of these salts increases with increasing temperature. In cases where the dissolution of a salt results in the release of heat, solubility decreases with increasing temperature.

We can most easily visualize the effect of temperature on the solubility of a salt using a graph called a **solubility–temperature curve**. To prepare a solubility–temperature curve, we plot experimental temperature on the abscissa, or x-axis, and salt concentration in a saturated solution on the ordinate, or y-axis. The solution concentration unit most commonly used for such plots is grams of salt per 100 mL of water. Sol-ubility–temperature curves for potassium bromide (KBr), potassium nitrate (KNO₃), potassium dichro-mate (K₂Cr₂O₇), and potassium chlorate (KClO₃) are shown in Figure 1. Using such curves, we can estimate the salt concentration of saturated solutions at temperatures in between those at which the solubility has been experimentally determined.

In this experiment, you will create a solubility–temperature curve for an unknown salt. To accomplish this, you will dissolve a known mass of the salt in various measured volumes of water. You will cool the increas-

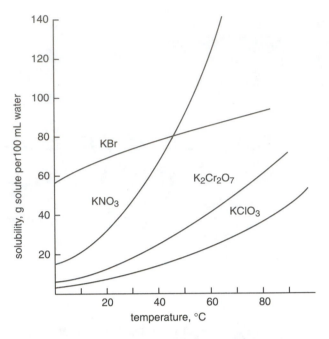

Figure 1   *Solubility-temperature curves for several salts in water*

ingly dilute solutions and determine the temperature at which the salt begins to crystallize out of solution.

The mass of salt and the volume of water associated with saturation temperature represent the concentration of a saturated solution at that temperature. You will use these data and Equation 1 to calculate the equivalent concentration in grams of salt per 100 mL water.

$$\text{concentration of saturated solution, g of salt per 100 mL water} = \left(\frac{\text{experimental mass of salt, g}}{\text{experimental volume of water, mL}} \right)(100) \quad \text{(Eq. 1)}$$

Finally, you will draw a solubility–temperature curve for your salt by plotting the grams of salt per 100 mL water against the related saturation temperature.

Procedure

Chemical Alert
Handle all solids and solutions with care, because of the possible presence of toxic or irritating substances.

Caution: Wear departmentally approved eye protection while doing this experiment.

I. Determining the Saturation Temperature of Your Original Salt Solution

1. Select a 2-hole rubber stopper that fits snugly into a 25 × 250-mm test tube.

> *Caution:* Use special precautions when inserting a thermometer and a stirring rod into a rubber stopper. Injury to your hands can result if excessive pressure on the thermometer or the rod causes either to break.
>
> A good procedure is to lubricate the thermometer and stopper with glycerine or water, cover both of your hands with a cloth, and avoid forcing the thermometer if it does not enter the stopper hole easily.
>
> Your laboratory instructor will demonstrate the technique you are to use for inserting the thermometer and glass rod into the rubber stopper.

2. Carefully insert a thermometer through one hole of the rubber stopper, positioning it so that the thermometer bulb will be about 1 cm from the bottom of the test tube. Carefully insert a 3-mm diameter glass stirring rod, which has a small loop at the bottom, through the other stopper hole. Position the glass rod so that it will extend from the test tube bottom to at least 5 cm above the stopper. Stir your solution in the test tube by moving this rod up and down.

3. Place an unlighted Bunsen burner on the base of a ring stand. Attach an iron ring to the ring stand at a level 5 cm above the top of the burner. Place a ceramic-centered wire gauze on the iron ring. Attach a second iron ring to the ring stand so that the second ring is 7–10 cm above and centered over the wire gauze.

> *Note:* In order to avoid the possibility of knocking the 400-mL beaker off the iron ring during the experiment, place the beaker through the second iron ring, as shown in Figure 2.

4. Use a 100-mL graduated cylinder to transfer approximately 300 mL of distilled or deionized water to a 400-mL beaker. Insert the beaker into the top ring on the stand (see Figure 2). Position the beaker so that the bottom of the beaker rests on the wire gauze on the bottom ring. Light the air–gas mixture from the

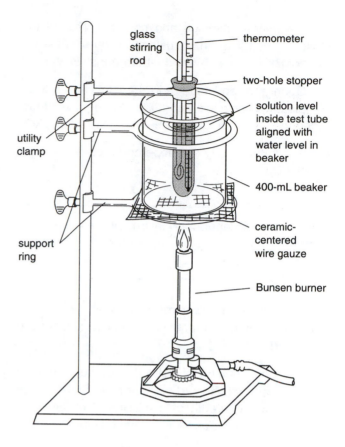

Figure 2 *Apparatus for determining saturation temperature*

Bunsen burner. Slowly begin to heat the beaker and water to about 80 °C, using your second thermometer to measure the water temperature.

> *Note:* Your laboratory instructor may prefer that you transfer your sample of unknown salt from a reagent bottle into a tared vial, instead of receiving it in an envelope. In this case, you will be given instructions for transferring and weighing your sample.

5. Obtain from your laboratory instructor an envelope containing the unknown salt you will study. Record the identification code of the unknown on your Data Sheet.

> *Note:* Your laboratory instructor will give you directions for using your balance and will tell you the number of significant digits to the right of the decimal point to use when recording your data.

6. Weigh the envelope and contents. Record this mass on your Data Sheet, using the number of significant digits specified by your laboratory instructor.

7. Open the envelope and transfer the salt as completely as possible into a 25 × 250-mm test tube. Stand the filled test tube in an empty 250-mL beaker.

8. Weigh the empty envelope and record this mass on your Data Sheet.

> **Note:** Your laboratory instructor will demonstrate a satisfactory procedure for filling, using, and reading a buret.

9. Fill a clean 50-mL buret with distilled water. Secure the buret in a buret clamp attached to a second ring stand.

10. Dispense 3.00 mL of distilled water from the buret into the test tube. Record on your Data Sheet, to the nearest two hundredth of a milliliter (0.02 mL), the exact volume of water added to the test tube.

11. Place the rubber stopper with thermometer and glass stirring rod in the test tube. Be sure the thermometer bulb is about 1 cm from the test tube bottom.

12. Clamp the test tube to the ring stand, using a utility clamp. (See Figure 2.) Position the test tube so that it is suspended in the hot-water bath. Adjust the test tube position until the liquid levels inside and outside the test tube are aligned.

> **Note:** Heat the solution in the test tube as rapidly as possible to avoid evaporation.
>
> Your laboratory instructor will inform you about the number of significant digits to the right of the decimal point to use when recording your temperature readings.

13. Stir the mixture in the test tube by slowly moving the stirring rod up and down in a uniform manner, while heating the test tube and contents to 80 °C.

14. If the salt is not completely dissolved when the solution temperature reaches 80 °C, add 0.50 mL of distilled water from the buret to the mixture in the test tube. Record on your Data Sheet the exact total volume of water added to the test tube. Reheat the test tube and its contents in the hot-water bath to 80 °C,

while continuing to stir the solution in a uniform manner.

15. If some salt still remains undissolved, repeat the procedure of adding 0.50 mL of distilled water to the mixture and reheating to 80 °C with mixing. Continue this process until you have added just enough distilled water to dissolve all of the salt at a temperature no higher than 80 °C. Be sure to record on your Data Sheet the exact total volume of water you added to the test tube to achieve complete dissolution of the salt.

16. After the salt has completely dissolved, carefully loosen the clamp holding the test tube and contents and lift them out of the hot-water bath. Reclamp the test tube and contents at a different position on the ring stand, so the test tube and contents are not immersed in the hot-water bath. Uniformly stir the test tube contents while allowing the solution to cool.

17. Carefully observe the solution. Record on your Data Sheet the exact temperature at which the first salt crystals appear in the solution. This is the saturation temperature for *this* solution.

If you are in doubt about the exact saturation temperature, replace the test tube and its contents in the hot-water bath and reheat them to 80 °C until all of the crystals dissolve. Repeat the cooling procedure, and record the saturation temperature.

Retain the solution in the test tube for use in Part II.

II. Determining the Saturation Temperature of Diluted Salt Solutions

18. Add 0.50 mL distilled water from the buret to the test tube and contents from Part I. Record on your Data Sheet the exact total volume of distilled water added to the test tube. Note that the total volume includes the amount already in the test tube from Part I.

19. Heat the test tube and its contents in the hot-water bath until all of the salt crystals dissolve.

20. Remove the test tube and its contents from the hot-water bath and allow them to cool until salt crystals appear. Stir during the cooling process as before. Record the saturation temperature on your Data Sheet.

21. Repeat Steps 18–20 to determine the saturation temperature of your salt solution after the addition of 0.50 mL of distilled water.

22. Continue determining the saturation temperature of your salt solution after each addition of a 0.50-mL portion of distilled water. Record on your Data Sheet the total volume of water added and the saturation temperature after each water addition. If 0.50-mL portions of distilled water do not significantly change the saturation temperature, add distilled water in 1.00-mL increments.

23. When the saturation temperature reaches room temperature, replace the hot-water bath with an ice-water bath. Add several more 0.50-mL portions of distilled water to the solution in the test tube. After each addition, cool the test tube and contents in the ice-water bath to obtain saturation temperature data that fall between room temperature and 0 °C.

Determine a total of eight different saturation temperatures.

24. Dispose of the solution in your test tube as directed by your laboratory instructor.

Caution: Wash your hands thoroughly with soap or detergent before leaving the laboratory.

Calculations

(Do the following calculations and record the results on your Data Sheet.)

1. Calculate the solubility, in grams of salt per 100 mL of water, of your salt at the original saturation temperature.

2. Recalculate the solubility of your salt at each experimental saturation temperature.

3. Prepare a solubility–temperature curve for your salt.

Post-Laboratory Questions

(Use the spaces provided for the answers and additional paper if necessary.)

1. What procedural change for this experiment would be necessary if the solubility of the salt were initially determined at 10 °C, followed by solubility determinations at 20 °C, 35 °C, 50 °C, 65 °C, and 80 °C?

4. It is important that you keep the test tube closed during heating to avoid solvent loss by evaporation.

(1) Briefly describe the effect such solvent loss would have on your initial calculation of the solubility of your salt.

2. Use the information in Figure 1 to determine the temperature at which 10.0 g KNO_3 dissolved in 12.5 mL of H_2O would be a saturated solution.

(2) Would this initial evaporation affect the calculated solubility of your salt at each subsequent experimental saturation temperature, or just at the initial temperature?

3. While transferring the weighed salt to the test tube during this experiment, a student spilled some. Will this salt loss result in calculated salt solubilities that are too high, too low, or unchanged? Briefly explain.

name   section   date

Data Sheet

I. Determining the Saturation Temperature of Your Original Salt Solution

unknown identification code   _____

mass of envelope and sample, g   _____

mass of envelope, g   _____

  mass of sample, g   _____

initial volume of water added, mL   _____

total volume of water added, mL   _____

saturation temperature, °C   _____

solubility, grams of salt per 100 mL of water   _____

II. Determining the Saturation Temperature of Diluted Salt Solutions

total volume of water added, mL	saturation temperature, °C	solubility, grams of salt per 100 mL water
_____	_____	_____
_____	_____	_____
_____	_____	_____
_____	_____	_____
_____	_____	_____
_____	_____	_____
_____	_____	_____
_____	_____	_____
_____	_____	_____

Pre-Laboratory Assignment

1. Briefly describe the hazards you should avoid when stirring a solution in a test tube suspended in a hot-water bath.

4. A student collected the following temperature–solubility data for ammonium chloride (NH_4Cl) dissolved in water.

mass of sample, g   2.33

total volume of water added, mL	saturation temperature, °C	solubility, grams of salt per 100 mL water
3.00	98.0	77.7
3.58	80.0	____
4.24	61.0	____
4.66	51.0	____
5.68	30.0	____
6.13	22.0	____
7.06	10.0	____
7.77	1.0	____

2. What is the difference between a saturated solution and an unsaturated solution?

(1)   Calculate the solubility of NH_4Cl, in grams of NH_4Cl per 100 mL water, for each saturation temperature. Enter these solubilities on the data table above.

3. Briefly explain why the solubility of most salts increases with increasing temperature.

(2)   Prepare a solubility–temperature curve for NH_4Cl on the graph paper on p. 12.
(3)   Estimate the solubility of NH_4Cl at 40 °C from the solubility–temperature curve you prepared in (2).

Enthalpy of Hydration

prepared by **H. A. Neidig**, Lebanon Valley College

Purpose of the Experiment

Determine the enthalpy of hydration of anhydrous magnesium sulfate.

Background Information

The **enthalpy of hydration** ($\Delta H_{hydration}$) of anhydrous sodium carbonate (Na_2CO_3) is the enthalpy change for converting 1 mol of anhydrous Na_2CO_3 to 1 mol of sodium carbonate decahydrate ($Na_2CO_3 \cdot 10\ H_2O$). The reaction for this process is shown in Equation 1.

$$Na_2CO_3(s) + 10\ H_2O(l) \rightarrow$$
$$Na_2CO_3 \cdot 10\ H_2O(s)\ \Delta H_1 \quad (Eq.\ 1)$$

Because the reaction takes place over a long period of time, we cannot easily collect temperature–time data to calculate ΔH_1.

However, we can obtain data to calculate the enthalpy change associated with dissolving anhydrous Na_2CO_3 and $Na_2CO_3 \cdot 10\ H_2O$ in water. These reactions are shown in Equations 2 and 3.

$$Na_2CO_3(s) + 80\ H_2O(l) \rightarrow Na_2CO_3(aq)\ \Delta H_2 \quad (Eq.\ 2)$$

$$Na_2CO_3 \cdot 10\ H_2O(s) + 70\ H_2O(l) \rightarrow$$
$$Na_2CO_3(aq)\ \Delta H_3 \quad (Eq.\ 3)$$

The **enthalpy of dissolution** is the enthalpy change resulting from dissolving 1 mol of a substance in a large amount of water; this process results in infinite dilution of the solute. Then, by using Hess' law, we can determine the enthalpy of hydration of anhydrous Na_2CO_3.

Hess' law, or the **law of constant heat summation**, states that at constant pressure the enthalpy change for a process is not dependent on the reaction pathway but is only dependent on the initial and final states of the system. The enthalpy changes of individual steps in a process can be added or subtracted to obtain the net enthalpy change for the overall process.

After reversing Equation 3 and changing the sign of ΔH_3, we obtain Equation 4, which, when added to Equation 2, gives us Equation 1.

$$Na_2CO_3(s) + 80\ H_2O(l) \rightarrow Na_2CO_3(aq)\ \Delta H_2 \quad (Eq.\ 2)$$
$$Na_2CO_3(aq) \rightarrow$$
$$70\ H_2O(l) + Na_2CO_3 \cdot 10\ H_2O(s)\ -\Delta H_3 \quad (Eq.\ 4)$$

$$Na_2CO_3(s) + 10\ H_2O(l) \rightarrow$$
$$Na_2CO_3 \cdot 10\ H_2O(s)\ \Delta H_1 \quad (Eq.\ 1)$$

The enthalpy of hydration, ΔH_1 equals $\Delta H_2 - \Delta H_3$. To calculate ΔH_1, we need to calculate ΔH_2 and ΔH_3 from experimental data.

Calculating the Enthalpy of Dissolution

To determine the enthalpy of dissolution of anhydrous Na_2CO_3, we conducted an experiment in which 5.19 g of Na_2CO_3 was dissolved in 75.0 g of distilled water. The temperature of the system increased by

---

Copyright © 1989 by Chemical Education Resources, Inc., P.O. Box 357, 220 S. Railroad, Palmyra, Pennsylvania 17078

No part of this laboratory program may be reproduced or transmitted in any form or by any means, electronic or mechanical, including photocopying, recording, or any information storage and retrieval system, without permission in writing from the publisher. Printed in the United States of America

3.80 °C. The heat capacity (or specific heat) of the resulting solution of Na_2CO_3 was 3.90 J g^{-1} deg^{-1}.

The heat transferred, Q, during the dissolution of anhydrous Na_2CO_3 is calculated, using Equation 5, where ΔT is the temperature change.

$Q = -$(mass of mixture, g) $\times$
(heat capacity of mixture, J g^{-1} deg^{-1})$(\Delta T$, °C)
(Eq. 5)

$Q = -$(80.2 g)(3.90 J g^{-1} deg^{-1})(3.80 °C) = -1189 J

Since ΔT is positive, heat was released to the system and the reaction is exothermic. Therefore, Q is negative for the dissolution of anyhdrous Na_2CO_3.

The number of moles of Na_2CO_3 used was

$$\text{number of moles} = \frac{\text{mass, g}}{\text{gram molecular mass, g mol}^{-1}}$$
(Eq. 6)

$$\text{number of moles of } Na_2CO_3 = \frac{5.19 \text{ g}}{106.00 \text{ g mol}^{-1}} = 4.90 \times 10^{-2} \text{ mol}$$

Assuming the heat transferred to the calorimeter is negligible, we can calculate the enthalpy of dissolution of anhydrous Na_2CO_3, ΔH_2, using Equation 7.

$$\Delta H_2 = \frac{Q, \text{ J}}{\text{number of moles of solute}}$$
(Eq. 7)

$$= \frac{-1188 \text{ J}}{4.90 \times 10^{-2} \text{ mol}} = -2.42 \times 10^4 \text{ J mol}^{-1}$$

$$= -2.42 \times 10^1 \text{ kJ mol}^{-1}$$

To determine the $\Delta H_{dissolution}$ of Na_2CO_3 • 10 H_2O, we conducted a second experiment in which 14.7 g of Na_2CO_3 • 10 H_2O was added to 65.0 g of distilled water. The temperature of the system decreased by 9.76 °C.

The heat transferred, Q_1, during the dissolution of Na_2CO_3 • 10 H_2O is calculated, using Equation 5.

$$Q_1 = -(79.7 \text{ g})(3.90 \text{ J g}^{-1} \text{ deg}^{-1})(-9.76 \text{ °C})$$

$$Q_1 = 3034 \text{ J}$$

Since ΔT is negative, heat was absorbed by the system and the reaction is endothermic. Therefore, Q_1 is positive for the dissolution of Na_2CO_3 • 10 H_2O.

The number of moles of Na_2CO_3 • 10 H_2O used is calculated, using Equation 6.

$$\text{number of moles of } Na_2CO_3 \bullet 10 \text{ } H_2O = \frac{14.7 \text{ g}}{286.16 \text{ g mol}^{-1}}$$

$$= 5.14 \times 10^{-2} \text{ mol}$$

Assuming the heat transferred from the calorimeter is negligible, we can calculate the enthalpy of dissolution of Na_2CO_3 • 10 H_2O, ΔH_3, using Equation 7.

$$\Delta H_3 = \frac{3034 \text{ J}}{5.14 \times 10^{-2} \text{ mol}}$$

$$= 5.90 \times 10^4 \text{ J mol}^{-1} = 5.90 \times 10^1 \text{ kJ mol}^{-1}$$

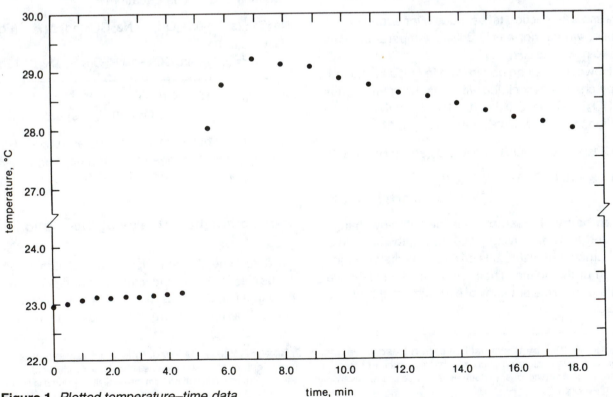

Figure 1 *Plotted temperature–time data*

time, min

Calculating the Enthalpy of Hydration

We can calculate the enthalpy of hydration (ΔH_1) from the $\Delta H_{dissolution}$ of $Na_2CO_3 \cdot 10\ H_2O$ (ΔH_3) and the $\Delta H_{dissolution}$ of anhydrous Na_2CO_3 (ΔH_2).

$$\begin{aligned}\Delta H_1 &= \Delta H_2 - \Delta H_3 \\ &= (-2.42 \times 10^1\ kJ\ mol^{-1}) - (5.90 \times 10^1\ kJ\ mol^{-1}) \\ &= -8.32 \times 10^1\ kJ\ mol^{-1} \qquad \text{(Eq. 8)}\end{aligned}$$

Thus, the enthalpy of hydration of anhydrous Na_2CO_3 is $-83.2\ kJ\ mol^{-1}$.

Calculating the Temperature Change

From a plot of your temperature–time data, you will find ΔT. Although heat is transferred essentially instantaneously at the time of mixing, thermometers cannot respond quickly enough to give an accurate indication of the temperature at the time of mixing and shortly thereafter. As a result, that portion of the temperature–time curve shows a temperature that is lower or higher than the actual temperature of the mixture at this point.

By plotting temperature–time data both prior to and after mixing, you can extrapolate the curve and determine the temperature at the time of mixing. In part, this procedure corrects for the deficiencies of your thermometers.

You will prepare a temperature–time graph by plotting the temperature, in degrees Celsius, on the ordinate (y-axis) and the time, in minutes, on the abscissa (x-axis). Note in Figure 1 that the temperature units on the y-axis are discontinued between 24.5 and 26.5 °C, where no data were collected. In this way, the temperature ranges where data were collected can be expanded, thus enhancing the accuracy of ΔT.

Draw a vertical line, perpendicular to the x-axis, through the point on the x-axis that represents the time of mixing. This line is called the **line of mixing.** Then, draw the best straight line representing the plotted points for the temperature of the water. Extend this line to intersect the line of mixing. This point of intersection represents the initial temperature of the water ($T_{initial}$).

Draw the best straight line representing the plotted points for the temperature of the reaction mixture. Extend this line to intersect the line of mixing. This point of intersection is the final temperature of the reaction mixture (T_{final}). Figure 2 illustrates this process.

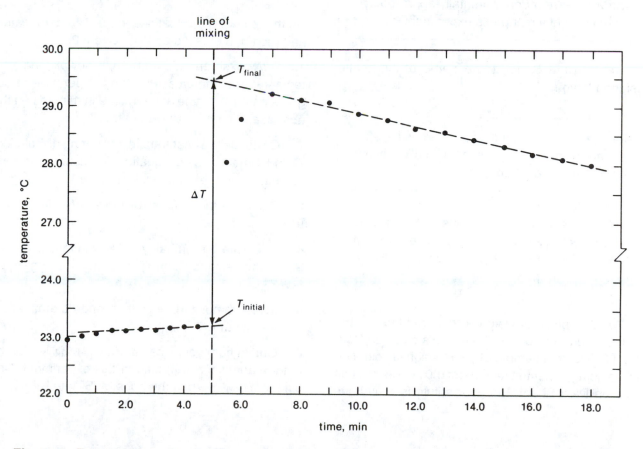

Figure 2 *Extrapolation of temperature–time data to find ΔT*

Determine ΔT for the process by subtracting $T_{initial}$ from T_{final}, as shown in Equation 9.

$$\Delta T = T_{final} - T_{initial} \qquad \text{(Eq. 9)}$$

In this experiment, you will dissolve a known mass of anhydrous magnesium sulfate ($MgSO_4$) or of hydrated magnesium sulfate ($MgSO_4 \cdot 7\ H_2O$) in 100.0 mL of water. You will collect temperature–time data while throughly stirring the solution.

You will find ΔT for the process and use it to calculate the enthalpy of dissolution of the solute assigned to you. After obtaining from your classmates the enthalpy of dissolution for the other magnesium salt used in this experiment, you will calculate the enthalpy of hydration of anhydrous $MgSO_4$.

Procedure

Chemical Alert

$MgSO_4$—hygroscopic and irritant
$MgSO_4 \cdot 7\ H_2O$—irritant

Caution: Wear departmentally approved eye protection while doing this experiment.

Record on Data Sheet 1 the name of the solute assigned to you.

Caution: While handling anhydrous $MgSO_4$, be careful to avoid having the solid absorb water from the atmosphere.

Note: The mass of anhydrous $MgSO_4$ should be between 7.00 and 8.00 g. The mass of powdered $MgSO_4 \cdot 7\ H_2O$ should be between 15.3 and 15.4 g.

Weigh a piece of weighing paper to the nearest tenth of a gram (0.01 g). Record this mass on Data Sheet 2. Weigh a sample of your assigned solute on the weighing paper to the nearest 0.01 g. Record this mass on Data Sheet 2. If your solute is anhydrous

$MgSO_4$, keep the solute covered until you add it to the water.

Place 100.0 mL of distilled or deionized water in a tared, 10-oz, pressed polystyrene cup. Measure and record on Data Sheet 1 temperature–time data for the water at 30-s intervals for a 4.5-min period. At the 5-min mark, rapidly transfer your weighed solute to the water in the polystyrene cup while stirring constantly with a glass stirring rod. Measure and record temperature–time data at 5.5 min and at 1-min intervals for an additional 14.5-min period. Measure and record on Data Sheet 2 the mass of the reaction mixture.

Do a second determination with new portions of solute and water. If time permits, do a third determination with new portions of solute and water.

Caution: Wash your hands thoroughly with soap or detergent before leaving the laboratory.

Calculations

Do the following calculations for each determination and record the results on Data Sheet 2.

1. Draw axes on the graph paper for plotting your temperature–time data. Plot the temperature in degrees on the ordinate and the time in minutes on the abscissa. Find ΔT, using Equation 9.

2. Calculate the heat transferred during dissolution, Q, in kJ mol^{-1}, using Equation 5. Assume the heat capacity of the reaction mixture is 3.84 J g^{-1} deg^{-1}.

3. Calculate the number of moles of solute used, using Equation 6.

4. Calculate the enthalpy of dissolution, $\Delta H_{dissolution}$, of the assigned solute, in kJ mol^{-1}, using Equation 7.

5. If you completed more than one determination, find the mean $\Delta H_{dissolution}$.

6. Obtain from your classmates $\Delta H_{dissolution}$ for the other magnesium salt used in this experiment. Calculate $\Delta H_{hydration}$ of anhydrous $MgSO_4$ in kJ mol^{-1}.

Post-Laboratory Questions

(Use the spaces provided for the answers and additional paper if necessary.)

1. A student was given a sample of an anhydrous salt to analyze, using the procedure described in this experiment. The student weighed the sample but decided to do the experiment the following day. The weighed sample was left uncovered on the bench overnight.

(1) What error(s) might result if the student used this salt sample in a determination the following day? Briefly explain.

(2) Would the student avoid any error(s) described in (1) by reweighing the salt sample on the following day, before doing the experiment?

(3) Briefly explain what effect the use of a glass beaker in place of the pressed polystyrene cup would have on the expected $\Delta H_{hydration}$.

2. A student doing this experiment decided to check the accuracy of the calibration of the thermometer. When placed in a beaker of boiling water, the thermometer indicated 103.5 °C. When placed in an ice bath, the indicated temperature was 3.5 °C. Would using this thermometer in this experiment cause any errors in the calculated $\Delta H_{dissolution}$? Briefly explain.

3. There is a large difference between $\Delta H_{dissolution}$ of an anhydrous salt dissolved in water and $\Delta H_{dissolution}$ of the hydrated form of the same salt dissolved in water. How do the enthalpy of dissolution, the crystal lattice energy, and the solvation energy of the two salts relate to these experimental observations?

name   section   date

Data Sheet 1

Temperature–Time Data

assigned solute _____

		determination	
	1	*2*	*3*
time, min	*temp, °C*	*temp, °C*	*temp, °C*
0.0	_____	_____	_____
0.5	_____	_____	_____
1.0	_____	_____	_____
1.5	_____	_____	_____
2.0	_____	_____	_____
2.5	_____	_____	_____
3.0	_____	_____	_____
3.5	_____	_____	_____
4.0	_____	_____	_____
4.5	_____	_____	_____
Mix solute and water			
5.5	_____	_____	_____
6.0	_____	_____	_____
7.0	_____	_____	_____
8.0	_____	_____	_____
9.0	_____	_____	_____
10.0	_____	_____	_____
11.0	_____	_____	_____
12.0	_____	_____	_____
13.0	_____	_____	_____
14.0	_____	_____	_____
15.0	_____	_____	_____
16.0	_____	_____	_____
17.0	_____	_____	_____
18.0	_____	_____	_____
19.0	_____	_____	_____
20.0	_____	_____	_____

Data Sheet 2

	determination		
	1	2	3
volume of water, mL			
mass of container + solute, g			
mass of container, g			
mass of solute, g			
number of moles of solute added			
initial temperature of water at time of mixing, $^\circ$C			
extrapolated final temperature of reaction mixture, $^\circ$C			
ΔT, $^\circ$C			
mass of calorimeter + reaction mixture, g			
mass of calorimeter, g			
mass of reaction mixture, g			
heat transferred during dissolution (Q), J			
$\Delta H_{dissolution}$, kJ mol^{-1}			
mean $\Delta H_{dissolution}$, kJ mol^{-1}			
mean $\Delta H_{dissolution}$ of other magnesium salt, kJ mol^{-1}			
$\Delta H_{hydration}$ of anhydrous $MgSO_4$, kJ mol^{-1}			

Pre-Laboratory Assignment

1. Read an authoritative source for a discussion of graphing techniques.

2. Why is it necessary for you to wash your hands thoroughly with soap or detergent before leaving the laboratory?

3. The following temperature–time data were collected when 14.00 g of anhydrous sodium acetate (GMM = 82.03 g mol^{-1}) was dissolved in 100.00 mL of water. The heat capacity of the reaction mixture was 4.045 J g^{-1} deg^{-1}.

time, min	temp, °C	time, min	time, °C
0.0	23.50	7.0	29.30
0.5	23.45	8.0	29.05
1.5	23.49	9.0	28.75
2.5	23.50	10.0	28.50
3.5	23.50	11.0	28.20
4.5	23.50	12.0	27.95
mix	——	13.0	27.65
5.5	28.30	14.0	27.35
6.0	29.00	15.0	27.10

(1) Prepare a graph by plotting the temperature on the *y*-axis (the ordinate) and the time on the *x*-axis (the abscissa).

(2) Calculate ΔT for the process, using Equation 9.

_____
answer

(3) Calculate the heat transferred, Q, using Equation 5.

_____
answer

(4) Calculate the number of moles of anhydrous sodium acetate dissolved, using Equation 6.

_____
answer

(5) Calculate $\Delta H_{dissolution}$ for anhydrous sodium acetate in kJ mol^{-1}, using Equation 7.

_____
answer

4. In a second experiment, 16.00 g of sodium acetate trihydrate (GMM = 136.08 g mol^{-1}) was dissolved in 100.00 mL of water. The initial temperature was 25.00 °C, and the final temperature was 20.06 °C. The heat capacity of the reaction mixture was 4.045 J g^{-1} mol^{-1}.

(1) Calculate ΔT, using Equation 9.

_____
answer

(2) Calculate the heat transferred, Q, using Equation 5.

_____
answer

(3) Calculate the number of moles of sodium acetate trihydrate dissolved, using Equation 6.

_____
answer

(4) Calculate $\Delta H_{dissolution}$ for sodium acetate trihydrate in kJ mol^{-1}, using Equation 7.

_____
answer

(5) Calculate $\Delta H_{hydration}$ for converting sodium acetate(s) sodium acetate trihydrate(aq).

_____
answer

Enthalpy of Neutralization

prepared by **H. A. Neidig**, Lebanon Valley College

Purpose of the Experiment

Determine the calorimeter constant for a calorimeter. Determine the enthalpy of neutralization for the reaction of a strong acid and a strong base.

Background Information

When substances react, the accompanying energy change frequently involves a transfer of heat. Many such reactions play important roles in our lives. Coal (C) reacts with the oxygen of the air and releases energy to heat our homes (see Equation 1). The burning of gaseous mixtures containing propane (C_3H_8) releases heat that can be used to cook a steak (see Equation 2). Commercial cold and hot packs provide instant heat transfer and are useful to athletic trainers or campers. In an activated cold pack, ammonium nitrate dissolves in water causing the solution to become cold (see Equation 3). In a hot pack, the addition of a seed crystal to a supersaturated solution of sodium thiosulfate results in the crystallization of some of the dissolved salt, causing the solution to become hot (see Equation 4).

$$C(s) + O_2(g) \rightarrow CO_2(g) \qquad \text{(Eq. 1)}$$

$$C_3H_8(g) + 5\,O_2(g) \rightarrow 3\,CO_2(g) + 4\,H_2O(g) \quad \text{(Eq. 2)}$$

$$NH_4NO_3(s) \xrightarrow{H_2O} NH_4^+(aq) + NO_3^-(aq) \quad \text{(Eq. 3)}$$

$$2\,Na^+(aq) + S_2O_3^{2-}(aq) \rightarrow Na_2S_2O_3(s) \quad \text{(Eq. 4)}$$

These examples illustrate some of the applications of **thermochemistry**, the study of heat changes and transfers associated with chemical reactions.

In thermochemical investigations, we study reactions that are carried out in well-insulated containers called **calorimeters**. We call the reactants and products contained in the calorimeter a **chemical system**. Everything else in the immediate vicinity of that system is the **surroundings**, including the air above the calorimeter, the water in the calorimeter, the thermometer, and the stirrer.

In these investigations, we compare: the **final state** of the system, the time when we stop making temperature measurements for a reaction; and the **initial state** of the system, the time when we begin to make temperature measurements. The absolute difference between the final temperature of the system (T_{final}) and the initial temperature of the system ($T_{initial}$) is the **temperature change** (ΔT) for the reaction (Equation 5).

$$\Delta T = T_{final} - T_{initial} \qquad \text{(Eq. 5)}$$

When two substances react in a sealed bomb calorimeter at constant volume, there is a change in their internal energies (ΔE). Thermodynamically, this change involves the difference between the in-

---

Copyright © 1988 by Chemical Education Resources, Inc., P.O. Box 357, 220 S. Railroad, Palmyra, Pennsylvania 17078
No part of this laboratory program may be reproduced or transmitted in any form or by any means, electronic or mechanical, including photocopying, recording, or any information storage and retrieval system, without permission in writing from the publisher. Printed in the United States of America

ternal energy of the final state of the system and the internal energy of the initial state of the system. We usually express this change as the difference between the internal energy of the products and that of the reactants at constant volume (see Equation 6).

$$\Delta E = E_{products} - E_{reactants} \qquad \text{(Eq. 6)}$$

However, most chemical reactions occurring in solution take place in calorimeters that are unsealed and open to the atmosphere. An example of such a container is a Styrofoam coffee-cup calorimeter. Under these conditions, the heat energy transferred is expressed in terms of **enthalpy** (H). The **change in enthalpy** (ΔH) is the difference between the enthalpy of the products and that of the reactants at constant pressure (see Equation 7). Calculations of enthalpy changes are based on temperature measurements made during a chemical reaction. In some cases, the difference between the internal energy change and the enthalpy change for a reaction is negligible.

$$\Delta H = H_{products} - H_{reactants} \qquad \text{(Eq. 7)}$$

When a chemical process or reaction occurs, the enthalpy of the system either increases of decreases. Figure 1 is an enthalpy diagram showing an enthalpy change. Notice that the enthalpy of the reactants is greater than that of the products. We conclude that heat has been transferred from the chemical system to the surroundings. In the reaction depicted in Figure 1, the enthalpy change is defined as negative, and we call this reaction an **exothermic reaction**.

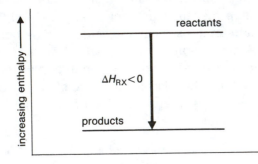

Figure 1 *Enthalpy diagram for an exothermic reaction*

In Figure 2, the enthalpy of the reactants is lower than that of the products. Heat has been transferred to the chemical system from the surroundings. In the reaction depicted in Figure 2, the enthalpy change is defined as positive, and we call this reaction an **endothermic reaction**.

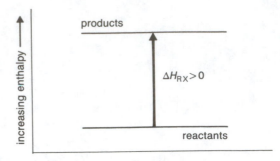

Figure 2 *Enthalpy diagram for an endothermic reaction*

Some of the heat transferred to the surroundings during an exothermic chemical reaction is absorbed by the contents of the calorimeter and some by the calorimeter itself. The heat absorbed by the solution in the calorimeter increases the temperature of this solution. Using this temperature increase, we can determine the exact amount of heat absorbed by the water. However, there is no easy way to measure the temperature change of the calorimeter itself. Therefore, we must determine indirectly the amount of heat absorbed by the calorimeter per degree rise in the temperature of the contents of the calorimeter. This amount of heat is called the **calorimeter constant** (C_{cal}) or the heat capacity of the calorimeter which is expressed in joules per degree (J deg^{-1}).

Determining the Calorimeter Constant

In Part I of this experiment, you will determine the calorimeter constant of your calorimeter, a Styrofoam coffee cup. You will measure the temperatures of two known volumes of water, one of which is hot and one of which is cold, for a period of 5 min. You will then mix the two samples of water and continue to collect temperature–time data for an additional 15 min.

From a plot of the temperature–time data, you will determine the ΔT for both the hot and the cold water. By using ΔT, the mass of water samples, and the specific heat of water, you will calculate the heat lost by the hot water and that gained by the cold water. You will use these data to determine the calorimeter constant of your calorimeter.

Although heat is transferred essentially instantaneously at the time of mixing, thermometers cannot respond quickly enough to give an accurate indication of the temperature at the time of mixing and shortly thereafter. As a result, that portion of the temperature–time curve shows a temperature that islower or higher than the actual temperature of the mixture at this point.

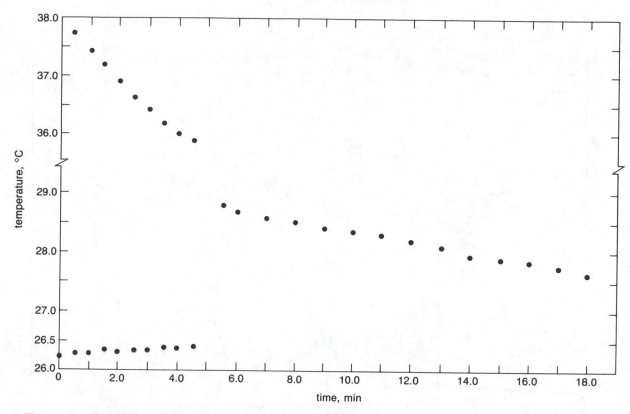

Figure 3   *Plotted temperature–time data for mixing hot and cold water*

By plotting temperature–time readings both prior to and after mixing, you can extrapolate the curve and determine the temperature at the time of mixing. In part, this procedure corrects for the deficiencies of your thermometers.

You will prepare a temperature–time graph by plotting the temperature, in degrees Celsius, on the ordinate (y-axis) and the time, in minutes, on the abscissa (x-axis). Note in Figure 3 that the temperature units on the y-axis are discontinued between 29.5 and 35.5 °C, where no data were collected. In this way, the temperature ranges where data were collected can be expanded, thus enhancing the accuracy of ΔT.

Draw a vertical line, perpendicular to the x-axis, through the point on the x-axis that represents the time of mixing. This line is called the **line of mixing**. Then, draw the best straight line representing the plotted points for the temperature of the sample of hot water and of the sample of cold water. Extend these lines to intersect the line of mixing. One point of intersection represents the initial temperature of the hot water ($T^{hot}_{initial}$) and the other, the initial temperature of the cold water ($T^{cold}_{initial}$).

Draw the best straight line representing the plotted points for the temperature of the reaction mixture. Extend this line to intersect the line of mixing. This point of intersection is the final temperature of the reaction mixture (T_{final}).

Determine ΔT for the cold water by subtracting $T^{cold}_{initial}$ from T_{final}. Figure 4 illustrates this procedure. In similar fashion, we can find ΔT for the hot water.

Example 1

We will calculate a calorimeter constant from laboratory data as an example. When 100.0 mL of water at 35.00 °C was added to a calorimeter containing 100.0 mL of water at 25.00 °C, the final temperature of the system was 29.63 °C.

First, we calculate the amount of heat lost by the hot water, using Equation 8a. The specific heat of water is the amount of heat ($4.184 \text{ J g}^{-1} \text{ K}^{-1}$) required to raise the temperature of 1 g of water 1 °C.

$$\begin{array}{l} \text{heat lost} \\ \text{by hot} \\ \text{water} \end{array} = \begin{pmatrix} \text{volume of} \\ \text{hot water,} \\ \text{mL} \end{pmatrix} \begin{pmatrix} \text{density of} \\ \text{water,} \\ \text{g mL}^{-1} \end{pmatrix} \times$$

$$\begin{pmatrix} \Delta T \text{ for} \\ \text{hot water,} \\ °C \end{pmatrix} \begin{pmatrix} \text{specific heat} \\ \text{of water,} \\ \text{J g}^{-1} \text{ K}^{-1} \end{pmatrix}$$

$$= (100.0 \text{ mL})(1.00 \text{ g mL}^{-1}) \times$$
$$(5.37 \text{ °C})(4.184 \text{ J g}^{-1} \text{ K}^{-1})$$

$$= \quad 2.25 \times 10^3 \qquad\qquad \text{(Eq. 8a)}$$

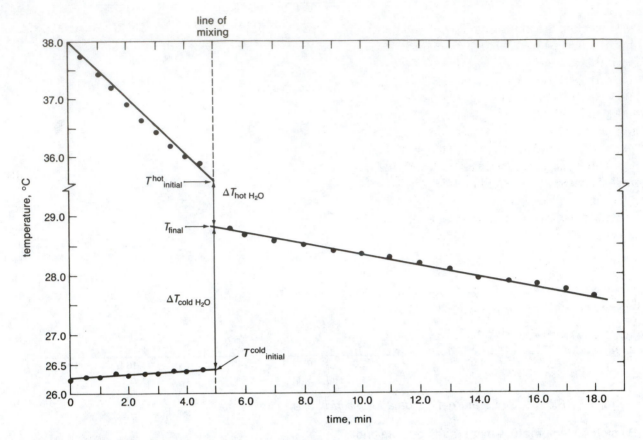

Figure 4 *Extrapolation of temperature–time data to find ΔT*

Note that the units °C and K cancel in Equation 8a because one degree on both the kelvin and the Celsius scales has the same magnitude. Therefore, throughout this module, we will use these units interchangeably.

In similar fashion, we calculate the amount of heat absorbed by the cold water, using Equation 8b.

$$
\begin{pmatrix} \text{heat gained} \\ \text{by cold} \\ \text{water} \end{pmatrix} = \begin{pmatrix} \text{volume of} \\ \text{cold water,} \\ \text{mL} \end{pmatrix} \begin{pmatrix} \text{density of} \\ \text{water,} \\ \text{g mL}^{-1} \end{pmatrix} \times
$$

$$
\begin{pmatrix} \Delta T \text{ for} \\ \text{cold water,} \\ °C \end{pmatrix} \begin{pmatrix} \text{specific heat} \\ \text{of water,} \\ \text{J g}^{-1} \text{ K}^{-1} \end{pmatrix}
$$

$$
= (100.0 \text{ mL})(1.00 \text{ g mL}^{-1})
$$
$$
\times (4.63 \text{ °C})(4.184 \text{ J g}^{-1} \text{ K}^{-1})
$$

$$
= 1.94 \times 10^3 \text{ J} \qquad \text{(Eq. 8b)}
$$

The hot water lost 2.25×10^3 J and the cold water gained 1.94×10^3 J. The difference of 3.1×10^2 J is the amount of heat absorbed by the calorimeter. We can determine the calorimeter constant, using Equation 9.

$$
C_{cal} = \frac{\begin{pmatrix} \text{head lost by} \\ \text{hot water, J} \end{pmatrix} - \begin{pmatrix} \text{heat gained by} \\ \text{cold water, J} \end{pmatrix}}{\Delta T \text{ of cold water}}
$$

$$
= \frac{(2.25 \times 10^3 \text{ J}) - (1.94 \times 10^3 \text{ J})}{4.63 \text{ °C}}
$$

$$
= 67.0 \text{ J °C}^{-1} = 67.0 \text{ J K}^{-1} \qquad \text{(Eq. 9)}
$$

Determining the Molar Enthalpy of Neutralization

In Part II of this experiment, you will determine the enthalpy of neutralization for the reaction of a strong acid with a strong base, NaOH. When a mole of acid reacts with a base, the enthalpy change is referred to as the molar enthalpy of neutralization of that acid. You will measure the temperature of a known volume of a standard solution of an acid and of a known volume of a standard NaOH solution. After pouring the acid into the NaOH solution, you will collect temperature data for a period of time. Using ΔT determined from your plotted temperature–time data, the volume and molarities of the reactants, and the density and specific heat of your reaction mixture, you will calculate ΔH_{neutzn} for the acid.

Example 2

We will calculate the molar enthalpy of neutralization for the reaction of hydrochloric acid and ammonia water from laboratory data as an example. When 100.0 mL of $0.9764M$ HCl was mixed with 100.0 mL of $0.9870M$ NH_3, the change in temperature of the system was 5.93 °C. The density of the reaction mixture was 1.101 g mL^{-1} and its specific heat was 3.97 J g^{-1} K^{-1}. We will assume that the enthalpies of dilution and mixing are negligible. The previously determined calorimeter constant, 67.0 J K^{-1}, will be used for these calculations.

First, we write the chemical equation for the reaction.

$$H_3O^+(aq) + NH_3(aq) \rightarrow NH_4^+(aq) + H_2O \quad \text{(Eq. 10)}$$

Then, we calculate the amount of heat absorbed by the reaction mixture, using Equation 11.

$$\text{heat absorbed by reaction mixture} =$$

$$\begin{pmatrix} \text{volume of} \\ \text{reaction} \\ \text{mixture, mL} \end{pmatrix} \begin{pmatrix} \text{density of} \\ \text{reaction} \\ \text{mixture, g mL}^{-1} \end{pmatrix} \times$$

$$(\Delta T, °C) \begin{pmatrix} \text{specific heat} \\ \text{of reaction} \\ \text{mixture, J g}^{-1} \text{ K}^{-1} \end{pmatrix}$$

$$= (200.0 \text{ mL})(1.101 \text{ g mL}^{-1})$$
$$\times (5.93 °C)(3.97 \text{ J g}^{-1} \text{ K}^{-1})$$

$$= 4.76 \times 10^3 \text{ J} \quad \text{(Eq. 11)}$$

Next, we calculate the amount of heat absorbed by the calorimeter, thermometer, and stirrer, using Equation 12.

$$\text{head absorbed by calorimeter} = \begin{pmatrix} \text{calorimeter} \\ \text{constant, J K}^{-1} \end{pmatrix} (\Delta T, °C)$$

$$= (67.0 \text{ J K}^{-1})(5.93 °C)$$

$$= 3.97 \times 10^2 \text{ J} \quad \text{(Eq. 12)}$$

We calculate the total heat absorbed from the reaction of 100.0 mL of $0.9764M$ HCl and 100.0 mL of $0.9870M$ NH_3, using Equation 13.

$$\text{total heat absorbed} = \begin{pmatrix} \text{heat absorbed by} \\ \text{reaction mixture, J} \end{pmatrix} + \begin{pmatrix} \text{heat absorbed by} \\ \text{calorimeter, J} \end{pmatrix}$$

$$= (4.76 \times 10^3 \text{ J}) + (3.97 \times 10^2 \text{ J})$$

$$= 5.16 \times 10^3 \text{ J} \quad \text{(Eq. 13)}$$

Because the reaction is exothermic, by convention we place a minus sign in front of 5.16×10^3 J to make it a negative number. Therefore, the total heat released by the reaction is -5.16×10^3 J.

Hydrochloric acid is the limiting reagent in this reaction. We calculate the number of moles of HCl reacting, using Equation 14.

$$\text{number of moles of HCl} = (\text{volume, L})(\text{molarity, mol L}^{-1})$$

$$= (1.000 \times 10^{-1} \text{ L})(0.9764 \text{ mol L}^{-1})$$

$$= 9.764 \times 10^{-2} \text{ mol} \quad \text{(Eq. 14)}$$

The molar enthalpy of neutralization of HCl is expressed in kJ mol^{-1}. We calculate the amount of heat that would be released if 1 mol of HCl had reacted with an excess of NH_3, using Equation 15.

$$\Delta H_{neutzn} = \frac{\text{total heat released, J}}{\text{number of moles of HCl}}$$

$$= \frac{-5.16 \times 10^3 \text{ J}}{9.764 \times 10^{-2} \text{ mol HCl}}$$

$$= -5.28 \times 10^4 \text{ J mol}^{-1} \text{ HCl}$$

$$= -52.8 \text{ kJ mol}^{-1} \text{ HCl} \quad \text{(Eq. 15)}$$

Procedure

Chemical Alert
hydrochloric acid—corrosive and toxic
phosphoric acid—corrosive and toxic
sodium hydroxide—corrosive and toxic
sulfuric acid—corrosive and toxic

Caution: Wear departmentally approved eye protection while doing this experiment.

I. Determining the Calorimeter Constant

Place 50.0 mL of room-temperature, distilled or de-ionized water in a clean, dry, 8-oz pressed-Styrofoam cup. Place a split rubber stopper on the upper end of a thermometer and suspend the thermometer by clamping the rubber stopper to a ring stand. Immerse the end of the thermometer in the water in the cup but do not allow the thermometer to touch the side or the bottom of the cup. Adjust the suspended thermometer so that the temperature can be read easily. For all temperature measurements, estimate the reading to the nearest 0.01 °C.

Place 50.0 mL of distilled or deionized water in a clean, dry 150-mL beaker and heat with occasional stirring to approximately 15 °C above room temperature. Place the beaker containing the hot water on a piece of insulated board. Suspend a thermometer in this sample of water as described above.

For a 5-min period, measure and record directly on Data Sheet 1 temperature–time data for the two samples of water at 1-min intervals. If two students are working as partners, one student should read the temperature while the other student immediately plots the temperature–time data on graph paper. The temperature, in degrees, should be plotted on the ordinate and the time, in minutes, on the abscissa. The scale units on the ordinate should be selected so that the range is from one degree below room temperature to fifteen degrees above room temperature. The temperature scale should be interrupted as discus-sed in the Background Information. The scale units on the abscissa should be selected so that the range is from 0 min to 20 min.

While stirring the sample of cold water in the calorimeter, rapidly pour the hot water from the beaker into the cold water. Be certain to transfer all of the hot water into the calorimeter. Continue to measure and record on Data Sheet 1 temperature–time data for an additional 15-min period at 1-min intervals.

Do a second determination with new portions of water.

II. Determining the Molar Enthalpy of Neutralization

Your laboratory instructor will assign you one of the following chemical systems: HCl–NaOH; H_2SO_4–NaOH; or H_3PO_4–NaOH. The concentration of the acids to be used are HCl, 2.00M; H_2SO_4, 1.00M; and H_3PO_4, 0.67M. The concentration of the NaOH solution is 2.05M. Record the actual molarity of the acid and base you use on Data Sheet 4.

Dispense 50.0 mL of the assigned acid at room temperature from a buret into a clean, dry, 8-oz, pressed-Styrofoam cup. Dispense 50.0 mL of 2.05M NaOH at room temperature from a buret into a clean, dry 150-mL beaker. Adjust the temperature of the NaOH solution by cooling or heating the beaker and contents with running water so that the temperature of the NaOH solution is the same as that of the acid. Suspend a clean, dry thermometer in each of the solutions as described previously under Part I of Procedure. For a 5-min period, measure and record directly on Data Sheet 3 temperature–time data for each of the two solutions at 30-s intervals.

If two students are working as partners, one student should read the temperature while the other student immediately plots the temperature–time data on graph paper. These data should be plotted as previously described under Part I of Procedure.

Note: Use a clean, dry glass stirring rod for each determination.

While carefully stirring the acid in the Styrofoam cup with a glass stirring rod, rapidly, but carefully, pour the NaOH solution into the acid. Continue to measure and record on Data Sheet 3 temperature–time data for a 15-min period at 1-min intervals.

Discard the reaction mixture following the directions of your laboratory instructor.

Do a second determination with new portions of the solution.

Caution: Wash your hands thoroughly with soap or detergent before leaving the laboratory.

Calculations

Do the following calculations for each determination and record the results on the appropriate Data Sheet.

I. Determining the Calorimeter Constant

1. If the temperature–time data were not plotted on graph paper as the measurements were made, plot and extrapolate these data to determine ΔT as described in Background Information.

2. Calculate the heat lost by the hot water, using Equation 8a.

3. Calculate the heat gained by the cold water, using Equation 8b.

4. Calculate the calorimeter constant, using Equation 9. Use the ΔT of the cold water.

5. Repeat calculations 1-4 for your second determination.

6. Calculate the mean calorimeter constant.

II. Determining the Molar Enthalpy of Neutralization

7. Write a chemical equation for the reaction of sodium hydroxide and the acid you used in this experiment.

8. If the temperature–time data were not plotted on graph paper as the measurements were made, plot the temperature in degrees Celsius on the ordinate and the time in minutes on the abscissa. The use of an interrupted temperature axis is a convenient way to plot data in an acceptable form. From the resulting curves, find the temperature change, ΔT, for the reaction. If the initial temperatures of the acid and base solutions were different, use the average temperature of the two solutions.

9. Calculate the heat absorbed by the reaction mixture, using Equation 11. Consult the reference table for the density and specific heat of your reaction mixtures. Assume the concentration of your reaction mixture to be 1.00M.

10. Calculate the amount of heat absorbed by the calorimeter, thermometer, and stirrer, using Equation 12.

11. Calculate the total heat absorbed, using Equation 13.

12. Calculate the number of moles of acid used, using Equation 14.

13. Calculate the molar enthalpy of neutralization, in kJ, for your acid, using Equation 15.

14. Repeat calculations 7-12 for your second determination.

15. Calculate the mean molar enthalpy of neutralization.

Reference Table   *Heat capacities and densities of reaction mixtures*[a]

solution	concentration, M	heat capacity, $J\ g^{-1}\ K^{-1}$	density, $g\ mL^{-1}$
sodium chloride	1.00	3.89	1.04
sodium hydrogen sulfate	1.00	3.80	1.09
sodium sulfate	1.00	3.76	1.12
sodium hydrogen phosphate	1.00	3.80	1.07
sodium dihydrogen phosphate	1.00	3.80	1.05
sodium phosphate	1.00	3.80	1.11
water	—	4.184	1.00

[a] *International Critical Tables*; McGraw-Hill: New York, 1933.

Post-Laboratory Questions

(Use the spaces provided for the answers and additional paper if necessary)

1. Obtain from your classmates or from your laboratory instructor, the molar enthalpies of neutralization of hydrochloric acid, sulfuric acid, and phosphoric acid when each reacts with sodium hydroxide.

(1)   Write a chemical equation for each of these neutralization reactions.

(2)   Compare the molar enthalpies of neutralization of hydrochloric acid, sulfuric acid, and phosphoric acid. Discuss the similarities and differences of these data.

2.   When a neutralization reaction was carried out using 100.0 mL of 0.7890M NH$_3$ water and 100.0 mL of 0.7940M acetic acid, ΔT was found to be 4.76 °C. The specific heat of the reaction mixture was 4.104 J g^{-1} K^{-1} and its density was 1.03 g mL^{-1}. The calorimeter constant was 3.36 J K^{-1}.

(1)   Calculate ΔH_{neutzn} for the reaction of NH$_3$ and acetic acid.

_____
answer

(2)   At the end of the experiment, it was discovered that the thermometer had not been cal-

ibrated. When it was calibrated, it was found that the thermometer read 0.50 °C low. What effect would this thermometer reading have on the reported ΔH_{neutzn} calculated in (2) above?

(3)   When the temperature–time data graph was reviewed, it was found that an error had been made in determining ΔT. Instead of 4.76 °C, ΔT was actually 4.70 °C. Based on this change only, calculate the correct ΔH_{neutzn} for the reaction of NH$_3$ and acetic acid.

_____
answer

(4)   Calculate the percent error for the correct ΔH_{neutzn} if a ΔT of 4.76 °C had been used.

_____
answer

3.   Consider the possibility of substituting a glass beaker for the Styrofoam cup for the calorimeter when doing the experiment described in this module.

(1)   Would you expect the calorimeter constant of the cup to be larger, smaller, or the same as that of the beaker? Briefly explain.

(2)   Would you expect the ΔH_{neutzn} to be larger, smaller, or the same when using the beaker instead of the cup? Briefly explain.

_____   _____   _____
name   section   date

Data Sheet 1

I. Determining the Calorimeter Constant

time, min	first determination temperature, °C			second determination temperature, °C		
	hot water	cold water	mixture	hot water	cold water	mixture
0.0	_____			_____		
0.5		_____			_____	
1.0	_____			_____		
1.5		_____			_____	
2.0	_____			_____		
2.5		_____			_____	
3.0	_____			_____		
3.5		_____			_____	
4.0	_____			_____		
4.5		_____			_____	

Mix hot and cold water

time, min	mixture (first)	mixture (second)
5.5	_____	_____
6.0	_____	_____
7.0	_____	_____
8.0	_____	_____
9.0	_____	_____
10.0	_____	_____
11.0	_____	_____
12.0	_____	_____
13.0	_____	_____
14.0	_____	_____
15.0	_____	_____
16.0	_____	_____
17.0	_____	_____
18.0	_____	_____
19.0	_____	_____
20.0	_____	_____

Data Sheet 2

I. Determining the Calorimeter Constant

	determination	
	first	second
$T_{initial}$ of hot water, ˚C	_____	_____
$T_{initial}$ of cold water, ˚C	_____	_____
maximum temperature of mixture, ˚C	_____	_____
ΔT for hot water, ˚C	_____	_____
ΔT for cold water, ˚C	_____	_____
volume of cold water, mL	_____	_____
volume of hot water, mL	_____	_____
calorimeter constant, J deg^{-1}	_____	_____

mean calorimeter constant, J deg^{-1}

Data Sheet 3

II. Determining the Molar Enthalpy of Neutralization

chemical system_____

	first determination temperature, °C			second determination temperature, °C		
time, min	base	acid	mixture	base	acid	mixture
0.0	_____			_____		
0.5		_____			_____	
1.0	_____			_____		
1.5		_____			_____	
2.0	_____			_____		
2.5		_____			_____	
3.0	_____			_____		
3.5		_____			_____	
4.0	_____			_____		
4.5		_____			_____	
Mix acid and base						
5.5			_____			_____
6.0			_____			_____
7.0			_____			_____
8.0			_____			_____
9.0			_____			_____
10.0			_____			_____
11.0			_____			_____
12.0			_____			_____
13.0			_____			_____
14.0			_____			_____
15.0			_____			_____
16.0			_____			_____
17.0			_____			_____
18.0			_____			_____
19.0			_____			_____
20.0			_____			_____

Data Sheet 4

II. Determining the Molar Enthalpy of Neutralization

chemical system _____

	determination	
	first	second
molarity of base, M	_____	_____
molarity of acid, M	_____	_____
volume of acid, mL	_____	_____
volume of base, mL	_____	_____
$T_{initial}$ of acid, °C	_____	_____
$T_{initial}$ of base, °C	_____	_____
T_{final} of mixture, °C	_____	_____
temperature change, °C	_____	_____
heat absorbed by the reaction mixture, J	_____	_____
heat absorbed by calorimeter, thermometer, and stirrer, J	_____	_____
total heat absorbed, J	_____	_____
number of moles of acid used	_____	_____
molar enthalpy of neutralization kJ per mole of acid	_____	_____
mean molar enthalpy of neutralization, kJ per mole of acid		

Pre-Laboratory Assignment

1. Read an authoritative source for a discussion of preparing and interpreting a graph, especially the extrapolation of a curve.

2. A student determined the calorimeter constant of the calorimeter, using the procedure described in this module. The student added 50.00 mL of cold water to 50.00 mL of heated, distilled water in a Styrofoam cup. The initial temperature of the cold water was 21.00 °C and of the hot water, 29.15 °C. The maximum temperature of the mixture was found to be 24.81°C. Assume the density of water is 1.00 g mL^{-1} and the specific heat is 4.814 J g^{-1} K^{-1}.

   (1) Determine the ΔT for the hot water and the cold water.

(2) Calculate the heat lost by the hot water.

_____
answer

(3) Calculate the heat gained by the cold water.

_____
answer

(4) Calculate the calorimeter constant, using the ΔT of the cold water.

hot water _____
              answer

cold water _____
              answer

_____
answer

3. The student then determined ΔH_{neutzn} for the reaction of sodium hydroxide and acetic acid, using the procedure described in this module. 100.0 mL of 0.8500M NaOH was added to 100.0 mL of 0.8404M acetic acid. Prior to and following the mixing of the acid and base solutions, the following temperature–time data were collected.

time, min	temperature, °C NaOH	HOAc	time, min	temperature, °C mixture
0	17.52		6	23.28
0.5		17.73	7	23.03
1.0	17.54		8	23.10
1.5		17.75	9	23.06
2.0	17.55		10	22.99
2.5		17.77	11	22.93
3.0	17.57		12	22.89
3.5		17.78	13	22.85
4.0	17.58		14	22.78
4.5		17.80	15	22.73
5.0	mixing		16	22.70
			17	22.64
			18	22.59

(1) Plot the temperature–time data and determine the mean temperature of the unmixed reagents.

_____
answer

(2) Determine ΔT from the graph.

_____
answer

(3) Calculate the heat absorbed by the reaction mixture. You may assume the density of the reaction mixture was 1.02 g mL^{-1} and the specific heat of the reaction mixture was 3.822 J g^{-1} K^{-1}.

_____
answer

(4) Calculate the amount of heat absorbed by the calorimeter, thermometer, and stirrer, using the calorimeter constant calculated in Question 2(4).

_____
answer

(5) Calculate the total heat absorbed.

_____
answer

(6) Find the total heat released by the reaction of 100.0 mL of 0.8500M NaOH and 100.0 mL of 0.8404M acetic acid.

_____
answer

(7) Identify the limiting reagent and briefly explain why it is limiting.

_____
answer

(8) Find ΔH_{neutzn} for the reaction.

_____
answer

Enthalpy of Formation of Ammonium Salts

prepared by **Philip H. Rieger**, Brown University
and **H. Anthony Neidig**, Lebanon Valley College

Purpose of the Experiment

The enthalpy of formation of solid ammonium chloride, ammonium nitrate, or ammonium sulfate will be calculated from experimental data and literature data.

Background Information

Enthalpy is a thermodynamic state function. This means that when a system moves from one state to another, by a chemical reaction for example, the enthalpy change, ΔH, is a function only of the initial and final states of the system, and is independent of the details of the process by which the change was accomplished.

Enthalpy data are usually tabulated in terms of enthalpies of formation, ΔH_f^0, defined as the enthalpy change occurring when a compound is prepared from the constituent elements in their most stable forms. By convention, most such enthalpies of formation refer to "standard states", i.e., 1 atm pressure for gases, $1 M$ ideal solutions for dissolved solutes, and a temperature of 298 K.

Because enthalpy is a state function, the standard enthalpy change ΔH^0 can be computed for any chemical reaction for which the ΔH_f^0 values are available for the reactants and products. For example, the enthalpy change for the conversion of acetylene to benzene in their standard states can be found from their enthalpies of formation.

$$3\ C_2H_2\ (g,\ 1\ atm,\ 298\ K) \rightarrow C_6H_6\ (l,\ 298\ K)$$

$$\text{(Eq. 1)}$$

Imagine following a path where C_2H_2 is first decomposed to its elements, and then the elements are recombined to produce C_6H_6. Reference tables give the standard enthalpy changes for the individual reactions:

$$6\ C(s) + 3\ H_2(g) \rightarrow C_6H_6(l), \quad \Delta H_f^0 = +49.0\ \text{kJ mol}^{-1}$$

$$\text{(Eq. 2)}$$

$$2\ C(s) + H_2(g) \rightarrow C_2H_2(g), \quad \Delta H_f^0 = 226.7\ \text{kJ mol}^{-1}$$

$$\text{(Eq. 3)}$$

If Equation 3 is multiplied by 3 and its direction reversed, Equation 4 is obtained:

$$3\ C_2H_2(g) \rightarrow 3\ H_2(g) + 6\ C(s), \Delta H^0 = -680.1\ \text{kJ mol}^{-1}$$

$$\text{(Eq. 4)}$$

When Equations 2 and 4 are added, the elements $C(s)$ and $H_2(g)$ cancel, and Equation 1 results. The standard enthalpy change is then simply the sum of the ΔH's for the two steps:

$$\Delta H^0 = 49.0 + (-680.1) = -631.1\ \text{kJ mol}^{-1}$$

In this experiment, enthalpy data will be collected which, combined with other data from the literature, will allow the calculation of the enthalpy of formation of one of the salts, NH_4Cl, NH_4NO_3, or $(NH_4)_2SO_4$. Consider

---

Copyright © 1978 by Chemical Education Resources, Inc., P.O. Box 357, 220 S. Railroad, Palmyra, Pennsylvania 17078
No part of this laboratory program may be reproduced or transmitted in any form or by any means, electronic or mechanical, including photocopying, recording, or any information storage and retrieval system, without permission in writing from the publisher. Printed in the United States of America

the case of ammonium chloride. Suppose that NH_4Cl is formed directly from the elements. It would be difficult to arrange experimental conditions such that the reaction of $N_2(g)$, $H_2(g)$, and $Cl_2(g)$ actually gave NH_4Cl and only NH_4Cl. Alternatively, consider the rather roundabout path represented by the following steps:

$$\tfrac{1}{2}\,N_2(g) + \tfrac{3}{2}\,H_2(g) + x\,H_2O(l) \rightarrow NH_3(aq), \quad \Delta H_1$$
$$\text{(Eq. 5)}$$

$$\tfrac{1}{2}\,H_2(g) + \tfrac{1}{2}\,Cl_2(g) + y\,H_2O(l) \rightarrow HCl(aq), \quad \Delta H_2$$
$$\text{(Eq. 6)}$$

$$NH_3(aq) + HCl(aq) \rightarrow NH_4Cl(aq), \quad \Delta H_3$$
$$\text{(Eq. 7)}$$

$$NH_4Cl(aq) \rightarrow NH_4Cl(s) + (x + y)H_2O(l), \quad \Delta H_4$$
$$\text{(Eq. 8)}$$

The sum of Equations 5–8 corresponds to the formation of $NH_4Cl(s)$ from its elements:

$$N_2(g) + 2\,H_2(g) + Cl_2(g) \rightarrow NH_4Cl(s) \quad \text{(Eq. 9)}$$

$$\Delta H_f^0 = \Delta H_1 + \Delta H_2 + \Delta H_3 + \Delta H_4 \quad \text{(Eq. 10)}$$

Thus, this roundabout path has the advantage that it gives ΔH_f^0 in terms of the enthalpies of formation of $NH_3(aq)$ and $HCl(aq)$, ΔH_1 and ΔH_2, which are available from the literature, and the remaining two ΔH's which will be calculated from the data of this experiment.

Calorimetry

According to thermodynamics, the heat absorbed by a system at constant pressure, Q_p, is equal to the change in the enthalpy of the system, ΔH. Enthalpy changes are associated with a variety of chemical or physical processes. For example, when a system is warmed so that its temperature rises, heat is absorbed and the enthalpy of the system increases. This enthalpy change is proportional to the temperature change, ΔT, and to the heat capacity of the system, C:

$$\Delta H = Q_p = C\,\Delta T \quad \text{(Eq. 11)}$$

Similarly, when a system undergoes a change in its physical or chemical state—melting, boiling, dissolution of a solid into a solvent to form a solution, or a chemical reaction—the enthalpy changes, even though the temperature may be held constant (by carrying out the process in a constant-temperature water bath, for example).

The enthalpy change accompanying a chemical reaction or a change in physical state is conveniently measured using an adiabatic calorimeter, which in its simplest form is just an insulated cup. An adiabatic system by definition does not exchange heat with the surroundings—it is thermally insulated. Thus, at constant pressure, the enthalpy of an adiabatic system is constant and the enthalpy change accompanying a chemical reaction (which would result in a heat transfer in a nonadiabatic system) must be balanced by an equal and opposite enthalpy change associated with a change in temperature. This is expressed by Equation 12, where ΔH_r is the enthalpy change for the reaction and ΔH_t is the enthalpy change accompanying the resulting temperature change:

$$\Delta H_r + \Delta H_t = \Delta H_{system} = 0 \quad \text{(Eq. 12)}$$

However, ΔH_t is given by

$$\Delta H_t = (C_s + C_c)\,\Delta T \quad \text{(Eq. 13)}$$

where C_s and C_c are the heat capacities of the solution and of the calorimeter, respectively, and ΔT is the temperature change. Combining Equations 12 and 13 gives

$$\Delta H_r = -(C_s + C_c)\,\Delta T \quad \text{(Eq. 14)}$$

The first step in using a calorimeter is to determine the effective heat capacity of the calorimeter, C_c, often called the calorimeter constant. This can be done by carrying out a reaction for which the enthalpy change, ΔH_r, and the heat capacity of the products, C_s, are known. If ΔT is measured, Equation 14 can be solved for C_c. Consider as an example the reaction

$$OH^-(aq) + H_3O^+(aq) \rightarrow 2\,H_2O(l)$$
$$\text{(Eq. 15)}$$

Careful measurements reported in the literature give $\Delta H = -58.3$ kJ mol^{-1} for this reaction when a $2.0M$ solution of NaOH is mixed with a $2.0M$ solution of HCl. The heat capacity and density of a $1.00M$ NaCl solution are given in Table 2. If 100 mL of $2.00M$ HCl is mixed with 100 mL of $2.02M$ NaOH in the calorimeter, the temperature rises by $T = 13.9°$. The calorimeter constant is determined as follows:

$$\Delta H_r =$$
$$\text{(moles } H_2O \text{ produced)(enthalpy change per mole)}$$

The number of moles of water produced is determined by the limiting reagent, in this case HCl; thus

$$\Delta H_r = (0.100 \text{ L})(2.00 \text{ mol L}^{-1})(-58.3 \text{ kJ mol}^{-1}) = 11.7 \text{ kJ}$$

C_s = (grams of solution)(heat capacity per gram)
   = (200 mL)(1.037 g mL^{-1})(3.90 J g^{-1} K^{-1})
   = 809 J K^{-1}

Substituting these numbers into Equation 14 gives

$$-11.7 \times 10^3 \text{ J} = -(809 + C_c)(13.9 \text{ K})$$

$$C_c = \frac{11.7 \times 10^3}{13.9} - 809 = 33 \text{ J K}^{-1}$$

The calorimeter constant is the effective heat capacity of the calorimeter, and so includes allowance for whatever heat is lost or gained during the reaction. Note that C_c is a small difference of two large numbers and so is rather imprecisely determined: if the uncertainty in ΔT is ±0.1°, then the uncertainty in C_c will be ±6 J K^{-1}.

The second step in using a calorimeter is to carry out a reaction for which the enthalpy change is to be determined. C_c is now known, ΔT is determined, and if C_s is known, it is only necessary to solve Equation 14 for ΔH_r and then convert the result to a molar basis.

In this experiment, you will use a simple adiabatic calorimeter to determine the enthalpy of neutralization of aqueous ammonia by hydrochloric acid, nitric acid, or sulfuric acid. You will also determine calorimetrically the enthalpy of dissolution of ammonium chloride, ammonium nitrate, or ammonium sulfate. The two measured enthalpies, along with the enthalpies of formation of aqueous ammonia and the appropriate aqueous acid, will be used to compute the enthalpy of formation of the ammonium salt.

Procedure

A series of three experiments will be performed.

Part A. Determination of the calorimeter constant by measuring the temperature change produced by the reaction of HCl and NaOH solutions.

Part B. Determination of the enthalpy of reaction of aqueous ammonia with a strong acid—HCl, HNO$_3$, or H$_2$SO$_4$—assigned by the instructor.

Part C. Determination of the enthalpy of solution of the ammonium salt of the acid used in Part B—NH$_4$Cl, NH$_4$NO$_3$, or (NH$_4$)$_2$SO$_4$.

In addition to reagents, the equipment needed is: a 6-oz pressed Polystyrene cup, a 150-mL beaker, two 50-mL burets or 25-mL pipets, a glass stirring rod or plastic spoon, and two thermometers which can be read to 0.1 °C.

Because temperature differences will be measured, the absolute calibration of the thermometers is not critical. However, they must give the same reading when at the same temperature. Check the thermometer calibrations by measuring the temperature of a beaker of water with both thermometers simultaneously. Estimate the temperature to 0.1°. If one thermometer reads lower than the other, label it and correct subsequent readings by adding the difference found in this step.

Solution temperature as a function of time will be recorded on the Data Sheet. In addition to recording the data, a temperature–time plot on graph paper should be prepared; this can be done while the data are being collected. The temperature should be plotted on the y axis and the time in minutes on the x axis. The ΔT observed will be between 7 and 14° for Parts A and B and between −1 and −6° for Part C. The scale units on the y axis should be selected so that all points can be plotted on scale; the scale units on the x axis should cover the range 0–20 minutes.

The procedure is essentially the same for all three parts:

1. Rinse the Polystyrene cup and beaker with distilled water and wipe dry with a clean piece of absorbent paper. Mount one thermometer in the cup using a split rubber stopper, clamp, and ring stand. Do not allow the thermometer to touch the bottom or sides of the cup.

2. Using a buret or pipet, measure solution 1 (see Table 1) into the cup and (except in Part C) measure solution 2 into the beaker. Record the concentrations of the solutions used on the Data Sheet.

Table 1 *Reagents to be used*

part	solution 1	solution 2
A	50.0 mL 2.00M HCl	50.0 mL 2.05M NaOH
B	50.0 mL acid (2.00M HCl, 2.00M HNO$_3$, or 1.00M H$_2$SO$_4$)	50.0 mL 2.05M NH$_3$
C	100.0 mL distilled water	*

*Calculate the concentration of the ammonium salt—NH$_4$Cl, NH$_4$NO$_2$, or (NH$_4$)$_2$SO$_4$—formed in the reaction of Part B. Determine the mass of salt in 100 mL of solution; weigh out that amount of salt and use in place of solution 2 in Part C.

3. Check the temperatures of both solutions. If they differ by more than 0.5°, warm or cool solution 2 (as necessary) by running tap water over the outside of the beaker. When the two solutions are within 0.5°, re-

cord the temperatures. Continue to record the temperatures of the solutions at 1-minute intervals for 5 minutes.

4. After five minutes of temperature readings pour solution 2 into the calorimeter cup. Immediately stir with the stirring rod or plastic spoon to mix the solution well. Record the time of mixing. (Never use a thermometer as a stirring rod!) Continue to record the temperature at 1-minute intervals for 10 minutes.

  Do two (if time permits, three) determinations of the temperature change for each part of the experiment.

Calculations

1. Find the temperature change ΔT from the temperature–time plots by extrapolating the plotted data to the time of mixing, reading the initial and final

temperatures, and taking the temperature change for each determination. See Figure 1 for an example.

2. Use Equation 14 to compute the calorimeter constant from the data for Part A. Necessary heat capacity and density data are given in Table 2 on p. 5.

3. Use the data from Part B and the calorimeter constant just determined to calculate the enthalpy of neutralization of NH_3 and the assigned acid.

4. Use the data from Part C to compute the enthalpy of solution of the ammonium salt.

5. Use the results for the enthalpy of neutralization and the enthalpy of dissolution, along with the literature values of ΔH_f for $NH_3(aq)$ and the assigned acid (given in Table 3, p. 5) to compute the enthalpy of formation of the ammonium salt.

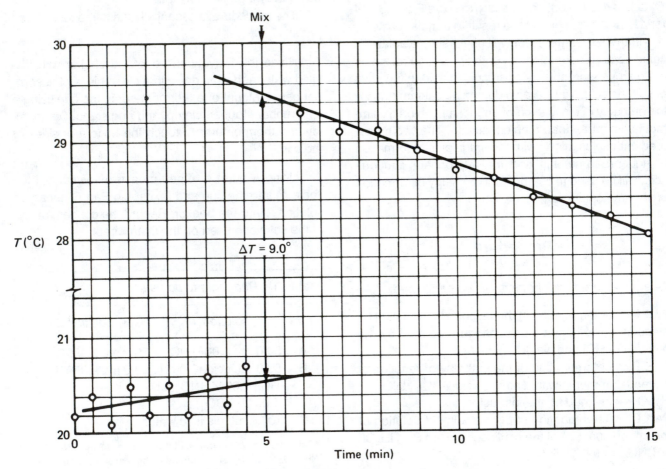

Figure 1 *Plot of temperature–time data showing determination of ΔT*

Table 2 *Molecular weights, heat capacities, and densities of solutions*

solute	molecular wt $g\,mol^{-1}$	concentration $mol\,L^{-1}$	heat capacity[a] $J\,g^{-1}\,K^{-1}$	density[a] $g\,mL^{-1}$
NH_4Cl	53.49	1.00	3.93	1.013
NH_4NO_3	80.04	1.00	3.90	1.029
$(NH_4)_2SO_4$	132.14	0.50	3.90	1.035
$NaCl$	58.44	1.00	3.90	1.037
H_2O	18.02	—	4.18	0.997

[a] at 25 °C

Table 3 *Enthalpies of formation at 25 °C[a]*

	$\Delta H_f\,kJ\,mol^{-1}$
$\tfrac{1}{2}\,N_2(g) + \tfrac{3}{2}\,H_2(g) + H_2O(l) \rightarrow NH_3(aq,\,2.0M)$	−80.7
$\tfrac{1}{2}\,H_2(g) + \tfrac{1}{2}\,Cl_2(g) + H_2O(l) \rightarrow HCl(aq,\,2.0M)$	−164.7
$\tfrac{1}{2}\,H_2(g) + \tfrac{1}{2}\,N_2(g) + \tfrac{3}{2}\,O_2(g) + H_2O(l) \rightarrow HNO_3(aq,\,2.0M)$	−206.0
$H_2(g) + S(s) + 2\,O_2(g) + H_2O(l) \rightarrow H_2SO_4(aq,\,1.0M)$	−884.6
$H_2(g) + \tfrac{1}{2}\,O_2(g) \rightarrow H_2O(l)$	−285.8

[a] "Selected Values of Chemical Thermodynamic Properties," Circular 500, National Bureau of Standards, Washington, D.C., 1961

Post-Laboratory Questions

(Use the space provided for the answer and additional paper if necessary.)

1. There are several possible sources of systematic error in this experiment. Evaluate the effect on the experimental results of each of the following possibilities:

    (a) Both thermometers read $0.5°$ lower than the true temperature.

    (b) The NaOH and NH_3 solutions used were 2% higher in concentration than stated on the labels.

    (c) The HCl, HNO_3, or H_2SO_4 concentration was 2% lower than stated on the label.

2. Random errors arise in this experiment primarily through the temperature measurements. For each of the three parts of the experiment, calculate the standard deviation in $\overline{\Delta T}$ by

$$\sigma_{\Delta T} = \left[\sum_i (\overline{\Delta T} - \Delta T_i)^2 / (n-1) \right]^{1/2}$$

where n is the number of runs and ΔT_i is the temperature change for the i th run. Use the estimated uncertainty in $\overline{\Delta T}_A$ to compute the uncertainty in C_c, the uncertainties in C_c and $\overline{\Delta T}_B$ to compute the uncertainty in ΔH(neutr.), and the uncertainties in C_c and $\overline{\Delta T}_C$ to compute the uncertainty in ΔH(dissol.). Finally, estimate the uncertainty in ΔH_f^0. In the propagation of error calculation, the following two simple formulas may be used:

for $z = x + y$ or $z = x - y$, $\sigma_z = (\sigma_x^2 + \sigma_y^2)^{1/2}$

for $z = xy$ or $z = x/y$, $\sigma z = z\{ (\sigma x/x)^2 + (\sigma y/y)^2 \}^{1/2}$

3. The enthalpy changes referred to by Equations 5 and 6 and given in Table 3 are not strictly standard enthalpies of formation since the products are not in standard states (though the reactants are). The measured enthalpies of neutralization and dissolution are not standard enthalpies either. Yet Equation 10 states that the standard enthalpy of formation is calculated by summing four non-standard enthalpy changes. Explain why this is so.

4. HCl is a strong acid and NaOH is a strong base; i.e., they are essentially completely ionized to H_3O^+ and Cl^- and to Na^+ and OH^-, respectively, in their solutions. Thus the neutralization reaction of HCl and NaOH is really just

$$H_3O^+(aq) + OH^-(aq) \rightarrow H_2O(l)$$

In this experiment, a strong acid and a weak base, NH^3, were used. Comparing the results with those for HCl and NaOH, estimate the enthalpy of ionization of ammonia,

$$NH_3(aq) + H_2O(l) \rightarrow NH_4^+(aq) + OH^-(aq)$$

From this result, would ammonia be expected to be a stronger or weaker base at $50\,°C$ than at $25\,°C$? Explain.

5. The enthalpy of neutralization of HCl with NaOH is -55.8 kJ mol^{-1} when very dilute solutions are mixed, but is -58.2 kJ mol^{-1} when $2M$ solutions are mixed, as in this experiment. The reaction is

$$H_3O^+(aq) + OH^-(aq) \rightarrow H_2O(l)$$

in both cases. Assuming that the enthalpy of water is nearly independent of electrolyte concentration, is the enthalpy per mole of the ions higher or lower in concentrated solutions?

Data Sheet 1

Part A: Calorimeter Constant

[HCl] = _____ M   [NaOH] = _____ M

temperature–time data

	run 1			*run 2*			*run 3*	
time, min	*temp. HCl, °C*	*temp. NaOH, °C*	*time, min*	*temp. HCl, °C*	*temp. NaOH, °C*	*time, min*	*temp. HCl, °C*	*temp. NaOH, °C*
0	_____		0	_____		0	_____	
0.5		_____	0.5		_____	0.5		_____
1	_____		1	_____		1	_____	
1.5		_____	1.5	_____		1.5		_____
2	_____		2	_____		2	_____	
2.5		_____	2.5		_____	2.5		_____
3	_____		3	_____		3	_____	
3.5		_____	3.5	_____		3.5		_____
4	_____		4	_____		4	_____	
4.5		_____	4.5		_____	4.5		_____
_____ mix			_____ mix			_____ mix		
6	_____		6	_____		6	_____	
7	_____		7	_____		7	_____	
8	_____		8	_____		8	_____	
9	_____		9	_____		9	_____	
10	_____		10	_____		10	_____	
11	_____		11	_____		11	_____	
12	_____		12	_____		12	_____	
13	_____		13	_____		13	_____	
14	_____		14	_____		14	_____	
15	_____		15	_____		15	_____	

calculation of calorimeter constant

	run 1	run 2	run 3
temperature change, ΔT, °C	_____	_____	_____
average temperature change, $\underline{\Delta T}$	_____		
mass of solution, g	_____		
solution heat capacity, C_s, J K^{-1}	_____		
ΔH_r, J	_____		
calorimeter constant, C_c, J K^{-1}	_____		

Data Sheet 2

Part B: Enthalpy of Neutralization

$$[\qquad] = \underline{\hspace{3cm}} M \qquad [NH_3] = \underline{\hspace{3cm}} M$$

temperature-time data

	run 1			run 2			run 3	
time, min	temp. HA, °C	temp. NH_3, °C	time, min	temp. HA, °C	temp. NH_3, °C	time, min	temp. HA, °C	temp. NH_3, °C
0	___		0	___		0	___	
0.5		___	0.5		___	0.5		___
1	___		1	___		1	___	
1.5		___	1.5		___	1.5		___
2	___		2	___		2	___	
2.5		___	2.5		___	2.5		___
3	___		3	___		3	___	
3.5		___	3.5		___	3.5		___
4	___		4	___		4	___	
4.5		___	4.5		___	4.5		___
___ mix			___ mix			___ mix		
6	___		6	___		6	___	
7	___		7	___		7	___	
8	___		8	___		8	___	
9	___		9	___		9	___	
10	___		10	___		10	___	
11	___		11	___		11	___	
12	___		12	___		12	___	
13	___		13	___		13	___	
14	___		14	___		14	___	
15	___		15	___		15	___	

calculation of enthalpy of neutralization

	run 1	run 2	run 3
temperature change, ΔT, °C	___	___	___
average temperature change, $\overline{\Delta T}$	___		
mass of solution, g	___		
solution heat capacity, C_s, J K^{-1}	___		
ΔH_r, J	___		
molar enthalpy of neutralization	___		

Data Sheet 3

Part C: Enthalpy of Dissolution

salt_____   mass_____g   moles_____

temperature–time data

run 1		run 2		run 3	
time, min	temp. H_2O, °C	time, min	temp. H_2O, °C	time, min	temp. H_2O, °C
0	_____	0	_____	0	_____
1	_____	1	_____	1	_____
2	_____	2	_____	2	_____
3	_____	3	_____	3	_____
4	_____	4	_____	4	_____
_____ mix		_____ mix		_____ mix	
6	_____	6	_____	6	_____
7	_____	7	_____	7	_____
8	_____	8	_____	8	_____
9	_____	9	_____	9	_____
10	_____	10	_____	10	_____
11	_____	11	_____	11	_____
12	_____	12	_____	12	_____
13	_____	13	_____	13	_____
14	_____	14	_____	14	_____
15	_____	15	_____	15	_____

calculation of enthalpy of dissolution

	run 1	run 2	run 3
temperature change, ΔT, °C	_____	_____	_____
average temperature change, $\overline{\Delta T}$	_____		
mass of solution, g	_____		
solution heat capacity, C_s, J K^{-1}	_____		
ΔH_r, J	_____		
molar enthalpy of dissolution	_____		

Pre-Laboratory Assignment

1. Read in your textbook the sections dealing with concepts of energy, enthalpy, and heat capacity, and the determination of enthalpy changes by calorimetry.

2. Standard enthalpies of formation are usually computed from calorimetric data from reactions which are easy to carry out in the laboratory. For many compounds, the enthalpy of the reaction with $O_2(g)$ (combustion) is commonly used, together with the standard enthalpies of formation of the combustion products, to compute ΔH_f^0. For example, sucrose, $C_{12}H_{22}O_{11}$, common table sugar, is found to have a standard enthalpy of combustion of -5640.9 kJ mol^{-1}. The standard enthalpies of formation of $CO_2(g)$ and $H_2O(l)$ are, respectively, -393.51 and -285.83 kJ mol^{-1}. Compute ΔH_f^0 for sucrose.

2. A student determined the enthalpy of solution of NaOH by dissolving 4.00 g of NaOH(s) in 180 mL of pure water (density 0.999 g mL^{-1}) at an initial temperature of 19.5 °C. Temperature–time data collected after mixing was extrapolated back to the time of mixing to obtain a temperature change, $\Delta T = +5.5°$. The density of the final solution was 1.013 mL^{-1} at 25°C, and the solution heat capacity was 4.08 J g^{-1} deg^{-1}. In a separate experiment, the calorimeter constant was found to be 21 J deg^{-1}.

(a) What was the molar concentration of the final NaOH solution at 25°C?

(b) What was the molar enthalpy of dissolution of NaOH in water?

Studying the Kinetics of the Solvolysis of 2-Chloro-2-methylpropane

prepared by **William H. Brown**, Beloit College

Purpose of the Experiment

Monitor the solvolysis reaction of 2-chloro-2-methylpropane in 2-propanol–water by analyzing one of the reaction products. Establish whether or not the reaction rate is directly proportional to the concentration of 2-chloro-2-methylpropane in the mixture.

Background Information

Alkyl halides are organic compounds composed of carbon, hydrogen, and halogen atoms. Under some conditions, the halogen atom in an alkyl halide can be replaced by another atom or group of atoms. When the halogen atom is replaced by a solvent molecule, we call the replacement reaction a **solvolysis** reaction.

Consider the reaction of 2-chloro-2-methylpropane, $(CH_3)_3CCl$ or *t*-butyl chloride, with water to produce hydrochloric acid (HCl) and a mixture of organic products, as shown in Equation 1a. Because 2-chloro-2-methylpropane is not soluble in water, the solvolysis reaction occurs only at the interface between the two liquids. The progress of a reaction occurring at an interface is difficult to monitor. Therefore, we add a third substance, 2-propanol, $(CH_3)_2CHOH$, or isopropyl alcohol, in which both 2-chloro-2-methylpropane and water are soluble. This addition allows the reaction to occur in a homogeneous solution. The presence of 2-propanol in the reaction mixture complicates the reaction, because 2-propanol also reacts with 2-chloro-2-methylpropane to produce HCl, as shown in Equation 1b. This complication is not a serious one, because each mole of chloride ion released from 2-chloro-2-methylpropane yields one mole of HCl as a product, as shown in Equations 1a and 1b, regardless of whether the reaction occurs with water or 2-propanol.

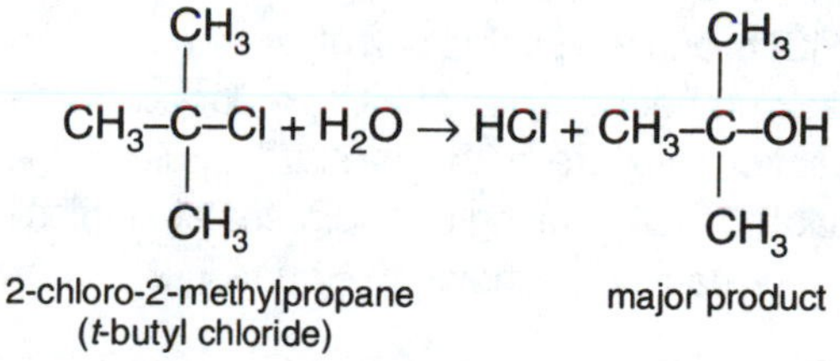

$$CH_3\text{–}\underset{\underset{CH_3}{|}}{\overset{\overset{CH_3}{|}}{C}}\text{–}Cl + H_2O \rightarrow HCl + CH_3\text{–}\underset{\underset{CH_3}{|}}{\overset{\overset{CH_3}{|}}{C}}\text{–}OH \qquad \text{(Eq. 1a)}$$

2-chloro-2-methylpropane   major product
(*t*-butyl chloride)

$$CH_3\text{–}\underset{\underset{CH_3}{|}}{\overset{\overset{CH_3}{|}}{C}}\text{–}Cl + CH_3\text{–}\underset{\underset{OH}{|}}{CH}\text{–}CH_3 \rightarrow$$

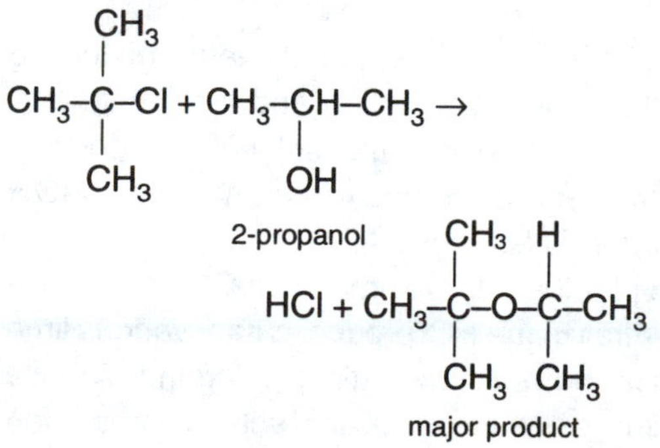

2-propanol

$$HCl + CH_3\text{–}\underset{\underset{CH_3}{|}}{\overset{\overset{CH_3}{|}}{C}}\text{–}O\text{–}\underset{\underset{CH_3}{|}}{\overset{\overset{H}{|}}{C}}\text{–}CH_3 \qquad \text{(Eq. 1b)}$$

major product

The rate at which 2-chloro-2-methylpropane reacts in a mixture of water and 2-propanol depends on several factors:

Copyright © 1993 by Chemical Education Resources, Inc., P.O. Box 357, 220 S. Railroad, Palmyra, Pennsylvania 17078
No part of this laboratory program may be reproduced or transmitted in any form or by any means, electronic or mechanical, including photocopying, recording, or any information storage and retrieval system, without permission in writing from the publisher. Printed in the United States of America

(1) the inherent properties of the alkyl halide;
(2) the alkyl halide concentration;
(3) the proportion of water and 2-propanol in the reaction mixture;
(4) the reaction temperature;
(5) the presence (or absence) of catalysts.

To observe the influence of one of these five variables, we must control the other four. In this experiment, we will control variables 1, 3, 4, and 5 above and vary the alkyl halide concentration. First, we will assume that the alkyl halide does not change its inherent properties (1) during the course of the reaction. By using a dilute 2-chloro-2-methylpropane solution, only a small fraction of the solvent will be consumed in the reaction, so the proportions of water and 2-propanol will remain practically constant (3). Also, we will control the reaction temperature (4) throughout the experiment, and we will assume that neither reactants nor products catalyze this reaction (5). Based on this experimental design, the only variable in this experiment is the 2-chloro-2-methylpropane concentration (2).

We can determine the 2-chloro-2-methylpropane concentration in the reaction mixture at the end of any given time interval by subtracting the amount of 2-chloro-2-methylpropane that has reacted during the time interval from the amount present at the beginning of the time interval. The number of moles of 2-chloro-2-methylpropane that reacted corresponds to the number of moles of HCl produced by this solvolysis reaction. To determine the number of moles of HCl produced, we titrate the reaction mixture with standard sodium hydroxide solution (NaOH), using phenolphthalein indicator. The titration reaction is shown in Equation 2.

$$H_3O^+(aq) + OH^-(aq) \rightarrow 2\ H_2O(l) \qquad \text{(Eq. 2)}$$

We must use the same 2-propanol–water mixture to prepare both the NaOH solution and the 2-chloro-2-methylpropane. Otherwise, the solvent composition will constantly vary as we add more and more NaOH solution to the reaction vessel.

We can plot the total volume of NaOH solution required to neutralize the HCl produced after various time intervals once the reaction begins. Using the plot, we can determine whether or not the solvolysis reaction proceeds at a constant rate.

We can also determine the instantaneous rate of this reaction. To illustrate what we mean by "instantaneous rate," consider the following analogy. Suppose it takes us 3 hr to drive 150 mi. Our average speed, or rate of travel, for the trip is 50 mi hr^{-1}. Yet, at any time during the trip, our instantaneous speed, as indicated by the speedometer, may be 20, 30, 60, or more mi hr^{-1}.

The instantaneous rate of the 2-chloro-2-methylpropane solvolysis reaction is difficult to measure directly using the titration method described. However, from our titration data, we can calculate the number of milliliters of NaOH neutralized per minute, which is equivalent to the reaction rate, at various stages of the reaction. To help visualize the relationship between reaction rate and elapsed time, we can plot the volume of NaOH neutralized as a function of time.

Most alkyl halide solvolysis reaction rates are directly proportional to the concentration of alkyl halide in solution. However, we must experimentally verify that the rate of solvolysis of 2-chloro-2-methylpropane is directly proportional to the concentration of 2-chloro-2-methylpropane in our chosen solvent mixture. Then, we can determine the precise mathematical relationship between the reaction rate and time. Assuming that there is a direct relationship between reaction rate and 2-chloro-2-methylpropane concentration, we can express that relationship in mathematical form, as shown in Equation 3.

$$\text{rate} = k\,[\text{2-chloro-2-methylpropane}] \qquad \text{(Eq. 3)}$$

The square brackets indicate molar concentration (mol L^{-1}) and k is the **specific rate constant** for the reaction. The magnitude of k depends on the inherent properties of the alkyl halide (1), and on the other variables that we are holding constant (3, 4, and 5) in this experiment. Before we can insert actual values into Equation 3, we must first define the following terms:

A = number of moles of 2-chloro-2-methylpropane present at the start of the reaction
P = number of moles of HCl formed at any time, t, after the start of the reaction
$A - P$ = number of moles of unreacted 2-chloro-2-methylpropane present at any time, t, after the start of the reaction

We substitute $(A - P)$ into Equation 3 to get Equation 4,

$$\text{rate} = k\,[A - P] \qquad \text{(Eq. 4)}$$

which shows the relationship between the solvolysis reaction rate and the concentration of 2-chloro-2-methylpropane remaining in the reaction mixture. Using the titration method, we cannot measure A and P directly. Instead, we measure the volume of NaOH solution required to neutralize the reaction mixture at any point during the reaction.

From Equation 4 we can develop Equation 5, which expresses the mathematical relationship between the NaOH solution volume and time.

$$\log\left(1 - \frac{V}{V_\infty}\right) = \frac{-kt}{2.3} \qquad \text{(Eq. 5)}$$

In Equation 5, V is the volume of NaOH neutralized at any elapsed time t, and V_∞ is the volume of NaOH neutralized after all the 2-chloro-2-methylpropane has reacted. V/V_∞ represents the fraction of the total amount of 2-chloro-2-methylpropane reacted at any time t, and $(1 - V/V_\infty)$ represents the fraction of 2-chloro-2-methylpropane remaining after any elapsed time t.

Equation 5 is in the form $y = mx + b$, which is a straight-line equation, in which

$$y = \log\left(1 - \frac{V}{V_\infty}\right) \qquad m = -\frac{k}{2.3} \qquad x = t \qquad b = 0$$

Now we can use our titration data and Equation 5 to test whether or not the solvolysis rate of 2-chloro-2-methylpropane in our chosen water and 2-propanol mixture is directly proportional to the concentration of unreacted 2-chloro-2-methylpropane. If our theory is correct, the plot of $\log(1 - V/V_\infty)$ against t should be a straight line with a slope of $-k/2.3$ and a y-intercept of 0, as shown in Figure 1.

In this experiment, you will study the reaction rate of 2-chloro-2-methylpropane with water and 2-propanol by titrating the HCl produced at various times during the reaction with standard NaOH solution, using phenolphthalein indicator. Phenolphthalein is colorless in acidic solution and pink in basic solution. You will then plot your data to demonstrate the dependence of the solvolysis rate on the concentration of unreacted 2-chloro-2-methylpropane.

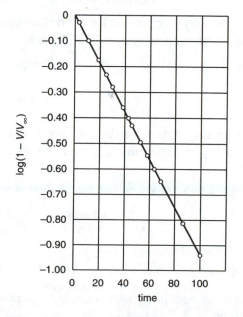

Figure 1　*The relationship between $\log(1 - V/V_\infty)$ and elapsed t for the solvolysis of 2-chloro-2-methylpropane in a 2-propanol–water mixture*

Procedure

Chemical Alert
2-chloro-2-methylpropane—flammable and irritant
2-propanol—flammable and irritant
0.350M sodium hydroxide—toxic and irritant

Caution: Wear departmentally approved eye protection when doing this experiment.

Caution: The reaction mixture is flammable. Extinguish all open flames in the laboratory before performing this experiment.

Measure 100 mL of the solvent mixture specified by your laboratory instructor in a clean, dry graduated cylinder. Transfer the mixture to a clean, dry 250-mL Erlenmeyer flask equipped with a tightly fitting rubber stopper. Stopper the flask to prevent contamination of the solvent mixture by diffusion of carbon (IV) oxide (CO_2) from the air. Record the composition of your solvent mixture on Data Sheet 1.

Note: If your laboratory instructor tells you to perform the experiment at room temperature, make sure that the solvent mixture, the 2-chloro-2-methylpropane, and the NaOH solution are all at room temperature before starting. If your instructor tells you to perform the experiment at a temperature above or below room temperature, place the Erlenmeyer flask containing the solvent in a water bath, and maintain the desired temperature by heating or cooling the bath as appropriate.

Record the reaction temperature on Data Sheet 1. Add 2-3 drops of phenolphthalein indicator to the Erlenmeyer flask and thoroughly mix by swirling the flask.

Caution: Sodium hydroxide solution is corrosive and toxic, and it can cause skin burns. Prevent contact with your eyes, skin, and clothing. Do not ingest the solution.

If you spill any NaOH solution, immediately notify your laboratory instructor.

Clean a 50-mL buret. Rinse the buret with 10 mL of distilled or deionized water, followed by two 10-mL portions of standard NaOH solution supplied by your laboratory instructor. Discard the NaOH rinses in the container specified by your laboratory instructor and labeled "Discarded NaOH Solution." Finally, fill the buret with the NaOH solution. Read the buret to the nearest 0.02 mL, and record this initial buret reading on Data Sheet 1. Record the exact molarity of the NaOH solution on Data Sheet 1.

> **Caution:** 2-Chloro-2-methylpropane is flammable and an irritant. Prevent contact with your eyes, skin, and clothing. Avoid inhaling the vapors and ingesting the compound.

Measure 2 mL of 2-chloro-2-methylpropane into a clean, dry 10-mL graduated cylinder. If you are performing the reaction at a temperature other than room temperature, carefully place the cylinder in your water bath to adjust the 2-chloro-2-methylpropane temperature.

To start the reaction, pour the 2-chloro-2-methylpropane into the Erlenmeyer flask. Record the exact time and volume of 2-chloro-2-methylpropane added on Data Sheet 1. Swirl the flask and its contents until the solution is completely mixed.

Immediately after you have mixed the solution containing the 2-chloro-2-methylpropane, add 1–2 mL of NaOH solution from the buret. Swirl the flask and its contents. The reaction mixture should now be pink. If it does not turn pink, continue adding 1–2 mL portions of NaOH solution *just until* the reaction mixture turns pink. Then carefully observe the solution in the flask, and note the exact time at which the pink fades completely. Record this time and the corresponding buret reading on Data Sheet 2.

Immediately after you record the buret reading and the time at which the pink fades, add another 1–2 mL of NaOH solution. The reaction mixture should again turn pink. Carefully observe the reaction mixture, and note the exact time at which the pink fades completely. Record this second time and the corresponding buret reading on Data Sheet 2.

Continue in this manner, adding 1–2 mL portions of NaOH solution and recording time and buret readings as frequently as possible without undue hurry. A smaller number of accurate measurements will be more useful than a larger number of erratic ones. If you miss the time at which a particular fading of the pink occurs, add more NaOH solution, mix, and observe the next fading.

> **Note:** The total volume of NaOH solution required for complete neutralization of the HCl formed may exceed 50 mL. When your buret reading approaches 45 mL, prepare to refill your buret with NaOH solution and to take a new initial buret reading. Record this new, initial buret reading on Data Sheet 1. *Do not let the level of NaOH solution in your buret drop below the 50-mL mark.*
>
> Your laboratory instructor will tell you how to determine when the reaction is 75% complete.

After the reaction is at least 75% complete, heat the Erlenmeyer flask and the mixture to 60 °C in a hot-water bath on a hot plate for about 15 min to bring the reaction to 100% completion. Then cool the solution to the temperature at which you carried out the reaction, and titrate the remaining HCl with NaOH solution until a faint pink persists for 30 s. Carefully perform this titration so that you obtain an accurate value of V_∞.

Pour the neutralized reaction mixture into the container specified by your laboratory instructor and labeled "Discarded Reaction Mixtures." Transfer any NaOH solution remaining in your buret into the "Discarded NaOH Solution" container. Rinse all glassware with tap water and then with distilled water. Pour the rinses into the drain.

> **Caution:** Wash your hands thoroughly with soap or detergent before leaving the laboratory.

Calculations

Do the following calculations for each determination, and record the results on your Data Sheets.

1. Use your experimental data to calculate the following:

$$t \qquad V \qquad \frac{V}{V_\infty} \qquad 1 - \frac{V}{V_\infty} \qquad \log\left(1 - V/V_\infty\right)$$

Record all calculated results with three significant digits. Enter your results in the appropriate columns on Data Sheet 2.

2. Prepare two graphs by plotting the following:

	on the vertical axis	on the horizontal axis
Graph 1–raw data	V	t
Graph 2–to test Eq.5	$\log\left(1 - V/V_\infty\right)$	t

3. Draw the best possible line through the points on each graph.

4. Calculate k for the solvolysis reaction you performed. Enter this value on Data Sheet 1.

Post-Laboratory Questions

(Use the spaces provided for the answers and additional paper if necessary.)

1. Briefly explain why the temperature of the reaction mixture must be held constant, or nearly constant, during the experiment.

2. Is your plot of log $(1 - V/V_\infty)$ against time a straight line? If so, what does this prove?

3. The half-life of an alkyl halide in a solvolysis reaction is the time required for half of the alkyl halide to react. Calculate $V_\infty/2$ and use this value and Graph 1 to determine the half-life of 2-chloro-2-methylpropane in the solvent mixture you used.

4. Using the half-life of 2-chloro-2-methylpropane you determined in Question 3, calculate the fraction of unreacted 2-chloro-2-methylpropane remaining after one half-life, after two half-lives, after three half-lives, and after four half-lives.

5. Estimate the rate of this reaction in terms of the volume of NaOH neutralized during 3-min intervals:
    (1)   at the start of the reaction (0–3 min)

    (2)   after one half-life

    (3)   after two half-lives

name   section   date

Data Sheet 1

solvent mixture, % water _____

solvent mixture, % 2-propanol (isopropyl alcohol) _____

reaction temperature, °C _____

volume of 2-chloro-2-methylpropane added, mL _____

concentration of NaOH solution, mol L^{-1} _____

initial buret reading, mL _____

final buret reading before refilling, mL _____

initial buret reading after refilling, mL _____

final buret reading after refilling, mL _____

total volume of NaOH added, mL _____

time, initial clock reading, min and s _____

volume of NaOH neutralized
   when reaction is 75% complete, mL _____

specific rate constant, k, min^{-1} _____

Data Sheet 2

time, clock reading	t (elapsed time, min)	buret reading, mL	V (volume of NaOH solution) added, mL	V/V_∞	$1 - (V/V_\infty)$	$\log(1 - V/V_\infty)$
_____	_____	_____	_____	_____	_____	_____
_____	_____	_____	_____	_____	_____	_____
_____	_____	_____	_____	_____	_____	_____
_____	_____	_____	_____	_____	_____	_____
_____	_____	_____	_____	_____	_____	_____
_____	_____	_____	_____	_____	_____	_____
_____	_____	_____	_____	_____	_____	_____
_____	_____	_____	_____	_____	_____	_____
_____	_____	_____	_____	_____	_____	_____
_____	_____	_____	_____	_____	_____	_____
_____	_____	_____	_____	_____	_____	_____
_____	_____	_____	_____	_____	_____	_____
_____	_____	_____	_____	_____	_____	_____
_____	_____	_____	_____	_____	_____	_____

Pre-Laboratory Assignment

1. (1) Briefly describe the hazard associated with the reaction mixture used in this experiment.

(3) Calculate the volume of $0.380M$ NaOH solution required to neutralize the total amount of HCl produced by the hydrolysis reaction.

(2) Because of this hazard, what safety precaution must you take when performing the experiment?

(4) Calculate the volume of $0.380M$ NaOH solution you will need to neutralize the HCl produced when the solvolysis reaction is 75% complete. Record this volume on Data Sheet 1.

2. This experiment calls for the use of 2 mL of 2-chloro-2-methylpropane. The molar mass of 2-chloro-2-methylpropane is 92.57 g mol^{-1}, and its density is 0.847 g mL^{-1}.

(1) Calculate the number of moles of 2-chloro-2-methylpropane used in this experiment.

(5) Will your buret contain sufficient NaOH solution to neutralize the HCl produced by the solvolysis reaction by the time the reaction is 75% complete? If not, you will have to refill the buret during the determination.

(2) Calculate the number of moles of HCl produced by the complete reaction of 2 mL of 2-chloro-2-methylpropane.

3. Briefly explain why the NaOH solution must be prepared by dissolving solid NaOH in the same solvent mixture as used in the reaction flask.

Kinetic Study of a Chemical Reaction

prepared by **Paul C. Moews, Jr.**, Royal Institution of Great Britain,
Ralph H. Petrucci, California State University, San Bernardino,
and **Judith C. Foster**, Bowdoin College

Purpose of The Experiment

Determine the rate law, activation energy, and the frequency factor for the reaction between peroxodisulfate ion and iodide ion. Determine the effect of copper(II) nitrate on the specific rate constant.

Background Information

General

Chemical kinetics is the branch of chemistry that deals with the rates and mechanisms of chemical reactions. The **rate** of a chemical reaction refers to the speed with which that reaction proceeds and is measured by the disappearance rate of one of the reactants or by the appearance rate of one of the products. Rates of chemical reactions are affected by changes in concentration, temperature, presence of a catalyst, and the nature of the reactants. The **mechanism** of a reaction is a series of one or more steps, each involving one or more reacting species, through which the reactants are converted to products. The rate of the overall reaction is determined by the rate of the slowest of these steps, which is called the **rate-limiting step**.

Reaction kinetics is based on a model of chemical reactions in which reactant species collide to form product species. The overall reaction rate is affected by those conditions that affect the frequency and strength of these collisons.

From the kinetic theory of gases, it may be seen that the collision frequency is proportional to concentration and the $T^{1/2}$ (Kelvin temperature). As concentration or temperature increases, there are more collisions, hence the reaction rate increases. It is reasonable to assume that increasing either concentration or temperature will increase the collision frequency and reaction rates in solutions also.

The concentration of a reacting species refers to the number of that species per unit volume of the reaction mixture. In general, the rate of a chemical reaction increases with an increase in the concentration of the reacting species. However, it is only by experiment that we can determine exactly how changes in the concentration of a particular reactant affect the rate of a chemical reaction. The exact mechanism of a chemical reaction cannot be predicted by theory alone.

The rate of a chemical reaction can be expressed in a mathematical form known as the **rate law**. The rate law for the hypothetical reaction $A + B + C \rightarrow D + E + F$ is usually of the form shown in Equation 1.

$$rate = k[A]^x[B]^y[C]^z \qquad \text{(Eq. 1)}$$

Copyright © 1987 by Chemical Education Resources, Inc., P.O. Box 357, 220 S. Railroad, Palmyra, Pennsylvania 17078
No part of this laboratory program may be reproduced or transmitted in any form or by any means, electronic or mechanical, including photocopying, recording, or any information storage and retrieval system, without permission in writing from the publisher. Printed in the United States of America

A,B,C, etc., represent the substances on whose concentration the rate of the chemical reaction depends. These concentrations generally involve the reactants but may also include one of the products if a reverse reaction occurs. Not all reactants have an effect on the rate of the reaction; hence, the rate law may not include all the reactants in the overall reaction. Even species that are not shown in the overall stoichiometric equation for the reaction may be present in the rate law. These species would be reactants or products in an intermediate step in the mechanism, the rate-limiting step. The symbol [ ] signifies the concentration in moles of reactant per liter of reaction mixture of the species in the brackets.

The specific rate constant, k, depends principally upon the nature of the reactants and the temperature at which the reaction occurs. In addition, for reactions between ions occurring in aqueous solution, the specific rate constant is affected by the total concentration of ions in solution.

The exponents x, y, z must be determined by experiment. They are often small positive integers but may be fractional and even negative in some cases. These exponents are further used to describe the **order** of the reaction. For example, in Equation 1, if $x = 1$, the reaction is said to be first order with respect to reactant A. If $y = 2$, the reaction is second order with respect to B. The sum of the exponents, $x + y + z$, gives the overall order of the reaction.

The rate constant k in Equation 1 is only constant as long as the reaction is carried out at the same temperature each time. As the temperature rises, the rate of collisions between reactant species increases; hence the reaction rate should increase. It is not unusual, however, for a temperature increase of ten degrees (say, from 298 to 308 K) to result in a doubling or tripling of the reaction rate. This dramatic increase in the rate cannot be accounted for simply by the increase in the rate of collision, which is proportional to $T^{1/2}$. A change from 298 to 308 K represents a 3% increase in temperature and only a 1.5% increase in the rate of collisions.

The Swedish chemist Svante Arrhenius (1859–1927) suggested that some minimum energy is needed for a reaction to be successful and result in product formation. This minimum energy is called the **activation energy**, E_a. The activation energy is constant for a given reaction occurring via a given reaction mechanism. Reacting species with a total kinetic energy of E_a or greater can successfully react. By considering the distribution of kinetic energies at two slightly different temperatures, we can see that the percent increase in

reactant species with kinetic energy greater than or equal to E_a is vastly greater than the percent increase in the square root of the temperature. This concept of an activation energy accounts for the observed large increase in rates over a relatively small increase in temperature. This concept also led to the development of the Arrhenius equation (Equation 2), relating k and T.

$$k = Ae^{-E_a/RT} \qquad \text{(Eq. 2)}$$

Equation 2 can be converted into Equation 3.

$$\log_{10} k = \log_{10} A - \left(\frac{E_a}{2.303\ R}\right)\left(\frac{1}{T}\right) \quad \text{(Eq. 3)}$$

In Equations 2 and 3, k is the specific rate constant; T is the kelvin temperature; A is a collision frequency factor constant for the reaction, related to the collision frequency and geometry of reacting species; R is the ideal gas constant, having a value of 8.314 J mol^{-1} K^{-1}; E_a is the activation energy of the reaction, expressed in J mol^{-1}. From Equation 3, we can see that the activation energy, E_a, is a proportionality factor between the $\log_{10} k$ and the inverse of the temperature (1/T, K^{-1}). For a reaction with a large E_a, the rate constant k will change more over a given change in temperature than will the k for a reaction with a small E_a.

The rates of some reactions can be increased by the presence of substances that do not undergo any net change during the reaction. These substances are known as **catalysts**. Because a catalyst is not consumed during a reaction, it must enter into the series of steps constituting the reaction mechanism in such a way that the catalyst is regenerated during the course of the reaction. The catalyst furnishes an alternative, less energetically demanding route for the conversion of reactants to products. This lower activation energy (E_a) means that at the same temperature, a greater portion of the reactant species have enough energy to react successfully, thus raising the rate constant and the rate. Very often, the new mechanism involves two new oxidation–reduction steps in which the catalyst is first oxidized and then reduced to its original form, or vice-versa.

The Chemical Reaction to Be Studied

In this experiment, we study the reaction of peroxodisulfate, also called persulfate ion, $S_2O_8^{2-}$, with iodide, I^-, as shown in Equation 4.

$$S_2O_8^{2-}(aq) + 3\ I^-(aq) \rightarrow 2\ SO_4^{2-}(aq) + I_3^-(aq)$$
$$\text{(Eq. 4)}$$

Starch solution is also present in each reaction mixture. The starch is a strong indicator of the triiodide ion, I_3^-, turning blue as soon as there is any buildup of I_3^-. Since it is produced immediately once the reagents are mixed, the starch would turn blue immediately.

A second reaction is used as a stopwatch to indicate when the peroxodisulfate–iodide reaction has reached a certain point (similar to using a stopwatch to time how long it takes a runner to go one mile). In this second reaction, the liberated iodine, in the form of triiodide ion, reacts with a known amount of thiosulfate ion, $S_2O_3^{2-}$, that is also present in the reaction mixture. The reaction is shown in Equation 5.

$$2\ S_2O_3^{2-}(aq) + I_3^-(aq) \rightarrow S_4O_6^{2-}(aq) + 3\ I^-(aq)$$
(Eq. 5)

The presence of thiosulfate ion in the reaction mixture serves a dual purpose. First, by reacting with the triiodide ion, I_3^-, as it is formed, the thiosulfate prevents the reverse reaction in Equation 4 from occurring. Thus the data obtained in the experiments will refer to the rate of the forward reaction in Equation 4 and will not be complicated by the reverse reaction involving sulfate ion and triiodide ion. Second, as long as thiosulfate ion is present, no free iodine in the form of triiodide ion can accumulate in the solution. However, as soon as all of the thiosulfate ion present reacts, a buildup of triiodide ion will occur. The presence of the free iodine, even in trace amounts, will be indicated by the dark blue color of the starch.

In each laboratory assignment in this investigation, reaction mixtures will contain the reactants $S_2O_8^{2-}$ and I^-, a constant amount of $S_2O_3^{2-}$, and starch solution. Because the same amount of thiosulfate is present in each experiment, the amount of I_3^- produced before the blue color change occurs will be the same in all experiments. Therefore, the peroxodisulfate–iodide reaction we are studying must proceed to the same point to produce this amount of I_3^-. The elapsed time between the moment of mixing of reactants and the blue color change is the reaction time.

This is the stopwatch function. If Run 2 has a longer reaction time than Run 1, then Run 2 has a lower reaction rate. (As in running a mile, a higher time means you are running slower.) The differing times recorded for each run to change color allow us to calculate the different rates for each laboratory assignment, using Equation 6.

$$\text{rate} = \frac{\text{constant}}{\text{time}}$$
(Eq. 6)

It is important to note that the thiosulfate–iodide and indicator reactions occur virtually instantaneously. Therefore, the time elapsed from mixing does in fact represent the reaction time for the thiosulfate–iodide reaction.

This **method of initial rates** is useful in the study of the kinetics of the peroxodisulfate–iodide reaction only because of a fortuitous circumstance. Peroxodisulfate ion is a very good oxidizing agent and in the reaction mixtures used, peroxodisulfate would be expected to oxidize thiosulfate ion as well as iodide ion. But, fortunately, under the conditions of this investigation, the reaction of peroxodisulfate with thiosulfate occurs only very slowly. Thus, the assumption can be made that in each laboratory assignment, thiosulfate is consumed only by reacting with iodine. The time required for the disappearance of thiosulfate can be related to the rate of the reaction being studied, as shown in Equation 6. If two competing reactions for thiosulfate were involved, such as one with iodide ion and the other with peroxodisulfate ion, the reaction system would be very much more complicated. In such a case, this specific method could not be used to study the kinetics of the reaction and a modified method would have to be devised to use initial rates to study such a system.

Determining the Rate Equation for the Reaction

To determine the order of the peroxodisulfate–iodide reaction with respect to the $[I^-]$ and $[S_2O_8^{2-}]$ requires that values of x and y in the rate law be established.

$$\text{rate} = k\,[\,I^-\,]^x[S_2O_8^{2-}]^y$$
(Eq. 7)

Consider two laboratory assignments, at the same temperature, in which the $[S_2O_8^{2-}]$ is held constant and the $[I^-]$ is decreased by a factor of 2. The measured rates and concentrations might be related as follows:

$$\text{rate}_1 = k\,[\,I^-\,]_1^x[S_2O_8^{2-}]^y$$
(Eq. 8)

$$\text{rate}_2 = k\,[\,I^-\,]_2^x[S_2O_8^{2-}]^y$$
(Eq. 9)

However, $[S_2O_8^{2-}]$ has the same value in Equation 8 and in Equation 9, k is constant because the temperature is the same, and we stated that $[\,I^-\,]_1 = 2^x\,[\,I^-\,]_2$. Thus, the ratio of the rates can be expressed as:

$$\frac{\text{rate}_1}{\text{rate}_2} = \frac{k\,2^x\,[\,I^-\,]_2^x\,[S_2O_8^{2-}]^y}{k\,[\,I^-\,]_2^x\,[S_2O_8^{2-}]^y} = 2^x$$
(Eq. 10)

If rate_1 is found by experiment to be twice as great as rate_2, $2^x = 2$ and $x = 1$; if rate_1 is four times rate_2, $2^x = 4$

and $x = 2$; and so on. Holding the $[I^-]$ constant and varying the $[S_2O_8^{2-}]$ results in a similar equation for y.

The specific rate constant, k, may be obtained by solving Equation 7, after the values of x and y have been determined. This leads to the expression given in Equation 11.

$$k = \frac{\text{rate}}{[I^-]^x [S_2O_8^{2-}]^y} \qquad \text{(Eq. 11)}$$

The terms $[I^-]$ and $[S_2O_8^{2-}]$ refer to the initial concentrations of the reactants for any given experiment. These are the concentrations that exist immediately after the solutions of reactants are mixed and before any reaction occurs.

A numerical value for the rate of the reaction to be used in Equation 11 is a bit more difficult to obtain but can be approximated in this way:

$$\text{rate} = -\frac{\Delta[S_2O_8^{2-}]}{(\Delta t, \text{sec})} \qquad \text{(Eq. 12)}$$

In Equation 12, $\Delta[S_2O_8^{2-}]$ represents the change in the $[S_2O_8^{2-}]$ that occurs during the time period, Δt, between mixing the reactants and the appearance of the blue iodine–starch color. The assumption made is that the rate of reaction remains constant during the time period measured.

The minus sign is incorporated in Equation 12 because the concentrations of the reactants decrease during the reaction (that is, $\Delta[S_2O_8^{2-}]$ is a negative quantity). This, then, makes the reaction rate and the specific rate constant positive quantities.

Equation 12 is used to calculate the rate for each laboratory assignment. In each assignment $\Delta[S_2O_8^{2-}]$ has the same value and only Δt will vary. The number of moles of $S_2O_8^{2-}$ consumed in the course of the reaction can be calculated from the number of moles of $S_2O_3^{2-}$ used in each experiment. Assume that in each experiment 1.0×10^{-2} L of $1.0 \times 10^{-2}M$ $Na_2S_2O_3$ is used. The number of moles of $S_2O_3^{2-}$ is found from

$$\begin{aligned}\text{number of moles} \\ \text{of } S_2O_3^{2-} \text{ used}\end{aligned} = (\text{molarity})(\text{volume, liters})$$
$$= (1.0 \times 10^{-2} \text{ mol L}^{-1})(1.0 \times 10^{-2} \text{ L})$$
$$= 1.0 \times 10^{-4} \text{ mol}$$

The first appearance of color occurs when all the $S_2O_3^{2-}$ has been consumed by reaction with I_3^-. Because two moles of $S_2O_3^{2-}$ are required for each mole of I_3^- and each mole of $S_2O_8^{2-}$ produces one mole of I_3^-, two moles of $S_2O_3^{2-}$ are consumed for each mole of $S_2O_8^{2-}$ consumed. Thus if 1.0×10^{-4} mol of $S_2O_3^{2-}$ are present in each laboratory assignment, one-half of

this number of moles of $S_2O_8^{2-}$ must have been consumed. The calculations may be summarized as follows:

$$\text{number of moles of } S_2O_8^{2-} \text{ consumed} =$$

$$\left(\begin{array}{c}\text{molarity} \\ \text{of } S_2O_3^{2-}\end{array}\right)\left(\begin{array}{c}\text{volume,} \\ L\end{array}\right)\left(\frac{1 \text{ mole } I_3^-}{2 \text{ mole } S_2O_3^{2-}}\right)\left(\frac{1 \text{ mole } S_2O_8^{2-}}{1 \text{ mole } I_3^-}\right)$$

The change in the number of moles of $S_2O_8^{2-}$ for each laboratory assignment is -5.0×10^{-5}. The minus sign is included because the number of moles of $S_2O_8^{2-}$ decreases during the course of the reaction.

Because $\Delta[S_2O_8^{2-}]$ is the change in $[S_2O_8^{2-}]$, the number of moles of $S_2O_8^{2-}$ consumed must be divided by the volume of the reaction mixture to establish $\Delta[S_2O_8^{2-}]$. In each laboratory assignment the reaction volume is the same. In this discussion it is assumed to be 6.5×10^{-2} L.

$$\Delta[S_2O_8^{2-}] = -\frac{5.0 \times 10^{-5} \text{ mol } S_2O_8^{2-}}{6.5 \times 10^{-2} \text{ L}}$$

Using this value for $\Delta[S_2O_8^{2-}]$ and the measured value of Δt, we can determine the rate of reaction for any laboratory assignment by using Equation 12.

$$\text{rate} = -\frac{\Delta[S_2O_8^{2-}]}{(\Delta t)} = -\frac{5.0 \times 10^{-5} \text{ mol } S_2O_8^{2-}}{(6.5 \times 10^{-2} \text{ L})(\Delta t, \text{sec})}$$
$$\text{(Eq. 12)}$$

Once the rate is obtained, Equation 10 is used to find x (and the corresponding equation to find y). Equation 11 may be used in conjunction with the determined order of the reaction (that is, with values of x and y) and the concentrations of the reactants to calculate the specific rate constant, k.

In a typical laboratory assignment, 25 mL of $2.0 \times 10^{-1}M$ KI and 25 mL of $2.0 \times 10^{-1}M$ $(NH_4)_2S_2O_8$ were added to 5.0 mL of 0.2% starch solution and 10 mL of $1.0 \times 10^{-2}M$ $Na_2S_2O_3$. The time elapsed from mixing to the appearance of the blue iodine–starch color was 24 s. The $[I^-]$ after the reactants were mixed, but before any reaction occurred, is found from

$$\begin{aligned}\text{number of moles} \\ \text{of KI used}\end{aligned} = (\text{molarity})(\text{volume, L})$$
$$= (0.20 \text{ mol}^{-1})(2.5 \times 10^{-2} \text{ L})$$
$$= 5.0 \times 10^{-3} \text{ mol}$$

$$\text{concentration of KI} = \frac{\text{number of moles of KI}}{\text{volume of reaction mixture, L}}$$
$$= \frac{5.0 \times 10^{-3} \text{ mol}}{0.065 \text{ L}} = 7.7 \times 10^{-2} M$$

A similar calculation for $(NH_4)_2S_2O_8$ gives

$$[(NH_4)_2S_2O_8] = 7.7 \times 10^{-2}M$$

The rate of the reaction is found from Equation 12.

$$\text{rate}_1 = -\frac{\Delta[S_2O_8^{2-}]}{(\Delta t)} = -\frac{-5.0 \times 10^{-5} \text{ mol } S_2O_8^{2-}}{(6.5 \times 10^{-2} \text{ L}) (24 \text{ s})}$$

$$= 3.2 \times 10^{-5} \text{ mol L}^{-1} \text{ s}^{-1} \quad \text{(Eq. 12)}$$

In a second laboratory assignment the $[S_2O_8^{2-}]$ was maintained at $7.7 \times 10^{-2}M$ and the $[I^-]$ was $3.9 \times 10^{-2}M$. The time elapsed to the color change was 48 s. The rate for this laboratory assignment becomes

$$\text{rate}_2 = -\frac{-5.0 \times 10^{-5} \text{ mol } S_2O_8^{2-}}{(6.5 \times 10^{-2} \text{ L}) (48 \text{ s})}$$

$$= 1.6 \times 10^{-5} \text{ mol L}^{-1} \text{ s}^{-1}$$

The ratio of the rates in the first and second laboratory assignments is

$$\frac{3.2 \times 10^{-5} \text{ mol L}^{-1} \text{ s}^{-1}}{1.6 \times 10^{-5} \text{ mol L}^{-1} \text{ s}^{-1}} = 2$$

From Equation 10,

$$\frac{\text{rate}_1}{\text{rate}_2} = 2^x$$

from which x is seen to be 1, so that the order with respect to the $[I^-]$ is **first**.

Assuming that the reaction is first order with respect to both the $[I^-]$ and $[S_2O_8^{2-}]$, the specific rate constant is then calculated from Equation 11.

$$k = \frac{\text{rate}}{[I^-]^x [S_2O_8^{2-}]^y} = \frac{3.2 \times 10^{-5} \text{ mol L}^{-1} \text{ s}^{-1}}{(7.7 \times 10^{-2}M) (7.7 \times 10^{-2}M)}$$

$$= 5.4 \times 10^{-3} \text{ L mol}^{-1} \text{ s}^{-1}$$

Calculating the Activation Energy and the Frequency Factor

The activation energy, E_a, and the collision frequency factor, A, for the reaction can be obtained from experimentally determined values for k at various temperatures. Using the specific rate constants calculated from experimental data for the laboratory assignments run at four different temperatures, we plot $\log_{10} k$ on the ordinate versus $1/T$ on the abscissa, where T is the kelvin temperature.

$$\log_{10} k = \log_{10} A - \left(\frac{E_a}{2.303 \, R}\right)\left(\frac{1}{T}\right) \quad \text{(Eq. 3)}$$

If the Arrhenius equation (Eq. 3) applies, then the plotted data should fall on a straight line graph whose slope is $-E_a / (2.303 \, R)$, where $R = 8.314$ J mol^{-1} K^{-1}.

The experimentally determined slope can be used to calculate the activation energy for this reaction. The intercept on the ordinate at $T^{-1} = 0$ is equal to $\log_{10} A$, where A is the frequency factor for the reaction and has the same units as the specific rate constant.

Procedure

Chemical Alert
ammonium peroxodisulfate—oxidant
potassium nitrate—oxidant and irritant
sodium thiosulfate—irritant

Note: Your laboratory instructor will indicate whether to use a graduated pipet, buret, or graduated cylinder for measuring each of the solutions to be used in this experiment.

I. Determining the Effect of the Concentration of the Reactants on the Reaction Rate

---

Caution: Wear departmentally approved eye protection during the experiment.

The amounts of reactants to be used in this part of the experiment are given in Tables 1 and 2. Your laboratory instructor will assign you to groups of four to do this investigation. Each of you will do the four assignments in Tables 1, 2, 3, or 4 in duplicate.

For each of the laboratory assignments you are asked to do, add the specified volumes of the 0.2% starch solution, $1.2 \times 10^{-2}M$ $Na_2S_2O_3$ solution, $2.0 \times 10^{-1}M$ KI solution, and $2.0 \times 10^{-1}M$ KNO$_3$ to a labeled, clean, dry 50-mL Erlenmeyer flask. Add the specified volume of $2.0 \times 10^{-1}M$ $(NH_4)_2S_2O_8$ and $2.0 \times 10^{-1}M$ $(NH_4)_2SO_4$ to a second labeled, clean, dry 50-mL Erlenmeyer flask.

Place these two flasks in a constant temperature bath, set at the specified temperature, and allow them to equilibrate. Record this temperature.

Note: Your laboratory instructor will indicate whether to use a wall clock, stopwatch, electric timer, or computer to make the time measurements.

Table 1   *Reaction volumes for determining the effect of [I⁻] on the reaction rate*

lab assign't	starch, mL	1.2×10^{-2} M $Na_2S_2O_3$, mL	0.20 M KI, mL	0.20 M KNO_3, mL	0.20 M $(NH_4)_2S_2O_8$, mL	0.20 M $(NH_4)_2SO_4$, mL
1	1.0	2.0	8.0	0.0	4.0	4.0
2	1.0	2.0	4.0	4.0	4.0	4.0
3	1.0	2.0	2.0	6.0	4.0	4.0
4	1.0	2.0	1.0	7.0	4.0	4.0

Table 2   *Reaction volumes for determining the effect of $[S_2O_8^{2-}]$ on the reaction rate*

lab assign't	starch, mL	1.2×10^{-2} M $Na_2S_2O_3$, mL	0.20 M KI, mL	0.20 M KNO_3, mL	0.20 M $(NH_4)_2S_2O_8$, mL	0.20 M $(NH_4)_2SO_4$, mL
5	1.0	2.0	4.0	4.0	8.0	0.0
6	1.0	2.0	4.0	4.0	4.0	4.0
7	1.0	2.0	4.0	4.0	2.0	6.0
8	1.0	2.0	4.0	4.0	1.0	7.0

Record the initial time of the reaction as soon as the one solution has been quickly added to the other. Rapidly mix the solutions by pouring the mixture from one flask to the other four times. Record the final time of the reaction when the mixture turns blue.

For each assignment, do a second determination at the same temperature. Between determinations, rinse the flasks thoroughly with distilled water and dry them. If the two reaction times do not agree within 10%, do a third determination.

II. Determining the Effect of Temperature on the Reaction Rate

In Laboratory Assignments 9, 10, 11, and 12, the concentrations of KI and $(NH_4)_2S_2O_8$ are held constant, but the temperature at which the reactants are mixed will be 10 °C below room temperature, room temperature, and 10 °C and 20 °C above room temperature. These assignments are given in Table 3.

Follow the procedure used in Part I to do two determinations of your designated laboratory assignments in Table 3. Record the initial time, final time, and the temperature of your constant temperature bath.

If the two reaction times do not agree within 10%, do a third determination.

III. Determining the Effect of Cu(NO₃)₂ on the Reaction Rate

Follow the procedure used in Part I to do two determinations of your designated laboratory assignments in Table 4. Add one drop of 2.0×10^{-2} M $Cu(NO_3)_2$ to the flask containing the starch, thiosulfate, and iodide solutions for each laboratory assignment. Record all data.

If the two reaction times do not agree within 10%, do a third determination.

> *Caution:* Wash your hands thoroughly with soap or detergent before leaving the laboratory.

Calculations

Exchange temperature and time data with other members of your group. Enter these data on your Data Sheet.

I. Determining the Effect of [I⁻] on the Reaction Rate

1. Calculate the concentration of KI in each reaction mixture after the reactants are mixed but before any reaction occurs.

Table 3 *Reaction volumes for determining the effect of temperature on the reaction rate*

lab assign't	temp., °C	starch, mL	$1.2 \times 10^{-2}M$ $Na_2S_2O_3$, mL	0.20M KI, mL	0.20M KNO_3, mL	0.20M $(NH_4)_2S_2O_8$, mL	0.20M $(NH_4)_2SO_4$, mL
9	room − 10°	1.0	2.0	4.0	4.0	2.0	6.0
10	room	1.0	2.0	4.0	4.0	2.0	6.0
11	room + 10°	1.0	2.0	4.0	4.0	2.0	6.0
12	room + 20°	1.0	2.0	4.0	4.0	2.0	6.0

Table 4 *Reaction volumes for determining the effect of copper(II) nitrate on the reaction rate*

lab assign't	starch, mL	$1.2 \times 10^{-2}M$ $Na_2S_2O_3$, mL	0.20M KI, mL	0.20M KNO_3, mL	0.20M $(NH_4)_2S_2O_8$, mL	0.20M $(NH_4)_2SO_4$, mL
13	1.0	2.0	8.0	0.0	4.0	4.0
14	1.0	2.0	4.0	4.0	4.0	4.0
15	1.0	2.0	2.0	6.0	4.0	4.0
16	1.0	2.0	1.0	7.0	4.0	4.0

$$\text{number of moles of KI used} = (\text{molarity})(\text{volume of KI used, L})$$

$$\text{concentration of KI after mixing} = \frac{\text{number of moles of KI}}{\text{volume of reaction mixture, L}}$$

2. Calculate the concentration of $(NH_4)_2S_2O_8$ in each reaction mixture after the reactants are mixed but before any reaction occurs.

$$\text{number of moles of } (NH_4)_2S_2O_8 \text{ used} = (\text{molarity})\left(\text{volume of } (NH_4)_2S_2O_8 \text{ used, L}\right)$$

$$\text{concentration of } (NH_4)_2S_2O_8 \text{ after mixing} = \frac{\text{number of moles of } (NH_4)_2S_2O_8}{\text{volume of reaction mixture, L}}$$

3. Calculate the rate of the reaction for each assignment.

$$\text{rate} = -\frac{\Delta[S_2O_8^{2-}]}{(\Delta t, \text{ sec})} \qquad \text{(Eq. 12)}$$

4. Find the order of the reaction with respect to I^- using the rates calculated for Laboratory Assignments 1 and 2, 2 and 3, 3 and 4. Assuming that $[I^-]_1/[I^-]_2 = 2$ and $[S_2O_8^{2-}]_1/[S_2O_8^{2-}]_2 = 1$,

$$\frac{\text{rate}_1}{\text{rate}_2} = 2^x \qquad \frac{\text{rate}_2}{\text{rate}_3} = 2^x \qquad \frac{\text{rate}_3}{\text{rate}_4} = 2^x \qquad \text{(Eq. 10)}$$

Find the average of x. Round to the nearest integer.

II. Determining the Effect of $[S_2O_8^{2-}]$ on the Reaction Rate

1. Calculate the concentration of KI in each reaction mixture after the reactants are mixed but before any reaction occurs.

$$\text{number of moles of KI used} = (\text{molarity})(\text{volume of KI used, L})$$

$$\text{concentration of KI after mixing} = \frac{\text{number of moles of KI}}{\text{volume of reaction mixture, L}}$$

2. Calculate the concentration of $(NH_4)_2S_2O_8$ in each reaction mixture after the reactants are mixed but before any reaction occurs.

$$\text{number of moles of } (NH_4)_2S_2O_8 \text{ used} = (\text{molarity})\left(\text{volume of } (NH_4)_2S_2O_8 \text{ used, L}\right)$$

$$\text{concentration of } (NH_4)_2S_2O_8 \text{ after mixing} = \frac{\text{number of moles of } (NH_4)_2S_2O_8}{\text{volume of reaction mixture, L}}$$

3. Calculate the rate of the reaction for each assignment.

$$\text{rate} = -\frac{\Delta[S_2O_8^{2-}]}{(\Delta t, \text{ sec})} \qquad \text{(Eq. 12)}$$

4. Find the order of the reaction with respect to $S_2O_8^{2-}$ using the rates calculated for Laboratory As-

signments 5 and 6, 6 and 7, 7 and 8. Assuming that $[I^-]_5/[I^-]_6 = 1$ and $[S_2O_8^{2-}]_5/[S_2O_8^{2-}]_6 = 2$,

$$\frac{rate_5}{rate_6} = 2^y \qquad \frac{rate_6}{rate_7} = 2^y \qquad \frac{rate_7}{rate_8} = 2^y \qquad \text{(Eq. 10)}$$

Find the average value of y. Round to the nearest integer.

Calculating the Specific Rate Constant

Calculate the specific rate constant, k, for each assignment in Part I, using the integer values determined for x and y.

$$k = \frac{rate}{[I^-]^x [S_2O_8^{2-}]^y} \qquad \text{(Eq. 11)}$$

where [I^-] = concentration of I^- in the reaction mixture after the reactants are mixed but before any reaction occurs.

[$S_2O_8^{2-}$] = concentration of $S_2O_8^{2-}$ in the reaction mixture after the reactants are mixed but before any reaction occurs.

III. Determining the Effect of Temperature on the Reaction Rate

1. Calculate the concentration of KI and $(NH_4)_2S_2O_8$ in the reaction mixture after the reactants are mixed but before any reaction occurs, as outlined in Part I.

2. Find the rate of the reaction for each temperature assignment.

$$rate = -\frac{\Delta[S_2O_8^{2-}]}{(\Delta t,\ sec)} \qquad \text{(Eq. 12)}$$

3. Calculate the specific rate constant for each temperature assignment as outlined under Calculating the Specific Rate Constant.

4. Plot the logarithm to the base 10 of the specific rate constant on the ordinate and $1/T$, where T is the kelvin temperature, on the abscissa. Draw the best straight line through the points. Using two points **on the line** (not necessarily two of your data points), calculate the slope of the line. From this calculated slope, use the point–slope method to establish the y-intercept. From the slope and the y-intercept, calculate E_a and A.

IV. Determining the Effect of $Cu(NO_3)_2$ on the Reaction Rate

1. Calculate the concentration of KI and $(NH_4)_2S_2O_8$ in the reaction mixture after the reactants are mixed but before any reaction occurs, as outlined in Part I.

2. Find the rate of the reaction for each assignment.

$$rate = -\frac{\Delta[S_2O_8^{2-}]}{(\Delta t,\ sec)} \qquad \text{(Eq. 12)}$$

3. Calculate the specific rate constant for each assignment.

4. Calculate the ratio of

$$\frac{k_{catalyzed}}{k_{uncatalyzed}}$$

for the four laboratory asignments

$$\frac{k_{13}}{k_1} \qquad \frac{k_{14}}{k_2} \qquad \frac{k_{15}}{k_3} \qquad \frac{k_{16}}{k_4}$$

Post-Laboratory Questions

(Use the space provided for the answer and additional paper if necessary.)

1. On graph paper, plot reaction temperature on the ordinate vs. (your reaction time)$^{-1}$ on the abscissa. What can you conclude about this reaction in relation to the rule of thumb that raising the temperature of a reaction by 10 °C doubles the rate?

(2) Sketch the Boltzmann distribution of the kinetic energy of molecules and show how that distribution changes with temperature. The fraction of molecules with enough energy to react successfully is equal to the area under each temperature curve from E_a to infinity. The rate constant is proportional to that area. Why does the specific rate constant increase with increasing temperature?

2. From the E_a and A obtained in this experiment, calculate the specific rate constant at 60 °C.

4. Compare the relative rates and rate constants calculated in Parts I and III. What is the effect of a trace of $Cu(NO_3)_2$ on the rate and rate constant?

3. E_a and A are constant and reasonably independent of temperature for a given reaction.

(1) From the Arrhenius equation, when temperature changes, what other variable changes?

Since Parts I and III are done at the same temperature, what can you conclude about E_a and A between Parts I and III (see Arrhenius equation)? Relate any effect on E_a to the Boltzmann distribution.

Data Sheet 1

I. Determining the effect of [I⁻]

	laboratory assignment number			
	1	2	3	4
concentration of reactants after mixing				
KI, M	_____	_____	_____	_____
$(NH_4)_2S_2O_8$, M	_____	_____	_____	_____
determination 1				
time of color change	_____	_____	_____	_____
initial time	_____	_____	_____	_____
reaction time, s	_____	_____	_____	_____
temperature, °C	_____	_____	_____	_____
determination 2				
time of color change	_____	_____	_____	_____
initial time	_____	_____	_____	_____
reaction time, s	_____	_____	_____	_____
temperature, °C	_____	_____	_____	_____
determination 3				
time of color change	_____	_____	_____	_____
initial time	_____	_____	_____	_____
reaction time, s	_____	_____	_____	_____
temperature, °C	_____	_____	_____	_____
average reaction time, s	_____	_____	_____	_____
rate of reaction, mol L⁻¹ s⁻¹	_____	_____	_____	_____
specific rate constant, k	_____	_____	_____	_____
order of reaction with respect to I⁻, x	_____	_____	_____	_____
average value of x		_____		

Data Sheet 2

I. Determining the Effect of $[S_2O_8{}^{2-}]$

	laboratory assignment number			
	5	6	7	8
concentration of reactants after mixing				
KI, *M*	_____	_____	_____	_____
$(NH_4)_2S_2O_8$, *M*	_____	_____	_____	_____
determination 1				
time of color change	_____	_____	_____	_____
initial time	_____	_____	_____	_____
reaction time, s	_____	_____	_____	_____
temperature, °C	_____	_____	_____	_____
determination 2				
time of color change	_____	_____	_____	_____
initial time	_____	_____	_____	_____
reaction time, s	_____	_____	_____	_____
temperature, °C	_____	_____	_____	_____
determination 3				
time of color change	_____	_____	_____	_____
initial time	_____	_____	_____	_____
reaction time, s	_____	_____	_____	_____
temperature, °C	_____	_____	_____	_____
average reaction time, s	_____	_____	_____	_____
rate of reaction, mol L^{-1} s^{-1}	_____	_____	_____	_____
specific rate constant, *k*	_____	_____	_____	_____
order of reaction with respect to $S_2O_8{}^{2-}$, *y*	_____	_____	_____	_____
average value of *y*		_____		

Data Sheet 3

II. Determining the Effect of Temperature

	laboratory assignment number			
	9	10	11	12
concentration of reactants after mixing				
KI, M	_____	_____	_____	_____
$(NH_4)_2S_2O_8$, M	_____	_____	_____	_____
determination 1				
time of color change	_____	_____	_____	_____
initial time	_____	_____	_____	_____
reaction time, s	_____	_____	_____	_____
temperature, °C	_____	_____	_____	_____
determination 2				
time of color change	_____	_____	_____	_____
initial time	_____	_____	_____	_____
reaction time, s	_____	_____	_____	_____
temperature, °C	_____	_____	_____	_____
determination 3				
time of color change	_____	_____	_____	_____
initial time	_____	_____	_____	_____
reaction time, s	_____	_____	_____	_____
temperature, °C	_____	_____	_____	_____
average reaction time, s	_____	_____	_____	_____
rate of reaction, mol L^{-1} s^{-1}	_____	_____	_____	_____
specific rate constant, k	_____	_____	_____	_____

II. Determining the Effect of Temperature (cont.)

laboratory assignment number

	9	10	11	12
$\log_{10} k$	_____	_____	_____	_____
temperature of reaction, °C	_____	_____	_____	_____
temperature of reaction, K	_____	_____	_____	_____
$1/T$, K^{-1}	_____	_____	_____	_____
slope of curve		_____		
energy of activation, J mol^{-1}		_____		
$\log_{10} A$		_____		
frequency factor, L mol^{-1} s^{-1}		_____		

Data Sheet 4

III.   Determining the Effect of $Cu(NO_3)_2$

	laboratory assignment number			
	13	14	15	16
concentration of reactants after mixing				
KI, *M*				
$(NH_4)_2S_2O_8$, *M*				
determination 1				
time of color change				
initial time				
reaction time, s				
temperature, °C				
determination 2				
time of color change				
initial time				
reaction time, s				
temperature, °C				
determination 3				
time of color change				
initial time				
reaction time, s				
temperature, °C				
average reaction time, s				
rate of reaction, mol L^{-1} s^{-1}				
specific rate constant, *k*				
k ratio ($k_{catalyzed}/k_{uncatalyzed}$)				

Pre-Laboratory Assignment

1. The following data were collected at 20 °C for the reaction:

$$acetone + H^+ + Br_2 \rightarrow bromoacetone$$

The time to produce a small but constant amount of bromoacetone was measured.

$[Br_2]$	[acetone]	$[H^+]$	time, s
0.010	0.25	0.10	35
0.010	0.13	0.10	71
0.020	0.25	0.10	36
0.020	0.13	0.40	18

What is the order of the reaction with respect to Br_2, acetone, and H^+?

2. If the small constant amount of bromoacetone produced corresponds to a concentration of $1.0 \times 10^{-4}\,M$, what is the rate constant and the rate law for this reaction?

3. What error will result if the KI buret (or pipet or cylinder) in Part I is rinsed with distilled water immediately before filling? Is this likely to result in an error in the determination of the order of the reaction with respect to $[I^-]$? Briefly explain.

4. What error occurred if the solution turns blue immediately on mixing? (**Hint**: What causes the delay in the change of color?)

5. For an activation energy of 40.0 kJ mol^{-1}, how much does the temperature have to change (from 20 $^{\circ}$C) to produce a doubling of the rate constant?

Standardizing a Hydrochloric Acid Solution

prepared by **Norman E. Griswold**, Nebraska Wesleyan University

---

Purpose of the Experiment

Determine the concentration of a 0.1M hydrochloric acid solution using pure sodium carbonate.

Background Information

Frequently, laboratory work involves the use of a solution of precisely known concentration. Such a solution is known as a **standard solution**. The process of determining its concentration is known as **standardizing** the solution. Standard solutions are used in titrations, a common form of analysis. **Titration** consists of the careful addition of one solution from a buret to another substance in a flask until all of the substance in the flask has reacted. The solution added from the buret is called the **titrant**. For every titration, there must be a way to determine when the titration reaction is complete. In acid–base titrations, this is accomplished by adding a small amount of a certain organic dye to the solution to be titrated. Several of these dyes, called **indicators**, are available. Indicators have one color in acidic solutions and another color in basic solutions. If the indicator is chosen correctly, the color change, called the **end point**, occurs upon addition of only a tiny amount of titrant.

If one of the reacting substances in a titration has been standardized, then data from the titration can be used to calculate the concentration of the other substance. Titration can be employed for acid–base reactions, oxidation–reduction reactions, complex formation reactions, and precipitation reactions. In this module, emphasis is placed on acid–base reactions. We will use solid sodium carbonate, a base, to standardize a hydrochloric acid solution by titration.

Selecting a Standard Substance

Once the reactants for a particular acid–base titration are chosen, one of the reactants must be standardized. The standardization can be accomplished in one of two ways.

First, one of the reactants may be a **primary standard**, a substance that is:

(1) available in very pure form,
(2) reasonably soluble,
(3) stable in the pure form and in solution,
(4) nonhygroscopic and easily dried, and
(5) a compound with molecular mass above 100.

A standard solution of a primary standard can be prepared by weighing the dry solute and making a solution of known volume. The mass of the solute and the volume of the solution can be used to calculate the exact concentration of the solution of a primary standard. The number of primary standards is quite limited

Copyright © 1988 by Chemical Education Resources, Inc., P.O. Box 357, 220 S. Railroad, Palmyra, Pennsylvania 17078
No part of this laboratory program may be reproduced or transmitted in any form or by any means, electronic or mechanical, including photocopying, recording, or any information storage and retrieval system, without permission in writing from the publisher. Printed in the United States of America

because few substances meet the criteria for primary standards. Two common primary standard bases are pure sodium carbonate and sodium borate decahydrate (borax). Potassium hydrogen phthalate and oxalic acid dihydrate are common primary standard acids.

Another way to standardize a solution is to determine its concentration by titration with a second solution that has been standardized. However, the standardization of the second solution probably was accomplished by reaction with a primary standard, so a primary standard is used either directly or indirectly in the preparation of standard solutions. Hydrochloric acid solutions may be standardized using the primary standard, sodium carbonate, or using a standardized sodium hydroxide solution. Sodium hydroxide solutions may be standardized using the primary standard, potassium hydrogen phthalate, or using a standardized hydrochloric acid solution. Standardization with a primary standard is best.

In this module, a hydrochloric acid solution is standardized using the primary standard, sodium carbonate. We will prepare a 0.1M HCl solution to be standardized. Then we will prepare solutions of the primary standard and titrate them with the prepared HCl solution so that we may determine the exact concentration of the HCl solution.

Preparing Solutions

A dilute HCl solution is prepared by dilution of a more concentrated solution of the substance. The commercially available concentrated acid is approximately 12M, but its concentration isn't sufficiently constant to permit its use for calculating the concentration of the dilute solution directly. For this reason, a solution of approximately the desired concentration of HCl is prepared and then standardized with a primary standard such as sodium carbonate.

A solution of the primary standard is prepared from the solid. The solid must be dried in an oven for at least an hour and cooled in a desiccator before use. Then a small sample is weighed with an analytical balance into a clean Erlenmeyer flask and dissolved in some distilled or deionized water. The mass of the solute is used in calculations, so it is unnecessary to know the volume or the concentration of this solution. After addition of an indicator, this solution is ready for titration with the solution to be standardized. Several more samples are prepared and titrated in the same way until you have at least three samples with consistent calculated results.

Storing Standard Solutions

Once solutions are standardized, it may be necessary to store them for some period of time before they are used for other applications. In this case, you must take some precautions. For example, solutions should not be stored in volumetric flasks. If the solution wets a ground-glass stopper, evaporation may cause the stopper to "freeze" into the top of the flask. This will make it difficult or impossible to remove the stopper. The solution should be transferred to a clean, dry bottle for storage. For any container, you must guard against evaporation of solvent. If the solvent evaporates, the actual concentration of the solute increases. It is also important to avoid contamination. Contamination is minimized by using only very clean equipment and by never returning any unused portion of standard solution to the storage container. These are the main precautions for storage of dilute HCl solutions and primary standard solutions. Certain solutions require additional precautions. For example, strong base solutions, such as sodium hydroxide, must be protected from carbon dioxide in the air.

Titrating and Detecting on End Point

The procedures for cleaning and using burets and pipets are given in Experiment 4. The titrant may splatter slightly as it is added from the buret. For this reason, it is wise to rinse the inner surface of the Erlenmeyer flask with distilled or deionized water before the end point is reached.

The approach of the titration end point is signaled by a very temporary appearance of the end-point color where the titrant first comes into contact with the solution in the flask. But this end-point color disappears when the flask is swirled. As the end point is approached more closely, these temporary flashes of color persist longer, and more swirling of the flask is needed to dispel them. As the color persists longer, smaller amounts of titrant should be added, always with constant swirling. Just prior to the end point, add the titrant no more than one drop at a time. The end point is reached when the end point color first appears faintly throughout the solution and persists even after swirling. For this experiment, the indicator bromcresol green is used. Bromcresol green is yellow in acid solution and blue in base solution. As it passes from one color to the other, it appears green. Green is the end-point color for this titration.

Titrating Na_2CO_3 with HCl

In this experiment, the dilute hydrochloric acid solution is referred to as the titrant because it is the solution dispensed from the buret. The titration reaction is shown in Equation 1.

$$2\ H_3O^+(aq) + CO_3^{2-}(aq) \rightleftarrows [H_2CO_3] \rightleftarrows$$
$$3\ H_2O(l) + CO_2(g) \quad \text{(Eq. 1)}$$

During the titration, the solution absorbs a significant amount of carbon dioxide produced by the reaction shown in Equation 1. The carbon dioxide is in equilibrium with carbonic acid, as shown in Equation 2.

$$CO_2(g) + H_2O(l) \rightleftarrows H_2CO_3(aq) \quad \text{(Eq. 2)}$$

Consequently, the titrant and the carbonic acid contribute to the acidity of the solution and produce a premature end point. To prevent this, the titration is interrupted just as the indicator begins to change color. The solution is boiled for a few minutes to expel the carbon dioxide produced by the titration reaction. This causes the carbonic acid concentration to become insignificant and the indicator changes back to the color it had in the sodium carbonate solution before the titration began. After the solution is cooled to room temperature, the titration is continued to completion.

The stoichiometry of the reaction in Equation 1 indicates that two moles of hydrochloric acid are required to react with one mole of sodium carbonate. This is represented mathematically in Equation 3.

number of moles of HCl =
$$2\ (\text{number of moles of } Na_2CO_3) \quad \text{(Eq. 3)}$$

We can make Equation 3 more useful if we use

$$\text{number of moles of HCl} = (M_{HCl})(V_{HCl}) \quad \text{(Eq. 4)}$$

where

$$M_{HCl} = \text{molarity of HCl in mol L}^{-1}$$
$$V_{HCl} = \text{volume of HCl in L}$$

The molecular mass of Na_2CO_3 is 106.0 g mol^{-1}, so we can use

$$\frac{\text{number of moles}}{\text{of } Na_2CO_3} = \frac{\text{mass of } Na_2CO_3,\ g}{106.0\ g\ mol^{-1}}$$
$$\text{(Eq. 5)}$$

and substitute Equations 4 and 5 into Equation 3 to give

$$(M_{HCl})\ (V_{HCl}) = \frac{2\ (\text{mass of } Na_2CO_3,\ g)}{106.0\ g\ mol^{-1}}$$
$$\text{(Eq. 6)}$$

The volume of HCl, in liters, is calculated using

$$V_{HCl} = (\text{mL HCl})\left(\frac{1\ L}{1000\ mL}\right) \quad \text{(Eq. 7)}$$

To calculate the molarity of the hydrochloric acid solution, Equation 6 is rearranged, giving Equation 8.

$$M_{HCl} = \frac{2(\text{mass of } Na_2CO_3,\ g)}{(106.0\ g\ mol^{-1})\ (V_{HCl},\ L)} \quad \text{(Eq. 8)}$$

Example. A 0.1338-g sample of solid Na_2CO_3 is dissolved in water and titrated with exactly 24.63 mL of dilute HCl solution. What is the concentration of HCl solution?

The volume of HCl must be converted to liters (Equation 7).

$$V_{HCl} = (24.63\ mL)\left(\frac{1\ L}{1000\ mL}\right) = 2.463 \times 10^{-2}\ L$$

Then, the concentration of the HCl solution is calculated, using Equation 8.

$$M_{HCl} = \frac{2(0.1338\ g)}{(106.0\ g\ mol^{-1})\ (2.463 \times 10^{-2}\ L)}$$
$$= 0.1025\ mol\ L^{-1}$$

Procedure

Chemical Alert
12*M* hydrochloric acid—very corrosive

Caution: Wear departmentally approved eye protection while doing this experiment.

I. Preparing Approximately 0.1*M* HCl

Caution: Concentrated HCl is a very corrosive substance. Prevent eye and skin contact. Avoid inhaling the vapors. Considerable heat is evolved when concentrated HCl is added to water.

Add about 500 mL of distilled or deionized water to a one-liter glass-stoppered bottle. Measure in a 10-mL graduated cylinder about 8.3 mL of concentrated hydrochloric acid (12*M*). Add the hydrochloric acid to the 500 mL of water in the glass-stoppered bottle. Dilute the solution to approximately one liter with distilled or deionized water. Stopper the bottle and swirl the contents thoroughly to mix the solution. Let the solution stand until to comes to room temperature.

Figure 1   *Handling a weighing bottle*

II.   Preparing Sodium Carbonate Solutions for Titration

> **Note:**   Do not put the weighing bottle in the oven with the stopper in the bottle.

Place approximately 2 g of the primary standard sodium carbonate in a weighing bottle. Dry the sample for 2 hr in an oven set at 110 °C. After drying, use crucible tongs to remove the hot weighing bottle from the oven. If crucible tongs are not available, fold a clean cloth or a strip of paper around the bottle as shown in Figure 1.

Never handle the weighing bottle with your bare fingers after it is dried. Allow the bottle to stand for about 2 min and then place the bottle in a desiccator for at least 30 min for further cooling.

Weigh the bottle and its contents to the nearest 0.1 mg (0.0001 g) on an analytical balance (refer to Experiment 3). This is the initial mass of the bottle and contents. Record this mass on your Data Sheet or in your laboratory notebook. Carefully pour about 0.10 to 0.15 g of sodium carbonate into a clean 250-mL Erlenmeyer flask. Weigh the weighing bottle and the remaining contents again to the nearest 0.1 mg. Record this final mass on your Data Sheet or in your laboratory notebook.

Dissolve the sample in the flask in about 40 mL of distilled or deionized water. Add three drops of bromcresol green indicator solution to the flask. The indicator should make the solution become blue. Carry this sample through the rest of the procedure before starting on the next sample of sodium carbonate.

III.   Standardizing the 0.1 *M* HCl solution

After cleaning the buret (see Experiment 4), rinse the buret with three 5-mL portions of the hydrochloric acid solution to be standardized. Make certain that the rinse solution comes in contact with the entire inner surface of the buret. Drain the rinse solution through the tip of

the buret and discard the rinse solution. Close the buret stopcock and fill the buret to a level above the top graduation mark with hydrochloric acid solution. Open the stopcock briefly to be certain that the buret tip is filled with solution. If the meniscus is still above the top graduation mark, open the stopcock and drain solution until the level reaches the calibrated portion of the buret. Read the buret carefully to the nearest 0.01 mL. Record this initial buret reading on your Data Sheet or in your laboratory notebook.

Place the Erlenmeyer flask containing the sodium carbonate sample and indicator under the buret. Lower the buret tip until it is below the lip of the flask, as shown in Figure 2. Swirl the flask with one hand and control the stopcock with the other hand. Titrate the sodium carbonate solution carefully, as described in Experiment 4, until the bromcresol green indicator just begins to change from blue to green throughout the solution. This is the premature end point described in the Background Information section of this module.

Place the flask and contents on a wire gauze supported on a ring attached to a ring stand. Boil the solution in the flask gently for a few minutes to remove dissolved carbon dioxide. The bromcresol green indicator should change back to a blue color.

Cool the flask and contents. Rinse the inner surface of the flask with a stream of distilled or deionized water from a wash bottle. Then complete the titration by adding the titrant dropwise from the buret. Read the final buret volume to the nearest 0.01 mL. Record this volume on your Data Sheet or in your laboratory notebook.

Weigh, dissolve, and titrate additional samples of sodium carbonate until you obtain three results for the calculated concentration of the HCl solution that agree within $5 \times 10^{-4} M$.

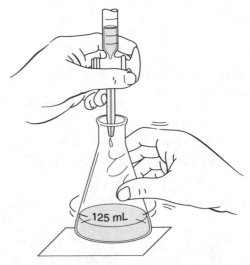

Figure 2   *Positioning the buret for titration*

When all of the titrations are completed, drain the buret and rinse it thoroughly with distilled or deionized water. Rinse all other containers and store for future use.

The standardized HCl can be used as a standard solution in titrations of base solutions. Stopper the bottle tightly and label it clearly with the identity, concentration, and date of standardization. If tightly stoppered, this solution will have a long shelf life.

Caution: Wash your hands thoroughly with soap or detergent before leaving the laboratory.

Calculations

(Do the following calculations for each determination. Record the results on your Data Sheet or in your laboratory notebook.)

1. Calculate the mass of Na_2CO_3 titrated.

mass of Na_2CO_3, g =
$$\text{(initial mass, g)} - \text{(final mass, g)} \quad \text{(Eq. 9)}$$

2. Calculate the volume, in liters, of the HCl used in the titration.

volume of HCl, mL =
$$\text{(final reading, mL)} - \text{(initial reading, mL)} \quad \text{(Eq. 10)}$$

Complete the calculation using Equation 7.

3. Calculate the molarity of the HCl solution to four significant digits, using Equation 8. If you do not have three results within $5 \times 10^{-4}M$, titrate additional samples of Na_2CO_3 until you do.

4. Use the molarities within $5 \times 10^{-4}M$ of each other to calculate the average molarity of the HCl solution. Label the bottle of dilute HCl solution with the average molarity and the date of standardization.

Post-Laboratory Questions

(Use the spaces provided for the answers and additional paper if necessary.)

1. For each of the following errors, indicate whether the calculated molarity of hydrochloric acid would be higher or lower than the real value or unaffected. Briefly explain your answer.

   (1) The sodium carbonate is not dried before use.

   (2) The sodium carbonate is dissolved in 80 mL of water (rather than the 40 mL indicated in the procedure).

   (3) The buret, wet with water, is not rinsed with HCl solution before filling.

   (4) The titration solution is not boiled at the appropriate step in the procedure.

2. Why should a primary standard have a high molecular mass?

3. When you finished the titration, you may have noticed that the green end-point color changed gradually to yellow after several minutes. What is a possible cause for this change?

Data Sheet

	first	second	determination third	fourth	fifth
final mass of container and Na_2CO_3, g	_____	_____	_____	_____	_____
initial mass of container, g	_____	_____	_____	_____	_____
mass of Na_2CO_3, g	_____	_____	_____	_____	_____
final buret reading, mL	_____	_____	_____	_____	_____
initial buret reading, mL	_____	_____	_____	_____	_____
volume of HCl, mL	_____	_____	_____	_____	_____
molarity of HCl solution, mol L^{-1}	_____	_____	_____	_____	_____
average molarity of HCl solution, mol L^{-1}			_____		

Pre-Laboratory Assignment

1. Review Experiments 3 and 4 for a discussion of volume and mass measurements and the use of a buret.

2. What are five criteria for a primary standard substance?

3. The exact concentration of a dilute HCl solution is determined by reaction with a primary standard. Why can't we just dilute the concentrated reagent carefully and calculate the concentration of the dilute solution?

4. What might happen if the stopper were left in the weighing bottle containing the primary standard when it is put in the hot oven?

5. (1) What is the name of the indicator used for this titration?

    (2) What color is the indicator in a sodium carbonate solution?

    (3) What color should appear at the end point?

6. Why is the titration interrupted and the solution boiled and cooled before the end point is reached?

7. A 0.1492-g sample of sodium carbonate is dissolved in distilled or deionized water. This solution is titrated with a dilute HCl solution. The initial buret reading is 0.06 mL and the final buret reading is 27.38 mL.

(1) What volume of HCl, in liters, is added?

(3) If the results of three additional titrations are 0.1039, 0.1033, and 0.1028*M*, what is the average molarity of the HCl solution?

_____
answer

_____
answer

(2) What is the molarity of the HCl solution?

_____
answer

Standardizing A Sodium Hydroxide Solution

prepared by **Norman E. Griswold**, Nebraska Wesleyan University

---

Purpose of the Experiment

Determine the concentration of a 0.1 M sodium hydroxide solution, using a standard hydrochloric acid solution.

Background Information

Frequently, laboratory work involves the use of a solution of precisely known concentration. Such a solution is known as a **standard solution**. The process of determining its concentration is known as **standardizing** the solution. Standard solutions are used in titrations, a common form of analysis. **Titration** consists of a careful addition of one solution from a buret to another substance in a flask until all of the substance in the flask has reacted. The solution added from the buret is called the **titrant.** For every titration, there must be a way to determine when the titration reaction is complete. In acid–base titrations, this is accomplished by adding a small amount of a certain organic dye to the solution to be titrated. Several of these dyes, called **indicators**, are available. Indicators have one color in acidic solutions and another color in basic solutions. If the indicator is chosen correctly, the color change, called the **end point**, occurs upon addition of only a tiny amount of titrant.

If one of the reacting substances in a titration has been standardized, then data from the titration can be used to calculate the concentration of the other substance. Titration can be employed for acid–base reac-

tions, oxidation–reduction reactions, complex formation reactions, and precipitation reactions. In this module, emphasis is placed on acid–base reactions. We will use a hydrochloric acid solution to standardize a sodium hydroxide solution by titration.

Selecting a Standard Substance

Once the reactants for a particular acid–base titration are chosen, one of the reactants must be standardized. The standardization can be accomplished in one of two ways.

First, one of the reactants may be a **primary standard**, a substance that is:

(1) available in very pure form,
(2) reasonably soluble,
(3) stable in the pure form and in solution,
(4) nonhygroscopic and easily dried, and
(5) a compound with molecular mass above 100.

A standard solution of a primary standard can be prepared by weighing the dry solute and making a solution of known volume. The mass of the solute and the volume of the solution can be used to calculate the exact concentration of the solution of a primary standard. The number of primary standards is quite limited

Copyright © 1988 by Chemical Education Resources, Inc., P.O. Box 357, 220 S. Railroad, Palmyra, Pennsylvania 17078
No part of this laboratory program may be reproduced or transmitted in any form or by any means, electronic or mechanical, including photo-copying, recording, or any information storage and retrieval system, without permission in writing from the publisher. Printed in the United States of America

because few substances meet the criteria for primary standards. Two common primary standard bases are pure sodium carbonate and sodium borate decahydrate (borax). Potassium hydrogen phthalate and oxalic acid dihydrate are common primary standard acids.

Another way to standardize a solution is to determine its concentration by titration with a second solution that has been standardized. However, the standardization of the second solution probably was accomplished by reaction with a primary standard, so a primary standard is used either directly or indirectly in the preparation of standard solutions. Hydrochloric acid solutions may be standardized using the primary standard, sodium carbonate, or using a standardized sodium hydroxide solution. Sodium hydroxide solutions may be standardized using the primary standard, potassium hydrogen phthalate, or using a standardized hydrochloric acid solution.

In this module, a sodium hydroxide solution is standardized using a standardized hydrochloric acid solution. We will prepare a 0.1M NaOH solution to be standardized. Then we will prepare solutions of the standard HCl and titrate them with the prepared NaOH solution so that we may determine the exact concentration of the NaOH solution.

Preparing Solutions

A dilute NaOH solution is prepared by dilution of a more concentrated solution of the substance. The concentrated base usually available is approximately 6M. It is prepared from the very hygroscopic solid NaOH. The concentration of this hygroscopic solid is not sufficiently constant to permit its use for calculating the concentration of the dilute solution directly. For this reason, a solution of approximately the desired concentration of NaOH is prepared and then standardized with another substance. Strong base solutions absorb carbon dioxide from the air. The carbon dioxide reacts with the base, and changes its concentration

$$CO_2(g) + 2\ OH^-(aq) \rightleftarrows CO_3^{2-}(aq) + H_2O(l) \qquad \text{(Eq. 1)}$$

The carbonate produced by this reaction may precipitate in very concentrated sodium hydroxide solutions, but should remain soluble in more dilute solutions. To prevent reaction of the base with carbon dioxide, distilled or deionized water is boiled to expel carbon dioxide before it is used to dilute a sodium hydroxide solution. If the 6M solution has solid carbonate apparent in the bottom of the container, the clear carbonate-free sodium hydroxide is decanted and diluted with freshly boiled and cooled distilled or deionized water.

A solution of standard HCl is prepared and standardized with sodium carbonate using the procedure in Experiment 29. The standardized solution is pipetted into an Erlenmeyer flask. After addition of an indicator, this solution is ready for titration with the solution to be standardized. Several more samples are prepared and titrated in the same way until you have at least three samples with consistent calculated results.

Storing Standard Solutions

Once solutions are standardized, it may be necessary to store them for some period of time before they are used for other applications. In this case, you must take some precautions. For example, solutions should not be stored in volumetric flasks. If the solution wets a ground-glass stopper, evaporation may cause the stopper to "freeze" into the top of the flask. This will make it difficult or impossible to remove the stopper. The solution should be transferred to a clean, dry bottle for storage. For any container, you must guard against evaporation of solvent. If the solvent evaporates, the actual concentration of the solute increases. It is also important to avoid contamination. Contamination is minimized by using only very clean equipment and by never returning any unused portion of standard solution to the storage container. These are the main precautions for storage of dilute HCl solutions and primary standard solutions.

A strong base solution such as sodium hydroxide requires additional precautions. First, sodium hydroxide solutions must be protected from carbon dioxide in the air. Second, sodium hydroxide gradually etches glass. Therefore, sodium hydroxide solutions should not be stored in glass containers but in screw-cap polyethylene bottles. Sodium hydroxide solutions have only a short shelf life.

Titrating and Detecting on End Point

The procedures for cleaning and using burets and pipets are given in Experiment 4. The titrant may splatter slightly as it is added from the buret. For this reason, it is wise to rinse the inner surface of the Erlenmeyer flask with distilled or deionized water before the end point is reached.

The approach of the titration end point is signaled by a very temporary appearance of the end-point color where the titrant first comes into contact with the solution in the flask. But this end-point color disappears when the flask is swirled. As the end point is approached more closely, these temporary flashes of color persist longer and more swirling of the flask is needed to dispel them. As the color persists longer, smaller amounts of titrant should be added, always with constant swirling. Just prior to the end point, add the titrant no more than one drop at a time. The end point is reached when the end-point color first appears faintly throughout the solution and persists even after swirling. For this experiment, the indicator phenolphthalein is used. Phenolphthalein is colorless in acid solution and red in base solution. As it passes from colorless to red, it appears pale pink. Pale pink is the end-point color for this titration.

Titrating HCl with NaOH

In this experiment, the dilute sodium hydroxide solution is referred to as the titrant, because it is the solution dispensed from the buret. The titration reaction is shown in Equation 2.

$$H_3O^+(aq) + OH^-(aq) \rightleftarrows 2\ H_2O(l) \quad \text{(Eq. 2)}$$

The stoichiometry of the reaction in Equation 2 indicates that one mole of hydrochloric acid is required to react with one mole of sodium hydroxide. This is represented mathematically in Equation 3.

number of moles of NaOH =
$$\text{number of moles of HCl} \quad \text{(Eq. 3)}$$

We can make Equation 3 more useful if we use

number of moles of NaOH =
$$(M_{NaOH})(V_{NaOH}) \quad \text{(Eq. 4)}$$

where

M_{NaOH} = molarity of NaOH in mol L^{-1}

V_{NaOH} = volume of NaOH in liters

Similarly we can use

$$\text{number of moles of HCl} = (M_{HCl})(V_{HCl}) \quad \text{(Eq. 5)}$$

and substitute Equations 4 and 5 into Equation 3 to give

$$(M_{NaOH})(V_{NaOH}) = (M_{HCl})(V_{HCl}) \quad \text{(Eq. 6)}$$

The volumes, in liters, are calculated using

$$V_{NaOH} = (\text{mL NaOH})\left(\frac{1\ L}{1000\ mL}\right) \quad \text{(Eq. 7a)}$$

$$V_{HCl} = (\text{mL HCl})\left(\frac{1\ L}{1000\ mL}\right) \quad \text{(Eq. 7b)}$$

To calculate the molarity of the sodium hydroxide solution, Equation 6 is rearranged, giving Equation 8.

$$M_{NaOH} = \frac{(M_{HCl})(V_{HCl},\ L)}{(V_{NaOH},\ L)} \quad \text{(Eq. 8)}$$

Example. A 25.00-mL sample of 0.1030M HCl is titrated with exactly 24.42 mL of dilute NaOH solution. What is the concentration of the NaOH solution? The volumes of NaOH and HCl must be converted to liters (Equation 7).

$$V_{NaOH} = (24.42\ mL)\left(\frac{1\ L}{1000\ mL}\right) = 2.442 \times 10^{-2}\ L$$

$$V_{HCl} = (25.00\ mL)\left(\frac{1\ L}{1000\ mL}\right) = 2.500 \times 10^{-2}\ L$$

Then, the concentration of the NaOH solution is calculated, using Equation 8.

$$(M_{NaOH}) = \frac{(0.1030M)(2.500 \times 10^{-2}\ L)}{(2.442 \times 10^{-2}\ L)}$$

$$= 0.1054\ \text{mol L}^{-1}$$

Procedure

> ### Chemical Alert
> 0.1M hydrochloric acid—corrosive
> 6M sodium hydroxide—very corrosive

> *Caution:* Wear departmentally approved eye protection while doing this experiment.

I. Preparing Approximately 0.1M NaOH

> *Caution:* Concentrated NaOH is very caustic. Handle with care. Prevent eye and skin contact. In case of a spill, notify your laboratory instructor immediately.

Place about 1200 mL of distilled or deionized water in a 2-L Erlenmeyer flask. Boil the water for about 3 min. Place a small watch glass over the top and then let the water cool. Measure in a 10-mL graduated cylinder about 17 mL of concentrated NaOH (6M). Pour the sodium hydroxide into a 1-L polyethylene bottle. Add about 1000 mL of the freshly boiled and cooled water to the bottle. Stopper the bottle and swirl the contents thoroughly to mix the solution. Let the solution stand until it comes to room temperature.

II. Preparing Hydrochloric Acid Solutions for Titration

Place approximately 125 mL of standard 0.1M HCl solution in a clean, dry 250-mL beaker.

> **Note:** The beaker must be dry or the water in the beaker will dilute the hydrochloric acid and change the concentration of the solution.

Record the exact molarity of the standard HCl solution on your Data Sheet or in your laboratory notebook.

Rinse a clean 25-mL volumetric pipet at least two times with 5-mL portions of the standard HCl solution (refer to Experiment 4 for the technique of cleaning and using volumetric pipets). Discard the rinse solutions, following the directions of your laboratory instructor.

Carefully pipet 25.00 mL of the acid solution into a 250-mL Erlenmeyer flask, holding the tip of the pipet against the inner surface of the flask to avoid splatter. Add about 40 mL of distilled or deionized water to the acid solution to give a total volume sufficient for easy swirling and observation of the color change. Add three drops of phenolphthalein indicator solution to the acid solution and swirl the flask to mix thoroughly. Phenolphthalein should be colorless in this solution. Carry this sample through the rest of the procedure before starting on the next sample of hydrochloric acid.

III. Standardizing the 0.1M NaOH Solution

After cleaning the buret (refer to Experiment 4), rinse the buret with three 5-mL portions of the sodium hydroxide solution to be standardized. Make certain that the rinse solution comes in contact with the entire inner surface of the buret. Drain the rinse solution through the tip of the buret and discard the rinse solution, following the directions of your laboratory instructor. Close the buret stopcock and fill the buret to a level above the top graduation mark with sodium hydroxide solution. Open the stopcock briefly to be certain that the buret tip is filled with solution. If the meniscus is still above the top graduation mark, open the stopcock and drain solution until the level reaches the calibrated portion of the buret. Read the buret carefully to the nearest 0.01 mL. Record this initial buret reading on your Data Sheet or in your labortory notebook.

Place the Erlenmeyer flask containing the hydrochloric acid sample and indicator under the buret.

Lower the buret tip until it is into the mouth of the flask as shown in Figure 1.

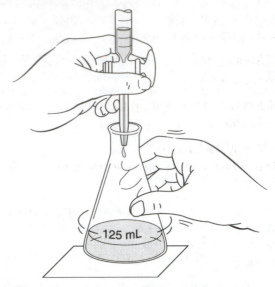

125 mL

Figure 1   *Positioning the buret for titration*

Titrate the hydrochloric acid solution carefully, (refer to Experiment 4). Swirl the flask with one hand and control the stopcock with the other hand. The phenolphthalein indicator should show flashes of pale pink as the end point is approached. When the pink begins to appear, rinse the inner surface of the flask with a stream of distilled or deionized water from a wash bottle. Then complete the titration by adding the titrant dropwise from the buret. Read the final buret volume to the nearest 0.01 mL. Record this volume on your Data Sheet or in your laboratory notebook.

Pipet, dilute, and titrate additional samples of hydrochloric acid until you obtain three results for the calculated concentration of the NaOH solution that agree within $5 \times 10^{-4}M$.

When all of the titrations are completed, drain the buret and rinse it thoroughly with distilled or deionized water. Rinse all other containers and store for future use.

The standardized NaOH can be used as a standard solution in titrations of acid solutions. Close the bottle tightly and label it clearly with the identity, concentration, and date of standardization. This solution has a short shelf life because of reaction with carbon dioxide, but this can be extended by keeping the bottle tightly capped.

> **Caution:** Wash your hands thoroughly with soap or detergent before leaving the laboratory.

Calculations

(Do the following calculations for each determination. Record the results on your Data Sheet or in your laboratory notebook.)

1. Calculate the volume, in liters, of HCl titrated, using Equation 7.

2. Calculate the volume, in liters, of the NaOH used in the titration.

volume of NaOH, mL =
   (final reading, mL) – (initial reading, mL)   (Eq. 9)

Complete the calculation using Equation 7.

3. Calculate the molarity of the NaOH solution to four significant digits, using Equation 8. If you do not have three results within $5 \times 10^{-4} M$, titrate additional samples of HCl until you do.

4. Use your molarity determinations that agree within $5 \times 10^{-4} M$ to calculate the average molarity of the NaOH solution. Label the bottle of dilute NaOH solution with the average molarity and the date of standardization.

Post-Laboratory Questions

(Use the spaces provided for the answers and additional paper if necessary.)

1. For each of the following errors, indicate whether the calculated molarity of NaOH would be higher or lower than the real value, or unaffected.
   (1) The water for the NaOH solution is not boiled and cooled before the solution is prepared.

   (2) The original HCl solution is obtained in a wet beaker.

   (3) The hydrochloric acid is diluted with 80 mL of water (rather than the 40 mL indicated in the procedure).

   (4) The buret, wet with water, is not rinsed with NaOH solution before filling.

2. Suppose you neglected to make the initial buret reading in Pre-Laboratory Assignment Question 6. What would the calculated molarity of NaOH be in this situation?

_____
answer

3. When you finished titrating, you may have noticed that the pale pink end-point color changed gradually to colorless after several minutes. What is a possible cause for this change?

Data Sheet

molarity of HCl _____

			determination		
	first	second	third	fourth	fifth
final buret reading, mL	_____	_____	_____	_____	_____
initial buret reading, mL	_____	_____	_____	_____	_____
volume of titrant, mL	_____	_____	_____	_____	_____
molarity of NaOH solution, mol L^{-1}	_____	_____	_____	_____	_____
average molarity of NaOH solution, mol L^{-1}			_____		

Pre-Laboratory Assignment

1. Review Experiment 4 for a discussion of titration and the use of a pipet and a buret.

2. What are two special precautions for storing sodium hydroxide solutions?

3. The exact concentration of a dilute NaOH solution is determined by reaction with another substance. Why can't we just dilute the concentrated reagent carefully and calculate the concentration of the dilute solution?

4. Why must standard HCl be put in a dry beaker?

5. (1) What is the name of the indicator used for this titration?

(2) What color is the indicator in a hydrochloric acid solution?

(3) What color should appear at the end point of this titration?

6. A 25.00-mL sample of 0.1010M hydrochloric acid is diluted with water. This solution is titrated with a dilute NaOH solution. The initial buret reading is 0.06 mL and the final buret reading is 23.58 mL.

(1) What are the volumes of NaOH and HCl, in liters?

NaOH, L _____
answer

HCl, L _____
answer

(2)   What is the molarity of the NaOH solution?

(3)   If the results of three additional titrations are 0.1066, 0.1072, and 0.1077M, what is the average molarity of the NaOH solution?

_____
answer

_____
answer

Evaluations of Commercial Antacids

prepared by **David N. Bailey**, Illinois Wesleyan University

---

Purpose of the Experiment

Determine the number of moles of hydrochloric acid required to neutralize a commercial antacid tablet. Calculate the weight and cost effectiveness of the tablet. Compare the cost effectiveness of different brands of antacids.

Background Information

Antacid preparations are sold over-the-counter for relieving heartburn or acid indigestion. These symptoms are due to an excess of stomach acid, conveniently thought of as hydrochloric acid. Antacids contain basic substances that function either by neutralizing the excess acid or by acting as buffers in the stomach. The types of bases commonly found in antacids are metallic hydroxides, carbonates, and hydrogen carbonates.

The neutralization of magnesium hydroxide, a typical metallic hydroxide, is shown in Equation 1.

$$Mg(OH)_2(s) + 2\,H_3O^+(aq) \rightleftarrows$$
$$Mg^{2+}(aq) + 4\,H_2O(l) \quad \text{(Eq. 1)}$$

In this reaction, two moles of hydronium ions react with two moles of hydroxide ions. Each mole of hydroxide ions neutralizes one mole of hydronium ions.

When calcium carbonate, a typical metallic carbonate, neutralizes hydrochloric acid, a two-step process occurs as shown in Equations 2 and 3.

$$CaCO_3(s) + 2\,H_3O^+(aq) \rightleftarrows$$
$$Ca^{2+}(aq) + H_2CO_3(aq) + 2\,H_2O(l) \quad \text{(Eq. 2)}$$

$$H_2CO_3(aq) \rightleftarrows CO_2(g) + H_2O(l) \quad \text{(Eq. 3)}$$

In the reaction in Equation 2, one mole of carbonate ions neutralizes two moles of hydronium ions.

An example of the buffering action of antacids is shown in Equations 4, 5, and 6. Either calcium carbonate or sodium hydrogen carbonate can be involved in a HCO_3^-/CO_3^{2-} buffer system. The important feature of a buffer is that it has components present to react with either added acids or bases. Equation 4 shows the reaction of the CO_3^{2-} ion with acid. Equation 5 shows the reaction of the HCO_3^- ion with a base. Either of these reactions causes the HCO_3^-/CO_3^{2-} ratio to change by only a small amount, resulting in a very minor change in pH. If all of the CO_3^{2-} ion is used by reaction with acid, or if $NaHCO_3$ is used as the antacid, then the HCO_3^- ion can react with additional acid, as shown in Equation 6.

$$CO_3^{2-}(aq) + H_3O^+(aq) \rightleftarrows$$
$$HCO_3^-(aq) + H_2O(l) \quad \text{(Eq. 4)}$$

$$HCO_3^-(aq) + OH^-(aq) \rightleftarrows$$
$$CO_3^{2-}(aq) + H_2O(l) \quad \text{(Eq. 5)}$$

$$HCO_3^-(aq) + H_3O^+(aq) \rightleftarrows$$
$$CO_2(g) + 2\,H_2O(l) \quad \text{(Eq. 6)}$$

One method of evaluating antacid preparations is to determine the amount of hydrochloric acid that will react

---

Copyright © 1986 by Chemical Education Resources, Inc., P.O. Box 357, 220 S. Railroad, Palmyra, Pennsylvania 17078
No part of this laboratory program may be reproduced or transmitted in any form or by any means, electronic or mechanical, including photocopying, recording, or any information storage and retrieval system, without permission in writing from the publisher. Printed in the United States of America

with a known amount of the antacid. The analytical procedure used is called **back-titration**. This procedure involves adding excess hydrochloric acid to an antacid tablet. The excess acid is then titrated with standard sodium hydroxide. By subtraction, the amount of acid used to neutralize the antacid can be determined.

The number of moles of HCl neutralized by a tablet is calculated using Equation 7:

$$N = (A - B) \qquad \text{(Eq. 7)}$$

where N = number of moles of HCl neutralized by a tablet
A = number of moles of HCl added to a tablet
B = number of moles of NaOH required for the back-titration of the excess HCl

The number of moles of HCl added to a tablet is found using Equation 8:

$$A = (V_{HCl})(M_{HCl}) \qquad \text{(Eq. 8)}$$

where A = number of moles of HCl added to a tablet
V_{HCl} = volume of HCl solution added to a tablet, in liters
M_{HCl} = molarity of HCl solution, in moles per liter

The number of moles of NaOH required in the back-titration is found using Equation 9:

$$B = (V_{NaOH})(M_{NaOH}) \qquad \text{(Eq. 9)}$$

where B = number of moles of NaOH required for the back-titration of the excess acid
V_{NaOH} = volume of NaOH solution required in the back-titration of the excess acid, in liters
M_{NaOH} = molarity of NaOH solution, in moles per liter

In order to compare different brands of antacids, the weight and cost effectiveness of each preparation are calculated. The weight effectiveness of an antacid is determined from Equation 10:

$$E = (N/W) \qquad \text{(Eq. 10)}$$

where E = weight effectiveness of a tablet, in number of moles of HCl neutralized per gram
N = number of moles of HCl neutralized by a tablet
W = mass of tablet, in grams

The cost effectiveness of a tablet is calculated using Equation 11:

$$C = (N/P) \qquad \text{(Eq. 11)}$$

where C = cost effectiveness of a tablet, in number of moles of HCl neutralized per penny
N = number of moles of HCl neutralized
P = cost of one tablet, in pennies

When you purchase an antacid, any consideration of weight or cost effectiveness must be made with caution. Some ingredients that contribute to cost effectiveness are harmful to persons with certain physical conditions.

In this experiment, you will dissolve a commercial antacid tablet in excess standard HCl solution. The excess HCl will be back-titrated with a standard NaOH solution.

From the titration data, you will calculate the number of moles of HCl required to neutralize the antacid tablet. The weight and cost effectiveness of the antacid will be calculated.

You will exchange data with your classmates who analyzed other brands of antacids. All of the preparations used in this investigation will be evaluated on the basis of their cost effectiveness.

Procedure

Caution: Wear departmentally approved eye-protection glasses while doing this experiment.

Note: Avoid touching the antacid tablets with your bare fingers. Any oil or dirt transferred from your fingers to a tablet will adversely affect the outcome of this experiment.

Obtain from your laboratory instructor three antacid tablets. Place them in an appropriate container. Record on the Data Sheet the brand name of your antacid, the number of tablets in the original container, and the cost of the container of tablets.

I. Using a Pipet

Note: In the next step, the beaker into which the standard solution will be placed must be clean and dry. If the beaker is not dry, the standard solution will be diluted. Consequently, the exact concentration of the solution will not be known.

Obtain approximately 125 mL of standard HCl solution in a clean, dry, labeled 250-mL beaker. Record on the Data Sheet the **exact** molarity of the standard HCl solution.

Rinse a clean 25-mL pipet by drawing about 5 mL of distilled water into the pipet using a rubber bulb.

Quickly disconnect the rubber bulb and place your index finger over the top opening of the pipet to prevent the water from draining out of it. Hold the pipet in a nearly horizontal position. Rotate the pipet to allow the water to contact all interior surfaces. Remove your finger briefly during this process to allow the water to enter the upper stem of the pipet. Allow the water to drain from the pipet through the tip.

In similar fashion, rinse the pipet with approximately 5 mL of the standard HCl solution. Discard the rinse solution into a waste container. Repeat this procedure twice with two separate 5-mL portions of the HCl solution.

> **Note:** After you discharge a solution from a pipet, there will be a small amount of liquid left in the tip of the pipet. Do not blow this liquid out of the pipet. The pipet is calibrated to deliver 25.00 mL of solution *excluding* the small amount remaining in the tip of the pipet.

Carefully pipet 25.00 mL of the standard HCl solution into a clean 250-mL Erlenmeyer flask. As you release the solution into the flask, hold the tip of the pipet against the side of the flask. Allow the solution to flow down the side of the flask to prevent splattering. After you deliver the solution from the pipet, continue to hold the tip of the pipet against the side of the flask for an additional 15 sec.

II. Analyzing an Antacid Preparation

> **Note:** In the next step, you will add a portion of distilled water to the acid in the flask. The solution will now have sufficient volume for the titration. The number of moles of acid in the solution will be the same after the addition of water as before. It is not necessary to know the new volume of the solution or its concentration.

Add approximately 40 mL of distilled water from a graduated cylinder to the acid in the flask.

> **Note:** In the next step, the beaker into which the standard NaOH solution will be placed must be clean and dry. If the beaker is not dry, the standard solution will be diluted and the exact concentration of the solution will not be known.

Obtain approximately 125 mL of standard NaOH solution in a clean, dry 250-mL beaker Record on the Data Sheet the *exact* molarity of the NaOH solution.

Place a small weighing dish or a weighing paper on the balance. Determine its mass to the nearest 1 mg. Record this mass on the Data Sheet. If the balance has an automatic taring feature, you may wish to use this feature and enter 0.000 for the mass of the container on the Data Sheet.

Place one of your antacid tablets in the weighing dish or on the weighing paper. Weigh the tablet and dish or paper to the nearest 1 mg. Record this mass on the Data Sheet.

> **Note:** In the next sequence of steps, the antacid–HCl solution is boiled to expel the CO_2. If any CO_2 is present during the subsequent titration, it will interfere with the determination of the end point.

Add the tablet to the HCl solution in the Erlenmeyer flask. Warm the flask and contents gently to dissolve the tablet. After the tablet has dissolved, boil the solution for 1–2 min. Cool the solution. Add 10 drops of bromphenol blue indicator solution. Mix the solution thoroughly by swirling the flask and contents.

> **Note:** If the solution in the flask is not yellow at this point, consult your laboratory instructor before proceeding to the next step.

Add 10 mL of distilled water to a clean 50-mL buret. Hold the buret in a nearly horizontal position. Rotate the buret in such a way that the water contacts all portions of the walls of the buret. Discharge the water through the tip of the buret into a waste container. Repeat this procedure twice with two separate 10-mL portions of water.

In similar fashion, rinse your buret with three separate 5-mL portions of the standard NaOH solution. Discharge the rinse solution through the tip of the buret into a waste container.

> **Note:** When you fill your buret in the following step, the tip of the buret must be filled with NaOH solution. Any air bubbles in the tip must be removed before you begin the titration. If they are not removed, the exact volume of solution used in the titration will be in question.

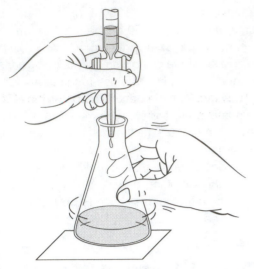

Figure 1   *Position of buret tip in mouth of flask*

Clamp the buret onto a ring stand. Close the stopcock. Fill the buret with the standard NaOH solution. Remove any air bubbles trapped in the tip of the buret. Allow the solution to flow through the stopcock into a waste container until the air is forced out of the tip. If necessary, lower the solution level in the buret until it is slightly below the 0.00-mL mark. Read the buret to the nearest 0.01 mL. Record this reading in the appropriate column on the Data sheet.

Touch the tip of the buret to the side of the waste container to remove any drop of solution that may be clinging to the tip. Place the tip of the buret about 2–3 cm below the rim of the Erlenmeyer flask, as shown in Figure 1.

Note: As the titration approaches the end point, you will notice momentary flashes of blue-green color where the NaOH solution hits the solution in the flask. When you are very close to the end point, these colored areas will grow larger, taking longer to change back to the original yellow color. At the end of the titration, you will be adding the titrant one drop at a time. The titration is complete when the solution in the flask becomes green and remains green for 15 sec after being completely mixed.

Swirl the Erlenmeyer flask with one hand. Control the stopcock of the buret with your other hand. Begin to titrate the antacid solution by slowly allowing the NaOH solution to drain into the flask. Continue to swirl the flask throughout the titration. Near the end of the titration, reduce the rate of addition of the titrant until only one drop at a time is added. At the end of the titration, read the buret to the nearest 0.01 mL. Record this reading in the appropriate column on the Data Sheet.

Do a second and third determination if time permits.

Note: The buret might not have to be refilled between determinations. Estimate the volume of NaOH needed for a second titration from the volume of the titrant used in the first titration. If there is sufficient volume remaining in the buret for the second titration, do not refill the buret. Always record the volume of the titrant in the buret before beginning a titration.

After completing the last determination, drain the solution from the buret into the waste container. Rinse the buret several times with distilled water. Rinse all other glassware used in this experiment with distilled water. Return the glassware to its proper place.

Calculations

(Note: Record all calculations on the Data sheet.)

1.   Calculate the number of moles of HCl added to the tablet in each determination (Equation 8).

2.   Calculate the mass of each antacid tablet.

3.   Calculate the number of moles of NaOH required to back-titrate the excess HCl in each determination (Equation 9).

4.   Calculate the number of moles of HCl neutralized per tablet in each determination (Equation 7).

5.   Calculate the average number of moles of HCl neutralized per tablet.

6.   Calculate the weight effectiveness of the antacid, which is the average number of moles of HCl neutralized per gram of tablet (Equation 10).

7.   Find the total cost of the number of tables in the container purchased.

8.   Calculate the cost of one tablet of the antacid.

9.   Calculate the cost effectiveness of the antacid, which is the average number of moles of HCl neutralized per penny (Equation 11).

10.   Exchange cost effectiveness data with your classmates to complete the Data sheet.

11.   Evaluate the antacids investigated by your laboratory section in terms of cost effectiveness. See Question 1 of the **Post-Laboratory Questions** in this experiment.

Post-Laboratory Questions

(Use the space provided for the answers and additional paper if necessary.)

1. Evaluate the antacids investigated by your laboratory section in terms of cost effectiveness.

2. A Brand X antacid tablet was analyzed by a student following the procedure used in this experiment. The back-titration required 5.85 mL of 1.015*M* NaOH to titrate the excess HCl added to the tablet. In the 25.00 mL of the HCl solution used to dissolve the tablet, there were 1.586×10^{-2} mol of HCl.

(1) Calculate the number of moles of acid neutralized by the tablet.

answer

(2) Calculate the concentration of the HCl solution in this experiment.

answer

name section date

DATA SHEET

Brand of antacid_____   Concentration of HCl, *M*_____

Cost of antacid $_____ per_____tablets

Cost of antacid $_____ per tablet   Concentration of NaOH, *M*_____

		determination	
	first	*second*	*third*
weight of weighing dish + tablet, g	_____	_____	_____
Weight of weighing dish, g	_____	_____	_____
Weight of tablet, g	_____	_____	_____
Initial buret reading, mL	_____	_____	_____
Final buret reading, mL	_____	_____	_____
Volume of titrant, mL	_____	_____	_____
Number of moles of HCl added	_____	_____	_____
Number of moles of NaOH added	_____	_____	_____
Number of moles of HCl neutralized per tablet	_____	_____	_____
Number of moles of HCl neutralized per gram	_____	_____	_____

Average number of moles of HCl neutralized per gram   _____

Weight effectiveness
(Average number of moles of HCl neutralized per tablet)   _____

Cost effectiveness
(Average number of moles of HCl neutralized per cost of tablet)   _____

Evaluation of Antacids					
brand of antacid					
cost effectiveness					

Pre-Laboratory Assignment

1. Review Experiment 4 for discussions of the proper techniques to be used in this experiment.

2. When a student analyzed an antacid tablet weighing 0.7743 g, 25.00 mL of 0.5900M HCl were used to dissolve the sample. The back-titration of the excess HCl required 5.04 mL of 1.015M NaOH.

    (1) Calculate the number of moles of HCl added to the tablet.

    (4) Calculate the number of moles of HCl that would be neutralized by a tablet of this antacid weighing 1.000 g.

---
answer

    (2) Calculate the number of moles of NaOH required to neutralize the excess HCl in the back-titration.

---
answer

    (5) Calculate the cost required to neutralize one mole of HCl if each tablet sells for $0.035.

---
answer

    (3) Calculate the number of moles of HCl neutralized by the tablet.

---
answer

---
answer

Determining the Percent Sodium Hypochlorite in Commercial Bleaching Solutions

prepared by **Enno Wolthuis**, Calvin College

Purpose of the Experiment

Determine the percent sodium hypochlorite in various commercial bleaching solutions by titration. Compare the cost effectiveness of different brands of commercial bleaching solutions.

Background Information

Commercial bleaching solutions are prepared by reacting chlorine (Cl_2) with a base. If the base is sodium hydroxide (NaOH), the product is sodium hypochlorite (NaOCl). Commercially available bleaching solutions usually contain NaOCl.

In aqueous solution, NaOCl dissociates into sodium ion (Na^+) and hypochlorite ion (OCl^-). Bleaching with OCl^- ion involves an oxidation-reduction reaction in which the Cl in the OCl^- ion, the oxidizing agent, is reduced to chloride ion (Cl^-). The reducing agent is usually the colored species or clothing stain being removed or oxidized by the bleach.

We can determine the OCl^- ion content of a bleaching solution by reacting a known mass or volume of the solution with excess reducing agent, such as iodide ion (I^-), in an acidic solution, as shown in Equation 1.

$$OCl^-(aq, colorless) + 2\ I^-(aq, colorless)$$
$$+ 2\ H_3O^+(aq) \rightarrow$$
$$I_2(aq, brown) + Cl^-(aq, colorless) + 3\ H_2O(l) \quad (Eq.\ 1)$$

The reaction in Equation 1 proceeds to completion. Visible evidence of reaction is the change in appearance of the solution from colorless to brown, due to the formation of iodine (I_2). In the presence of excess I^- ion, the amount of I_2 formed is a measure of the amount of OCl^- ion reacting.

Then, we determine the amount of I_2 formed by titrating the I_2 with a standard sodium thiosulfate solution ($Na_2S_2O_3$). The titration reaction for this experiment is shown in Equation 2.

$$I_2(aq, brown) + 2\ S_2O_3^{2-}(aq, colorless) \rightarrow$$
$$2\ I^-(aq, colorless) + S_4O_6^{2-}(aq, colorless) \quad (Eq.\ 2)$$

Thiosulfate ion ($S_2O_3^{2-}$) is a reducing agent that reacts quantitatively with I_2. The titration reaction is complete when the I_2 formed from the reaction of OCl^- ion with I^- ion is reconverted to I^- ion by $S_2O_3^{2-}$ ion.

As the titration proceeds, the I_2 concentration in the solution decreases. This causes the solution color to change from brown to pale yellow near the end of the titration. The end point occurs when all the I_2 has reacted and the solution is colorless. Because the change from pale yellow to colorless is not very distinct, establishing the end point on the basis of this final color change is difficult. We can make the end point more distinct by adding a small amount of starch solution to the titration mixture when the solution turns pale yellow. The unreacted I_2 combines with the starch, forming a

---

Copyright © 1992 by Chemical Education Resources, Inc., P.O. Box 357, 220 S. Railroad, Palmyra, Pennsylvania 17078
No part of this laboratory program may be reproduced or transmitted in any form or by any means, electronic or mechanical, including photocopying, recording, or any information storage and retrieval system, without permission in writing from the publisher. Printed in the United States of America

deep blue complex. Additional $S_2O_3^{2-}$ ion reacts with the complexed I_2, causing a breakdown of the complex. Disappearance of the blue color signals the end point.

We determine the concentration of OCl^- ion in bleaching solution from the volume of solution titrated, the volume and concentration of titrant used, and the stoichiometry of Equations 1 and 2. We can determine the number of moles of $S_2O_3^{2-}$ ion required for titration from the volume and concentration of titrant used and Equation 3.

number of moles of $S_2O_3^{2-}$ ion required, mol =

$$\left(\begin{array}{c}\text{volume of } Na_2S_2O_3 \\ \text{solution, mL}\end{array}\right)\left(\dfrac{1\ L}{1000\ mL}\right)$$

$$\left(\dfrac{\begin{array}{c}\text{number of moles} \\ \text{of } Na_2S_2O_3\end{array}}{\begin{array}{c}\text{volume of} \\ \text{solution, L}\end{array}}\right)\left(\dfrac{1\ mol\ S_2O_3^{2-}}{1\ mol\ Na_2S_2O_3}\right) \quad \text{(Eq. 3)}$$

We then calculate the number of moles of I_2 produced from the reaction of OCl^- ion, with I^- ion using, the stoichiometry of Equation 2 and Equation 4.

number of moles of I_2 produced, mol =

$$\left(\begin{array}{c}\text{number of moles} \\ \text{of } S_2O_3^{2-}\ \text{ion} \\ \text{required, mol}\end{array}\right)\left(\dfrac{1\ mol\ I_2}{2\ mol\ S_2O_3^{2-}\ \text{ion}}\right) \quad \text{(Eq. 4)}$$

We calculate the number of moles of OCl^- ion present in the titrated bleaching solution, using the stoichiometry of Equation 1 and Equation 5.

number of moles of OCl^- ion present, mol =

$$\left(\begin{array}{c}\text{number of} \\ \text{moles of } I_2 \\ \text{produced, mol}\end{array}\right)\left(\dfrac{1\ mol\ OCl^-\ \text{ion}}{1\ mol\ I_2}\right) \quad \text{(Eq. 5)}$$

Finally, we calculate the mass of NaOCl present in the titrated bleaching solution, using Equation 6.

$$\begin{array}{c}\text{mass of NaOCl in} \\ \text{bleaching solution, g}\end{array} = \left(\begin{array}{c}\text{number of moles of} \\ OCl^-\ \text{ion present, mol}\end{array}\right)$$

$$\left(\dfrac{1\ mol\ NaOCl}{1\ mol\ OCl^-\ \text{ion}}\right)\left(\dfrac{74.44\ g\ NaOCl}{1\ mol\ NaOCl}\right) \quad \text{(Eq. 6)}$$

We can calculate the volume of commercial bleaching solution titrated, using Equation 7.

$$\begin{array}{c}\text{volume of commercial} \\ \text{bleaching solution} \\ \text{titrated, mL}\end{array} = \left(\begin{array}{c}\text{volume of diluted} \\ \text{bleaching solution} \\ \text{titrated, mL}\end{array}\right)$$

$$\left(\dfrac{\begin{array}{c}\text{volume of commercial bleaching} \\ \text{solution diluted, mL}\end{array}}{\begin{array}{c}\text{total volume of} \\ \text{diluted bleaching solution, mL}\end{array}}\right) \quad \text{(Eq. 7)}$$

We determine the mass of commercial bleaching solution titrated from the density of the bleaching solution and the volume titrated. The strength of a bleaching solution is usually designated as percent NaOCl, as determined in Equation 8.

percent NaOCl in bleaching solution, % =

$$\left(\dfrac{\text{mass of NaOCl, g}}{\text{mass of bleaching solution, g}}\right)(100\%) \quad \text{(Eq. 8)}$$

In this experiment, you will determine the percent NaOCl in a commercial bleaching solution. The experimental error when measuring small volumes of liquids is usually greater than the error when measuring larger volumes. To improve the accuracy of data obtained from this experiment, you will dilute 10 mL of commercial bleaching solution to 100-mL using distilled or deionized water. You will titrate 25-mL portions, called **aliquots**, of the diluted bleaching solution. Done this way, your volume measurements will be more accurate than if you titrate small volumes of undiluted commercial bleaching solution.

To perform the titration, you will dissolve a measured volume of diluted bleaching solution in water and add excess potassium iodide solution (KI), a source of I^- ion. Then you will acidify the reaction mixture and determine the amount of I_2 formed by titrating with standard $Na_2S_2O_3$ solution. You will do duplicate determinations to assure reproducible results.

From the titration data, you will calculate the mass of OCl^- ion in your sample. Based on the density of the commercial bleaching solution and the volume titrated, you will express the OCl^- ion content as percent NaOCl. Finally, you will calculate the cost of the volume of your commercial bleaching solution required to supply 100 g of NaOCl. You will compare this cost with the costs of other brands to determine which commercial bleaching solution is the most cost effective.

Procedure

Chemical Alert
bleaching solution—corrosive
$6M$ hydrochloric acid—toxic and corrosive
10% potassium iodide—irritant
$0.1M$ sodium thiosulfate—irritant

Caution: Wear departmentally approved eye protection while doing this experiment.

Note: Discard titration mixtures, unused reagent solutions, and rinses into the appropriate 400-mL beakers labeled "Discarded $Na_2S_2O_3$ Solution and Titration Mixtures" and "Discarded Bleaching Solution."

Obtain 150 mL of $Na_2S_2O_3$ solution in a clean, **dry**, labeled 250-mL Erlenmeyer flask. Stopper the flask with a **new** cork, and keep the flask stoppered when not in use. Record the exact molarity of the $Na_2S_2O_3$ on your Data Sheet.

Rinse a clean 50-mL buret with a small amount of your standardized $Na_2S_2O_3$ solution. Fill the buret with the solution to a point above the top calibration.

Clamp the buret to a ring stand and place a 400-mL discard beaker under the buret tip. Slowly turn the stopcock to drain titrant until the meniscus is at, or slightly below, the 0-mL calibration. Read and record the initial buret reading on your Data Sheet.

Caution: Commercial bleaching solution is corrosive and should be handled with care.

Obtain 50 mL of a commercial bleaching solution in a clean, **dry**, labeled 125-mL Erlenmeyer flask. Stopper the flask with a **new** cork, and keep the flask stoppered when not in use. Record the brand name, the cost per gallon, and the density of the bleaching solution on your Data Sheet.

Caution: **Never** use mouth suction to draw a liquid or a solution into a pipet. Always use a rubber suction bulb to draw liquid into a pipet.

Rinse a clean 10-mL volumetric pipet with several milliliters of your commercial bleaching solution. Use the pipet to measure 10.00 mL of the solution and transfer it into a clean 100-mL volumetric flask. Fill the flask to the calibration mark with distilled or deionized water to prepare your diluted bleaching solution. Stopper the flask and invert 15 times to mix the solution.

Use a 100-mL graduated cylinder to transfer 100 mL distilled water and 10 mL 10% KI solution to a clean, labeled 250-mL Erlenmeyer flask.

Rinse a clean 25-mL volumetric pipet with the diluted bleaching solution. Use the pipet to transfer a 25-mL sample of the diluted bleaching solution to the Erlenmeyer flask.

Caution: $6M$ HCl solution is toxic, corrosive, and can cause burns. Handle with care.

Note: The solution should be titrated **immediately** after adding acid. Additional I_2 will be liberated by the reaction of aqueous KI with O_2 in the air.

Use a 10-mL graduated cylinder to add 4 mL of $6M$ HCl solution to the sample. Thoroughly mix by swirling.

Note: Starch decomposes in the presence of acid. Premature addition of starch solution will result in erroneous results due to decomposition of the starch and irreversible attachment of I_2 to the starch. Do not add starch until the color of the titration mixture is pale yellow.

While swirling the flask and its contents, add $Na_2S_2O_3$ solution from the buret to the solution in a fairly rapid stream until the solution turns a light yellow. Add 1 mL of starch solution. Continue dropwise addition of titrant, while swirling, until the blue color disappears. Record your final buret reading on your Data Sheet.

Refill the buret and repeat the titration with a second sample of your diluted bleach solution, recording all data on your Data Sheet.

Caution: Wash your hands thoroughly with soap or detergent before leaving the laboratory.

Calculations

Do the following calculations for each determination and record the results on your Data Sheet.

1. Find the volume, in milliliters, of $Na_2S_2O_3$ solution used to titrate the diluted bleaching solution.

2. Calculate the number of moles of $S_2O_3^{2-}$ ion required for the titration.

3. Calculate the number of moles of I_2 produced in the titration mixture.

4. Calculate the number of moles of OCl^- ion present in the titrated diluted bleaching solution.

5. Calculate the mass of NaOCl present in the titrated diluted bleaching solution.

6. Determine the volume of commercial bleaching solution titrated.

7. Calculate the mass of commercial bleaching solution titrated.

8. Determine the percent NaOCl in your commercial bleaching solution.

9. Repeat these calculations using the data from your second determination.

10. Find the mean percent NaOCl in your commercial bleaching solution.

11. Find the mass of one gallon of your commercial bleaching solution, using Equation 9.

$$\begin{array}{c}\text{mass of 1 gal} \\ \text{bleaching solution, g}\end{array} = \left(\begin{array}{c}\text{1 gal} \\ \text{bleaching solution}\end{array}\right)$$

$$\left(\begin{array}{c}\dfrac{3.785 \times 10^3 \text{ mL}}{\text{bleaching solution}} \\ \dfrac{}{\text{1 gal}} \\ \text{bleaching solution}\end{array}\right)\left(\begin{array}{c}\text{mass of} \\ \dfrac{\text{bleaching solution, g}}{1.00 \text{ mL}} \\ \text{bleaching solution, mL}\end{array}\right)$$

$$\text{(Eq. 9)}$$

12. Find the cost of 100 g of your commercial bleaching solution, using Equation 10.

$$\begin{array}{c}\text{cost of 100 g} \\ \text{bleaching solution, \$}\end{array} = \left(\begin{array}{c}\text{100 g} \\ \text{bleaching solution}\end{array}\right)$$

$$\left(\begin{array}{c}\dfrac{\text{1 gal}}{\text{bleaching solution}} \\ \dfrac{}{\text{mass bleaching}} \\ \text{solution, g}\end{array}\right)\left(\begin{array}{c}\dfrac{\text{cost, \$}}{\text{1 gal bleaching}} \\ \text{solution, gal}\end{array}\right) \quad \text{(Eq. 10)}$$

13. Find the cost of the amount of commercial bleaching solution required to supply 100 g of NaOCl, using Equation 11.

$$\begin{array}{c}\text{cost of bleaching solution} \\ \text{required to supply 100 g NaOCl, \$}\end{array} = \left(\begin{array}{c}\text{100 g} \\ \text{NaOCl}\end{array}\right)$$

$$\left(\begin{array}{c}\dfrac{\text{100 g}}{\text{bleaching solution}} \\ \dfrac{}{\text{mass of}} \\ \text{NaOCl, g}\end{array}\right)\left(\begin{array}{c}\dfrac{\text{cost, \$}}{\text{100 g bleaching}} \\ \text{solution, g}\end{array}\right) \quad \text{(Eq. 11)}$$

Post-Laboratory Questions

(Use the spaces provided for the answers and additional paper if necessary.)

1. Graduated cylinders are not as precisely calibrated as are burets or volumetric pipets. Briefly explain why it is acceptable to measure the KI and HCl solutions used in the titration with graduated cylinders rather than with pipets or burets.

2. Would the following procedural errors result in an incorrectly high or low calculated percent NaOCl in commercial bleaching solution? Briefly explain.

(1) A student failed to allow the volumetric pipet to drain completely when transferring the diluted bleaching solution to the Erlenmeyer flask.

(2) A student blew the last drops of solution from the pipet into the volumetric flask when transferring commercial bleaching solution to the flask.

(3) A student began a titration with an air bubble in the buret tip. The bubble came out of the tip after 5 mL of $Na_2S_2O_3$ solution had been released.

3. An overly efficient student simultaneously prepared two titration mixtures, consisting of diluted bleaching solution, KI solution, and HCl solution. The student found that data from the two titrations yielded significantly different percents NaOCl in the commer-

cial bleaching solution. Which determination would give the higher % NaOCl? Briefly explain.

4. Based on your titration data and calculations for determination 1:

(1) Calculate the volume of $Na_2S_2O_3$ solution required had you transferred commercial bleaching solution instead of diluted bleaching solution to Erlenmeyer flask 1.

(2) Assuming you used only the glassware supplied for the original experiment, briefly comment on the procedural change necessary to titrate 25.00 mL of commercial bleaching solution.

(3) Would you anticipate a greater error in the results of an analysis done in this way than is the case with the standard procedure? Briefly explain.

5. Commercial bleaching solutions found on store shelves are usually labeled: "Contains at least 5.25% sodium hypochlorite." Analysis will often show a lower OCl^- ion content. Briefly explain.

name _____

section _____ date _____

Data Sheet

brand name of commercial bleaching solution   _____

cost of commercial bleaching solution per gallon   _____

density of commercial bleaching solution, g mL^{-1} _____

volume of commercial bleaching solution diluted to 100 mL, mL _____

	determination	
	1	2
molarity of $Na_2S_2O_3$ solution, *M*	_____	_____
volume of diluted bleaching solution titrated, mL	_____	_____
final buret reading, mL	_____	_____
initial buret reading, mL	_____	_____
volume of $Na_2S_2O_3$ solution used, mL	_____	_____
number of moles of $S_2O_3^{2-}$ ion required for titration, mol	_____	_____
number of moles of I_2 produced in the titration mixture, mol	_____	_____
number of moles of OCl^- ion in diluted bleaching solution titrated, mol	_____	_____
mass of NaOCl present in diluted bleaching solution titrated, g	_____	_____
volume of commercial bleaching solution titrated, mL	_____	_____
mass of commercial bleaching solution titrated, g	_____	_____
percent NaOCl in commercial bleaching solution, %	_____	_____
mean percent NaOCl in commercial bleaching solution, %		_____
mass of 1 gal commercial bleaching solution, g		_____
cost of 100 g commercial bleaching solution, $		_____
cost of the amount of commercial bleaching solution required to supply 100 g of NaOCl, $		_____

Pre-Laboratory Assignment

1. Describe the hazards you should be aware of when working with 6M HCl solution.

2. Briefly explain what is meant by the following terms as they apply to this experiment.
(1) oxidation-reduction reaction

(2) standard solution

(3) meniscus

(4) aliquot

3. A student followed the procedure of this experiment to determine the percent NaOCl in a commercial bleaching solution that was found in the basement of an abandoned house. The student diluted 50.00 mL of the commercial bleaching solution to 250 mL in a volumetric flask, and titrated a 20-mL aliquot of the diluted

bleaching solution. The titration required 35.46 mL of 0.1052M Na$_2$S$_2$O$_3$ solution. A faded price label on the gallon bottle read $0.79. The density of the bleaching solution was 1.10 g mL^{-1}.
(1) Calculate the number of moles of S$_2$O$_3^{2-}$ ion required for the titration.

(2) Calculate the number of moles of I$_2$ produced in the titration mixture.

(3) Calculate the number of moles of OCl$^-$ ion present in the diluted bleaching solution titrated.

(4) Calculate the mass of NaOCl present in the diluted bleaching solution titrated.

(5) Determine the volume of commercial bleaching solution present in the diluted bleaching solution titrated.

(6)   Calculate the mass of commercial bleaching solution titrated.

(7)   Determine the percent NaOCl in the commercial bleaching solution.

(8)   Calculate the mass of one gallon of the commercial bleaching solution.

(9)   Calculate the cost of 100 g of the commercial bleaching solution.

(10) Determine the cost of the amount of commercial bleaching solution required to supply 100 g of NaOCl.

Quantitatively Determining the Acid Content of Fruit Juices

prepared by **Andrew W. Zanella**, Claremont McKenna, Pitzer, and Scripps Colleges

---

Purpose of the Experiment

Determine the acid content of various fruit juices by titration with standard sodium hydroxide solution.

Background Information

The sour taste of many fruit juices is due in large part to the presence of acids. Citric acid, $C_3H_5O(CO_2H)_3$, is one of several acids present in these juices. Citric acid reacts with sodium hydroxide (NaOH), a base, as shown in Equation 1.

$$C_3H_5O(CO_2H)_3(aq) + 3\ NaOH(aq) \rightarrow$$
$$C_3H_5O(CO_2)_3Na_3(aq) + 3\ H_2O(l) \quad \text{(Eq. 1)}$$

Equation 1 describes a **neutralization reaction**, in which an acid and base react to form water and an ionic compound called a **salt**. In the neutralization of $C_3H_5O(CO_2H)_3$, the salt is sodium citrate, $Na_3C_3H_5O(CO_2)_3$. Although various acids are found in different amounts in different fruit juices, for the purposes of this experiment, we will assume that the acid content of these juices consists entirely of $C_3H_5O(CO_2H)_3$.

We can determine the amount of acid in a given volume of fruit juice by titrating the juice with a standard NaOH solution. A **standard solution** is a solution of known concentration. We frequently express the concentration of a standard solution in terms of molarity.

The **molarity** (mol L^{-1}, or M) of a solution is the number of moles of solute per liter of solution, as expressed by Equation 2.

$$\text{molarity, } M = \frac{\text{number of moles of solute}}{\text{volume of solution, L}} \quad \text{(Eq. 2)}$$

We measure the volume of NaOH solution required for the neutralization by titration. **Titration** is the measurement of the volume of a standard solution required to completely react with a measured volume or mass of the substance being analyzed. We add the standard solution from a calibrated glass tube called a **buret**. Before beginning the titration, we add an indicator to the titration mixture. An **indicator** is a substance that changes color at the point when the titration reaction is complete. In this experiment, you will use phenolphthalein as an indicator when you titrate fruit juice with NaOH solution. Phenolphthalein is a complex organic dye that is colorless in acidic solutions and pink in solutions that are slightly alkaline, or basic.

Assume that we want to determine the acidity ($C_3H_5O(CO_2H)_3$ content) of an orange juice sample. We find that 39.62 mL of 0.106M NaOH solution are required to titrate a 10.0-mL sample of orange juice. We determine the number of moles of NaOH required to

---

Copyright © 1993 by Chemical Education Resources, Inc., P.O. Box 357, 220 S. Railroad, Palmyra, Pennsylvania 17078
No part of this laboratory program may be reproduced or transmitted in any form or by any means, electronic or mechanical, including photocopying, recording, or any information storage and retrieval system, without permission in writing from the publisher. Printed in the United States of America

neutralize the $C_3H_5O(CO_2H)_3$ from the concentration and volume of NaOH solution used in the titration, and a rearrangement of Equation 2, shown as Equation 3. Note that in part of Equation 3 the NaOH volume is converted from milliliters to liters.

$$\begin{pmatrix} \text{number of} \\ \text{moles of NaOH} \\ \text{required, mol} \end{pmatrix} = \begin{pmatrix} \text{volume of} \\ \text{NaOH solution} \\ \text{required, mL} \end{pmatrix} \begin{pmatrix} 1\ L \\ \hline 1000\ mL \end{pmatrix}$$

$$\begin{pmatrix} \text{concentration} \\ \text{of NaOH} \\ \text{solution, mol L}^{-1} \end{pmatrix} \qquad \text{(Eq. 3)}$$

$$= 39.62\ mL \left(\frac{1\ L}{1000\ mL} \right) \left(\frac{0.106\ mol\ NaOH}{1\ L\ solution} \right)$$

$$= 4.20 \times 10^{-3}\ mol\ NaOH$$

We determine the number of moles of $C_3H_5O(CO_2H)_3$ in the titrated juice sample using Equation 4 and the stoichiometry of the titration reaction given in Equation 1.

number of moles of $C_3H_5O(CO_2H)_3$, mol =

$$\begin{pmatrix} \text{number of moles} \\ \text{of NaOH} \end{pmatrix} \begin{pmatrix} 1\ mol\ C_3H_5O(CO_2H)_3 \\ \hline 3\ mol\ NaOH \end{pmatrix} \quad \text{(Eq. 4)}$$

$$= (4.20 \times 10^{-3}\ mol\ NaOH) \begin{pmatrix} 1\ mol\ C_3H_5O(CO_2H)_3 \\ \hline 3\ mol\ NaOH \end{pmatrix}$$

$$= 1.40 \times 10^{-3}\ mol\ C_3H_5O(CO_2H)_3$$

We compute the mass of $C_3H_5O(CO_2H)_3$ in the titrated juice sample using Equation 5. The molar mass of $C_3H_5O(CO_2H)_3$ is 192.12 g mol^{-1}.

mass of $C_3H_5O(CO_2H)_3$ in sample, g =

$$\begin{pmatrix} \text{number of} \\ \text{moles of} \\ C_3H_5O(CO_2H)_3 \end{pmatrix} \begin{pmatrix} 192.12\ g\ C_3H_5O(CO_2H)_3 \\ \hline 1\ mol\ of\ C_3H_5O(CO_2H)_3 \end{pmatrix} \quad \text{(Eq. 5)}$$

$$= \begin{pmatrix} 1.40 \times 10^{-3}\ mol \\ C_3H_5O(CO_2H)_3 \end{pmatrix} \begin{pmatrix} 192.12\ g\ C_3H_5O(CO_2H)_3 \\ \hline 1\ mol\ of\ C_3H_5O(CO_2H)_3 \end{pmatrix}$$

$$= 0.269\ g\ C_3H_5O(CO_2H)_3$$

For convenience in making comparisons between different juices, we determine the mass of $C_3H_5O(CO_2H)_3$ present in 1 mL of juice, using Equation 6.

mass of $C_3H_5O(CO_2H)_3$ in 1 mL juice, g mL^{-1} =

$$\frac{\text{mass of } C_3H_5O(CO_2H)_3 \text{ in sample, g}}{\text{volume of juice titrated, mL}} \quad \text{(Eq. 6)}$$

$$= \frac{0.269\ g\ C_3H_5O(CO_2H)_3}{10.0\ mL\ juice}$$

$$= 2.69 \times 10^{-2}\ g\ C_3H_5O(CO_2H)_3\ mL^{-1}$$

When evaluating the result of this calculation, we need to recall that we assumed that the only acid present in the juice was $C_3H_5O(CO_2H)_3$. This assumption is useful for comparative purposes. We would need to devise and execute a much more complex analytical scheme if we wanted to determine the exact amounts and identities of the individual acids actually present in the juice sample.

In this experiment, you will determine the number of grams of acid, assuming it is entirely $C_3H_5O(CO_2H)_3$, present in 1 mL of a fruit juice by titrating the juice sample with standard NaOH solution. If a variety of juice samples are available for analysis, you will be able to compare the acidities of these juices by comparing your results with those of your classmates.

Procedure

Chemical Alert

0.1*M* sodium hydroxide—toxic and corrosive

Caution: Wear departmentally approved eye protection while doing this experiment.

I. Preparing the Fruit Juice for Titration

Note: The numbers appearing in parentheses indicate the specific lines on your Data Sheet on which the indicated data should be entered.

---

1. Obtain 75 mL of fruit juice from your laboratory instructor in a clean, dry 150-mL beaker. Record the type of fruit juice and the code identification of the juice sample on your Data Sheet (1,2).

Note: If the juice has substantial amounts of pulp floating in it, filter the juice by pouring it from the beaker into another clean, dry 150-mL beaker through some glass wool placed in a conical funnel. Otherwise, the pulp tends to obscure the titration end point.

Note: Your laboratory instructor will tell you whether or not you should titrate any volume of juice **other than** that specified in Step 3. If it is necessary for you to dilute your juice prior to titration, your instructor will give you information concerning the appropriate calculations.

2. Label two clean, 125-mL Erlenmeyer flasks "1" and "2."

Note: Your laboratory instructor may suggest that you dispense your juice from a dispensing buret or a 20-mL pipet. If so, you will deliver 20.00 mL of sample, rather than 20.0 mL as mentioned in Step 3, due to the greater precision of these types of glassware.

3. Measure 20.0 mL of your juice from the beaker into a 25-mL graduated cylinder. Transfer the juice into Erlenmyer flask 1. Record on your Data Sheet the volume of juice transferred (3).

Rinse the graduated cylinder twice, using 5 mL of distilled or deionized water each time. Transfer the rinses into Erlenmeyer flask 1.

4. Use the procedure in Step 3 to transfer a 20.0-mL juice sample to Erlenmeyer flask 2.

5. Add three drops of phenolphthalein indicator solution to the solution in each Erlenmeyer flask. Gently swirl each flask and its contents to thoroughly mix each solution.

II. Cleaning and Filling the Buret

Note: Your laboratory instructor will demonstrate and describe an acceptable technique for adding solutions to your buret. One such method involves placing a funnel in the top of the buret and pouring the solution into the funnel.

If you use a funnel, be sure to rinse the inner surface of the funnel with the solution you will be adding to the buret, prior to placing the funnel in the buret.

6. Holding the buret vertically, close the stopcock and fill the buret with tap water. Open the stopcock, and drain the water through the buret tip into the drain.

If any water drops remain on the inner surface of the barrel, the buret is dirty and must be cleaned.

7. If the buret needs cleaning, dip a buret brush into a warm detergent solution in a 150-mL beaker. Scrub the inside buret wall. **Do not push the end of the brush below the 50-mL calibration line.**

8. Close the stopcock, and add approximately 10 mL of tap water to the buret. Open the stopcock, and drain the water through the buret tip into the drain. Repeat this procedure until you have rinsed all of the detergent from the buret.

9. Close the stopcock, and add 10 mL of distilled water to the buret. Clamp the buret to a support stand, and let the buret stand for a few minutes. Consult your laboratory instructor if the buret leaks.

10. Remove the buret from the support stand. Hold the buret almost horizontally. Carefully rotate the buret so that water contacts the entire inner surface.

11. Drain the water through the buret tip into the sink.

12. Repeat Step 10 with two additional 10-mL volumes of distilled water.

III. Titrating the Fruit Juice

Note: If you have a standard NaOH solution from a previous experiment, use this solution for the following titration. If not, use the standard NaOH solution supplied by your laboratory instructor.

Caution: NaOH solution is corrosive and toxic, and it can cause skin burns. Prevent contact with your eyes, skin, and clothing. Do not ingest the solution.

If you spill any NaOH solution, immediately notify your laboratory instructor.

13. Obtain about 75 mL of standard NaOH solution in a clean, **dry** 125-mL Erlenmeyer flask. Stopper the flask with a new cork, and keep the flask stoppered when not in use. Record the exact molarity of the NaOH solution on your Data Sheet (4).

14. Rinse the buret with 5 mL of your standard NaOH solution. Hold the buret nearly horizontally. Rotate the buret so that the NaOH solution contacts the entire

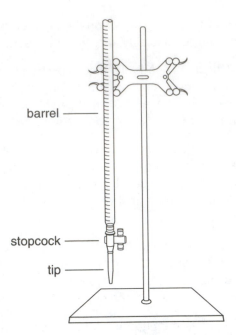

Figure 1　*Titration apparatus*

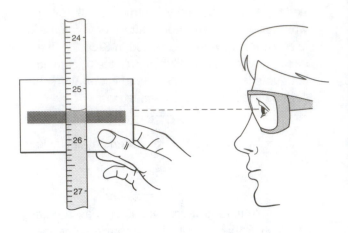

Figure 2　*Reading a buret*

inner surface. Drain the solution through the buret tip into a 400-mL beaker labeled "Discarded Rinses and Titration Mixtures."

15. Repeat the buret rinsing with two additional 5-mL volumes of NaOH solution.

16. Clamp the buret to the support stand, as shown in Figure 1.

17. Place the labeled 400-mL beaker directly under the buret tip. Close the buret stopcock. Lower the buret until the tip extends about 3–4 cm into the beaker and rests against the inside wall.

18. Rinse the inner surface of your short-stem funnel twice, using 5 mL of NaOH solution each time. Collect the rinses in your discard beaker. Place the funnel in the open top of your buret.

> **Note:** Make sure that the buret tip is filled with NaOH solution. There should not be any air bubbles in the solution between the stopcock and tip or in the solution in the buret barrel.

19. Close the stopcock. Fill the buret with NaOH solution to a level above the 0-mL calibration near the top of the buret. Remove the funnel from the buret, and place it in a clean 150-mL beaker.

20. Eliminate any air bubbles in the buret tip by carefully but rapidly rotating the stopcock a few times. Collect the small amount of drained NaOH solution in the

"Discarded Rinses" beaker. Then, slowly drain NaOH solution into the beaker until the bottom of the meniscus is at, or slightly below, the 0-mL calibration.

21. Lift the beaker and touch the buret tip with the wet inner side wall of the beaker *above* the solution surface to remove the drop of NaOH solution that may be clinging to the tip.

> **Note:** A 50-mL buret is calibrated in units of 0.1 mL, but measurements to the nearest 0.02 mL can be reproducibly estimated. Estimate the liquid level if it is between calibration marks, and record every reading to the nearest 0.02 mL.
>
> When reading the meniscus in the buret, you may find it helpful to hold a white card marked with a dark stripe directly behind and with the stripe slightly below the meniscus, as shown in Figure 2. Your line of sight must be level with the bottom of the meniscus.

22. Read the meniscus to the nearest 0.02 mL. Record this initial reading on your Data Sheet (6).

23. Place Erlenmeyer flask 1 under the buret tip. Lower the buret so that the tip extends about 3–4 cm into the mouth of the flask, as shown in Figure 3.

> **Note:** If you are right-handed, gently swirl the flask with your right hand and control the stopcock with your left hand, as shown in Figure 3. If you are left-handed, swirl the flask with your left hand.

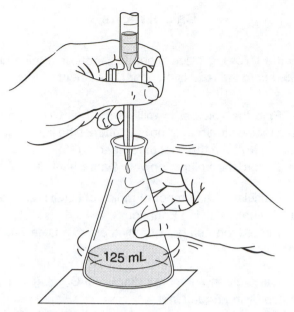

Figure 3 *Positioning the buret for titration, and manipulating the stopcock*

> **Note:** As the titration proceeds, you will observe a pink coloration at the point where NaOH solution contacts the juice solution. As you approach the end point of the titration, the pink will begin to momentarily flash through the entire solution. At this point, begin adding the NaOH solution dropwise.
>
> Because of the color of the fruit juice, the end point may be difficult to observe. Consult your laboratory instructor if you encounter difficulty detecting the end point.

24. Add 1- to 2-mL volumes of NaOH solution from the buret to the juice sample, while gently swirling the flask.

> **Note:** Stop titrating when pink persists throughout the solution for 30 s after you have thoroughly swirled the flask. When this condition occurs, you have reached the end point.

25. Take a final buret reading after the titration is complete. Record this reading to the nearest 0.02 mL on your Data Sheet (5).

26. Refill the buret with your NaOH solution, following the procedure in Steps 19–21.

27. Titrate the juice sample in Erlenmeyer flask 2, using the procedure in Steps 22–25. Record your data under column 2 on your Data Sheet (3, 4, 5, 6).

28. Consult your laboratory instructor to determine whether or not you need to perform a third determination. If a third titration is necessary, repeat Steps 3, 5, and 19–25, using a third juice sample. Record these titration data in the right margin of your Data Sheet.

After completing the third determination, ask your laboratory instructor to check and initial your Data Sheet (8).

29. Discard the NaOH solution remaining in your buret in the "Discarded Rinses and Titration Mixtures" beaker.

Thoroughly rinse your buret twice with 10 mL of tap water each time. Pour the rinses into the discard beaker. Then, rinse your buret twice with 10 mL of distilled water each time. Pour the rinses into the drain.

30. Empty the titration mixtures from Erlenmeyer flasks 1 and 2 into the "Discarded Rinses" beaker. Rinse each flask, following the procedure in Step 29.

31. Pour the remaining juice into the discard beaker. Rinse the juice beaker following the procedure in Step 29.

32. Unless your laboratory instructor indicates otherwise, pour any remaining standard NaOH solution into the discard beaker. Rinse the 125-mL flask, following the procedure in Step 29. Discard rinses into the discard beaker.

IV. Treating the "Discarded Rinses and Titration Mixtures" for Disposal

33. Add 2 drops of phenolphthalein indicator solution to the solution in the "Discarded Rinses and Titration Mixtures" beaker. Note the color of the solution in the beaker after adding the indicator. Record the color on your Data Sheet. (9)

> **Note:** Your laboratory instructor will provide 1*M* HCl and 1*M* NaOH solutions prepared specifically for neutralizing "Discarded Rinses and Titration Mixtures." If these solutions are not dispensed from dropping bottles, your laboratory instructor will demonstrate and describe a satisfactory method for dispensing the solutions.

Caution: 1*M* HCl is a corrosive, toxic solution that can cause skin irritation. Prevent contact with your eyes, skin, and clothing. Do not ingest the solution.

If you spill any HCl solution, immediately notify your laboratory instructor.

34. If the solution prepared in Step 33 is colorless, proceed to Step 35.

If the solution prepared in Step 33 is pink, add one drop of 1*M* HCl to the "Discarded Rinses and Titration Mixtures" beaker. Stir the solution with a glass stirring rod. Continue adding HCl solution dropwise, stirring after each addition, until the solution in the beaker *just* turns colorless.

Pour the colorless, neutralized solution into the drain, and dilute with a large amount of running water. Omit Step 35.

Caution: 1*M* NaOH is corrosive and toxic, and it can cause skin burns. Prevent contact with your eyes, skin, and clothing. Do not ingest the solution.

If you spill any NaOH solution, immediately notify your laboratory instructor.

35. If the solution prepared in Step 33 is colorless, add one drop of 1*M* NaOH solution to the "Discarded Rinses and Titration Mixtures" beaker. Stir with a glass stirring rod. Continue adding NaOH solution dropwise, stirring after each addition, until the solution in the beaker *just* turns light pink.

Pour the light pink, neutralized solution into the drain, diluting with a large amount of running water. Rinse the beaker with tap water and with distilled water. Discard the rinses into the drain.

Caution: Wash your hands thoroughly with soap or detergent before leaving the laboratory.

Calculations

Do the following calculations for each determination and record the results on your Data Sheet.

1.   Find the volume, in milliliters, of NaOH solution used for the titration. To do so, subtract the initial buret reading (6) from the final buret reading (5).

Record this volume on your Data Sheet (7).

2.   Calculate the number of moles of NaOH required for the titration, using Equation 3.

Record the number of moles on your Data Sheet (11).

3.   Calculate the number of moles of $C_3H_5O(CO_2H)_3$ titrated, using Equation 4.

Record the number of moles of $C_3H_5O(CO_2H)_3$ on your Data Sheet (12).

4.   Calculate the mass of $C_3H_5O(CO_2H)_3$ present in the juice sample, using Equation 5.

Record this mass on your Data Sheet (13).

5.   Calculate the mass of $C_3H_5O(CO_2H)_3$ present in 1 mL of juice, using Equation 6.

Record this mass on your Data Sheet (14).

6.   Repeat these calculations, using the data from your second determination (7, 11–14).

7.   Find the mean mass of $C_3H_5O(CO_2H)_3$ present in 1 mL of juice, using Equation 7.

mean mass of $C_3H_5O(CO_2H)_3$ in 1 mL of juice, g =

$$\frac{\text{results from Calculation 5 for determination 1} + \text{results from Calculation 5 for determination 2}}{2} \quad \text{(Eq. 7)}$$

Record the mean mass of $C_3H_5O(CO_2H)_3$ per 1 mL of juice on your Data Sheet (15).

Post-Laboratory Questions

(Use the spaces provided for the answers and additional paper if necessary.)

1. If available, compare your results with those of other class members who analyzed different juices. List the juices in order of increasing acidity.

type of juice   g $C_3H_5O(CO_2H)_3$ mL^{-1}

_____   _____

_____   _____

_____   _____

_____   _____

2. Briefly explain why it is essential that the flask in which you obtain the standard NaOH solution be completely dry, while the flask into which you pour the measured juice sample need not be dry.

3. A procedural change in this experiment would be required if a student wanted to determine the acidity of tomato juice by titrating a juice sample with NaOH solution. Briefly explain.

4. Briefly explain why you would probably obtain inaccurate results if you used the titration data you collected in this experiment to calculate the actual percent $C_3H_5O(CO_2H)_3$ in a juice sample.

name _____   section _____   date _____

Data Sheet

Experimental Data

I.   Preparing the Fruit Juice for Titration

(1)   type of fruit juice _____

(2)   code identification of fruit juice sample _____

	determination	
	1	2

(3)   volume of fruit juice transferred, mL _____   _____

III.   Titrating the Fruit Juice

(4)   molarity of NaOH solution, M _____   _____

(5)   final buret reading, mL _____   _____

(6)   initial buret reading, mL _____   _____

(7)   volume of NaOH used, mL _____   _____

(8)   acceptable titration results

_____

laboratory instructor's initials

IV.   Treating the "Discarded Rinses and Titration Mixtures" for Disposal

(9)   color of "Discarded Rinses and Titration Mixtures"
solution after addition of phenolphthalein indicator solution   _____

(10)   name and molarity of solution used to neutralize
"Discarded Rinses and Titration Mixtures" solution   _____

Treatment of Data

(11)   number of moles of NaOH required, mol   _____   _____

(12)   number of moles of $C_3H_5O(CO_2H)_3$ titrated, mol   _____   _____

(13)   mass of $C_3H_5O(CO_2H)_3$ in sample, g   _____   _____

(14)   mass of $C_3H_5O(CO_2H)_3$
       per mL of undiluted juice, g   _____   _____

(15)   mean mass of $C_3H_5O(CO_2H)_3$ per mL of undiluted juice, g   _____

Pre-Laboratory Assignment

1. Briefly explain why you should not drink any undiluted juice that has been brought into the laboratory.

2. Briefly explain the meaning of the following terms as they relate to this experiment.

(1) titration

(2) standard solution

(3) indicator

3. Briefly describe the procedure you should follow if your fruit juice sample contains excess pulp.

4. A student followed the procedure in this experiment to determine the number of grams of $C_3H_5O(CO_2H)_3$ per 1 mL of an apple juice sample. The titration of 20.0 mL of the undiluted juice required 12.84 mL of $9.580 \times 10^{-2} M$ NaOH solution.

(1) Calculate the number of moles of NaOH required for the titration.

(2) Calculate the number of moles of $C_3H_5O(CO_2H)_3$ titrated.

(3) Calculate the mass of $C_3H_5O(CO_2H)_3$ present in the juice sample.

(4) Calculate the mass of $C_3H_5O(CO_2H)_3$ present in 1 mL of apple juice.

Determining the Dissociation Constant of a Weak Acid Using pH Measurements

prepared by **Donald C. Raney**, Cañada College and
M. L. Gillette, Indiana University/Kokomo

Purpose of the Experiment

Determine the dissociation constant for an unknown weak acid using pH measurements.

Background Information

According to the Brønsted-Lowry theory, an **acid** is a substance that donates a hydrogen ion to an acceptor substance, called a **base**. A typical acid-base reaction is that of hydrochloric acid (HCl) with water, shown in Equation 1.

$$HCl(aq) + H_2O(l) \rightleftarrows H_3O^+(aq) + Cl^-(aq) \quad \text{(Eq. 1)}$$

In this reaction, HCl, the acid, donates a hydrogen ion (H^+, a proton) to a water molecule, the base. The products of the reaction are a hydronium ion (H_3O^+) and a chloride ion (Cl^-).

Another acid-base reaction is shown in Equation 2.

$$CH_3CO_2H(aq) + H_2O(l) \rightleftarrows$$
$$H_3O^+(aq) + CH_3COO^-(aq) \quad \text{(Eq. 2)}$$

In this reaction, an acetic acid molecule (CH_3CO_2H) becomes an acetate ion (CH_3COO^-) when it donates a proton to a water molecule, changing the latter into a H_3O^+ ion.

These reactions appear to be quite similar. In both, water is the base, and H_3O^+ ion is a product. There is one major difference between the two reactions: Whereas the dissociation of HCl essentially proceeds to completion, the dissociation of CH_3CO_2H occurs only to the extent of about 1%.

The relative strength of acids is determined by the extent to which they dissociate in water. The dissociation of **strong acids**, such as $HClO_4$, HNO_3, H_2SO_4, HCl, HBr, and HI, in water is virtually complete. Other acids that dissociate to a much lesser extent are considered **weak acids**. The extent of dissociation of any weak acid in water is indicated by the magnitude of its acid dissociation constant (K_a), which is the equilibrium constant for the dissociation reaction for that acid.

The acid dissociation constant expression for CH_3CO_2H is shown in Equation 3.

$$K_a = \frac{[H_3O^+][CH_3COO^-]}{[CH_3CO_2H]} \quad \text{(Eq. 3)}$$

The square brackets, [ ], indicate the molar concentration (mol L^{-1}) of the species in the brackets. Although water appears as a reactant in Equation 2, its concentration remains essentially constant at $55.5 M$ in dilute acid solutions. Therefore, the concentration of water is incorporated into the constant K_a. Because all equilibrium constants are temperature dependent, acid dissociation constants are reported with a temperature reference. For example, K_a for CH_3CO_2H is 1.7×10^{-5} at 25 $^{\circ}C$. The acid dissociation constants for some other weak acids at 25 $^{\circ}C$ are listed in Table 1.

Copyright © 1989 by Chemical Education Resources, Inc., P.O. Box 357, 220 S. Railroad, Palmyra, Pennsylvania 17078

No part of this laboratory program may be reproduced or transmitted in any form or by any means, electronic or mechanical, including photocopying, recording, or any information storage and retrieval system, without permission in writing from the publisher. Printed in the United States of America

Table 1 *Acid dissociation constants for some weak acids at 25 °C*

acetic acid	1.7×10^{-5}
boric acid	5.8×10^{-10}
chloroacetic acid	1.4×10^{-3}
crotonic acid	2.0×10^{-5}
formic acid	1.8×10^{-4}
hydrogen carbonate ion	4.8×10^{-11}
dihydrogen phosphate ion	6.2×10^{-8}
hydrogen phosphate ion	1.7×10^{-12}
hydrogen sulfite ion	5.6×10^{-8}
propionic acid	1.4×10^{-5}

Calculating Equilibrium Concentrations from K_a

Acid dissociation constants can be used to calculate H_3O^+ ion concentration in aqueous solutions of weak acids, if the molarity of the solution is known. We can, for example, calculate H_3O^+ ion concentration of a $1.0 \times 10^{-2}\,M$ solution of CH_3CO_2H at 25 °C, using Equation 3. The H_3O^+ ion concentration of this solution is $4.2 \times 10^{-4}\,M$. On the other hand, the H_3O^+ ion concentration of $1.0 \times 10^{-2}\,M$ HCl is $1.0 \times 10^{-2}\,M$. Thus you can see the large difference in the extent of dissociation of weak and strong acids.

Calculating K_a from pH Measurements

If we use the general formula HAn to stand for a typical weak acid, we can represent its dissociation using Equation 4.

$$HAn(aq) + H_2O(l) \rightleftarrows H_3O^+(aq) + An^-(aq) \quad \text{(Eq. 4)}$$

The acid dissociation constant expression for HAn is shown in Equation 5.

$$K_a = \frac{[H_3O^+][An^-]}{[HAn]} \quad \text{(Eq. 5)}$$

We can calculate the K_a for HAn if we know the concentrations of the dissociated and undissociated species in Equation 5. We can determine the H_3O^+ ion equilibrium concentration by measuring the pH of the solution and using Equation 6.

$$pH = -\log [H_3O^+] \quad \text{(Eq. 6)}$$

The An^- ion equilibrium concentration is equal to the H_3O^+ ion equilibrium concentration, according to Equation 4. If we know the original HAn concentration,

we can calculate the undissociated HAn equilibrium concentration by difference. Then we can calculate K_a, using Equation 5. When measuring solution pH for this purpose, be as accurate as possible. A small error in pH measurement will result in a large error in the calculated K_a.

Calculating K_a from Neutralization Data

Another method for determining K_a involves using pH measurements made following the partial neutralization of a weak acid with hydroxide ion (OH^-). A general equation for this reaction is shown in Equation 7.

$$HAn(aq) + OH^-(aq) \rightleftarrows (l) + An^-(aq) \quad \text{(Eq. 7)}$$

A solution containing H_3O^+ ion, An^- ion, and undissociated HAn is prepared by partially neutralizing an aqueous solution of a weak acid with aqueous sodium hydroxide (NaOH) solution. Due to the partial neutralization of HAn, the H_3O^+ ion and An^- ion concentrations are not equal. Because the reaction of OH^- ion with HAn goes essentially to completion, the number of moles of An^- ions formed is equal to the number of moles of NaOH used. Because the An^- ion concentration is so large and K_a is small, any additional dissociation of HAn shown in Equation 4 is suppressed. We can assume, therefore, that all unneutralized HAn remains in the undissociated form.

We can calculate the HAn and An^- ion equilibrium concentrations if we know the volume and concentration of HAn originally present and the volume and concentration of NaOH solution added.

The molarity (M) of a solution is defined by Equation 8.

$$\text{molarity, } M = \frac{\text{number of moles of solute}}{\text{volume of solution, L}} \quad \text{(Eq. 8)}$$

We can rearrange Equation 8 into Equation 9, which we can use to determine the initial number of moles of each reactant.

initial number of moles of solute =
$$(\text{molarity, moles L}^{-1})(\text{volume, L}) \quad \text{(Eq. 9)}$$

The number of moles of HAn at equilibrium can be determined, using Equation 10.

equilibrium number of moles of HAn		initial number of moles of HAn		initial number of moles of OH^-
	=		−	

$$\text{(Eq. 10)}$$

Then we can calculate the molar equilibrium concentration of each substance, using Equation 8.

From a pH measurement of the partially neutralized solution, we can determine the H_3O^+ ion equilibrium concentration, using Equation 6. We can substitute these concentrations into Equation 5 and calculate the K_a of HAn. Because K_a is constant for any specific weak acid, we can carry out this procedure several times, using different amounts of NaOH. From our calculated acid dissociation constants, we can determine a mean K_a.

The following example illustrates this method for determining K_a. Three solutions were prepared by adding 10.0, 20.0, and 30.0 mL of 0.200M NaOH solution to three 20.0 mL samples of 0.500M HAn solution. These mixtures were diluted with distilled water to a final volume of 250.0 mL. The pH of each of these solutions was measured. The resulting data are summarized in Table 2.

Table 2 *Volume of reactants and solutions and their pH*

	solution 1	solution 2	solution 3
volume of HAn, mL	20.0	20.0	20.0
volume of NaOH, mL	10.0	20.0	30.0
final volume, mL	250.0	250.0	250.0
pH	5.00	5.43	5.78

We can use Equation 9 to determine the initial number of moles of HAn and OH^- ion for Solution 1, as shown in Equations 11 and 12.

initial number of moles of HAn $= (0.500 \text{ mol L}^{-1})(2.00 \times 10^{-2} \text{ L})$

$= 1.00 \times 10^{-2} \text{ mol}$ (Eq. 11)

initial number of moles of OH^- $= (0.200 \text{ mol L}^{-1})(1.00 \times 10^{-2} \text{ L})$

$= 2.00 \times 10^{-3} \text{ mol}$ (Eq. 12)

Equation 13 shows the number of moles of HAn present in the mixture at equilibrium, as determined by substituting the concentrations into Equation 10.

number of moles of HAn at equilibrium $= (1.00 \times 10^{-2} \text{ mol HAn})$

$- (2.00 \times 10^{-3} \text{ mol OH}^-)$

$= 8.0 \times 10^{-3} \text{ mol HAn})$ (Eq. 13)

Using Equation 8, we can calculate the molar equilibrium concentrations of HAn and An^- ion, as shown in Equation 14 and 15, respectively.

molar equilibrium concentration of HAn $= \dfrac{8.0 \times 10^{-3} \text{ mol}}{0.250 \text{ L}}$

$= 3.2 \times 10^{-2} \text{ M}$ (Eq. 14)

molar equilibrium concentration of An^- $= \dfrac{2.00 \times 10^{-3} \text{ mol}}{0.250 \text{ L}}$

$= 8.00 \times 10^{-3} \text{ M}$ (Eq. 15)

We can then calculate the H_3O^+ ion equilibrium concentration from the measured pH, using Equation 6.

The resulting initial and equilibrium number of moles of HAn and OH^- ion and the equilibrium concentrations of HAn, An^- ion, and H_3O^+ ion are given for each solution in Table 3.

Table 3 *Initial and equilibrium number of moles and equilibrium concentrations*

initial number of moles of	solution 1	solution 2	solution 3
HAn(aq)	1.00×10^{-2}	1.00×10^{-2}	1.00×10^{-2}
OH^-(aq)	2.00×10^{-3}	4.00×10^{-3}	6.00×10^{-3}
number of moles at equilbirium of			
HAn(aq)	8.0×10^{-3}	6.0×10^{-3}	4.0×10^{-3}
An^-(aq)	2.0×10^{-3}	4.0×10^{-3}	6.0×10^{-3}
equilibrium concentration, mol L^{-1}			
HAn(aq)	3.2×10^{-2}	2.4×10^{-2}	1.6×10^{-2}
An^-(aq)	8.0×10^{-3}	1.6×10^{-2}	2.4×10^{-2}
H_3O^+(aq)	1.0×10^{-5}	3.7×10^{-6}	1.7×10^{-6}

We can calculate the K_a for HAn in Solution 1 using Equation 5 and the data in Table 3.

$$K_a = \frac{[H_3O^+][An^-]}{[HAn]} \qquad \text{(Eq. 5)}$$

$$= \frac{(1.0 \times 10^{-5})(8.0 \times 10^{-3})}{(3.2 \times 10^{-2})} \qquad \text{(Eq. 16)}$$

$$= 2.5 \times 10^{-6}$$

Similarly, we can calculate the K_a for HAn from the data for Solutions 2 and 3. Using the three values, we can determine a mean K_a.

Determining K_a from a Graph of Neutralization Data

We can determine a K_a for HAn by graphing the data used in the preceeding method. We know that we can calculate the number of moles of HAn present at equilibrium in any solution by subtracting the number of moles of OH^- ion added from the number of moles of HAn initially present. If we symbolize the number of moles of HAn initially present as B, we can rewrite Equation 5 as Equation 17.

$$K_a = \frac{[H_3O^+][An^-]}{[B] - [An^-]} \qquad \text{(Eq. 17)}$$

We can rearrange Equation 17 in several steps to give Equation 18.

$$[H_3O^+] = \frac{K_a([B] - [An^-])}{[An^-]} = \frac{K_a[B]}{[An^-]} - K_a$$

$$\frac{[H_3O^+]}{K_a[B]} = \frac{1}{[An^-]} - \frac{1}{[B]} \qquad \text{(Eq. 18)}$$

$$\frac{1}{[An^-]} = \left(\frac{1}{K_a[B]}\right)[H_3O^+] + \frac{1}{[B]}$$

Equation 18 is in the form of a straight line equation, $y = mx + b$, where $y = 1/[An^-]$ and $x = [H_3O^+]$. We can plot $[H_3O^+]$ on the x–axis (the abscissa) and $1/[An^-]$ on the y-axis (the ordinate) and draw a best straight line through the points. The slope of this line will be $1/(K_a[B])$ and the x-intercept where $x = 0$ will be equal to $1/[B]$.

To determine K_a graphically from the data for Solutions 1, 2, and 3, we use the H_3O^+ ion concentrations and the reciprocal of the An^- ion equilibrium concentrations listed in Table 3. These data are given in Table 4 and plotted in Figure 1.

We can find the slope of the straight line connecting the data points by first determining the coordinates of two widely separated points on the line. The ordinates of two such points in Figure 1 are 45 and 95, while the abscissae are 2.0×10^{-6} and 7.0×10^{-6}, respectively.

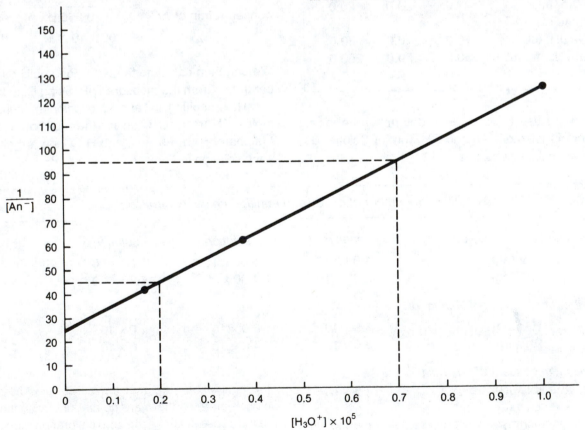

Figure 1   *Graphically determining K_a*

Table 4 *Equilibrium [H₃O⁺] and reciprocals of equilibrium [An⁻]*

solution number	[H₃O⁺]	$\frac{1}{[An^-]}$	K_a
1	1.0×10^{-5}	1.3×10^{2}	2.5×10^{-6}
2	3.7×10^{-6}	6.3×10^{1}	2.5×10^{-6}
3	1.7×10^{-6}	4.2×10^{1}	2.6×10^{-6}
			mean $K_a = 2.5 \times 10^{-6}$

We can calculate the slope of the line, using Equation 19.

$$\text{slope} = \frac{\Delta y}{\Delta x} = \frac{(95 - 45)}{(7.0 \times 10^{-6}) - (2.0 \times 10^{-6})} \quad \text{(Eq. 19)}$$

$$= 1.0 \times 10^{7}$$

In Equation 18, we saw the relation between the dissociation constant and the slope.

$$\text{slope} = \frac{1}{(K_a[B])}$$

We can rewrite this relationship as Equation 20

$$K_a = \frac{1}{(\text{slope})[B]} \quad \text{(Eq. 20)}$$

where $[B] = [HAn]$ prior to any dissociation or neutralization and is calculated as follows, using Equation 21.

$$[B] = \frac{\text{initial number of moles of HAn(aq)}}{\text{total volume of solution, L}} \quad \text{(Eq. 21)}$$

$$= \frac{1.00 \times 10^{-2} \text{ mol}}{2.50 \times 10^{-1} \text{ L}} = 4.00 \times 10^{-2}$$

If we insert this value of B into Equation 20, we can determine K_a as follows:

$$K_a = \frac{1}{(1.0 \times 10^{7})(4.0 \times 10^{-2})}$$

$$= 2.5 \times 10^{-6}$$

Because this and the preceeding methods of determining K_a result in the exact same value, either of these methods can be used to determine the K_a of an unknown weak acid.

Doing this Experiment

In this experiment, you will prepare several aqueous solutions of a weak acid. The solutions will be composed of equal volumes of the acid, but will contain various amounts of a standard NaOH solution. You will add distilled or deionzied water to each solution so that the final volume of all of the solutions will be identical. You will measure the pH of each solution, using a pH meter. From the pH and the volumes of acid and NaOH solution, you will determine the concentrations of H_3O^+ ion, of undissociated acid, and of the anion of the dissociated acid in each solution. You will use these data to calculate the K_a of your unknown acid. Finally, you will compare these K_a values to the K_a you will determine from a graphical treatment of the same data.

Procedure

Chemical Alert

0.500*M* NaOH—corrosive and toxic

Caution: Wear departmentally approved eye protection while doing this experiment.

1. Clean two 50-mL burets and a 250-mL volumetric flask. Rinse each buret and the volumetric flask three times with separate 10-mL portions of distilled or deionized water.

2. Obtain 180 mL of a 1.00*M* unknown weak acid solution in a clean, dry 250-mL Erlenmeyer flask. Record the number of your unknown and the exact molarity of the solution on Data Sheet 1.

3. Rinse one of the clean burets with a 5-mL portion of the unknown weak acid solution. Be sure that the rinse solution contacts as much of the inner surface of the buret as possible. Drain the rinse solution through the tip of the buret into a small beaker. Collect this solution and other solutions and waste liquids in this beaker, and discard them according to the directions of your laboratory instructor. Repeat this procedure twice with two additional 5-mL portions of the weak acid solution.

4. Close the buret stopcock. Fill the buret to a level above the zero calibration mark with the unknown weak acid solution. Allow any air bubbles trapped in the solution to rise to the surface. Briefly open the stopcock to be sure that the buret tip is filled with solution and that there are no air bubbles in the tip. If the meniscus is above the zero calibration mark, open the stopcock and drain the solution through the buret tip into the small beaker, containing the waste solutions, until the level reaches the calibrated portion of the buret. The level does not have to align exactly with the zero calibration mark, as long as the actual initial volume is read and recorded. Touch the buret tip to a wet glass surface to remove the hanging droplet. Read the buret to the nearest 0.01 mL, and record this volume on Data Sheet 1.

> **Caution:** 0.500M NaOH is a corrosive and toxic solution that can cause burns. Prevent contact with your eyes, skin, and clothing.

5. Obtain 140 mL of an approximately 0.500M NaOH solution in another clean, dry 250-mL Erlenmeyer flask. Record the exact molarity of the solution on Data Sheet 1.

6. Following the same procedure you used for the unknown acid solution, rinse and fill the second buret with the NaOH solution. Record your initial buret reading to the nearest 0.01 mL on Data Sheet 1.

7. Prepare Solution 1 by adding 40.00 mL of the unknown acid solution from the buret to a clean 250-mL volumetric flask. Record your final buret reading to the nearest 0.01 mL on Data Sheet 1.

Add 12.50 mL of the NaOH solution from the second buret to the acid solution in the volumetric flask. Record your final buret reading to the nearest 0.01 mL on Data Sheet 1.

Table 5   *Laboratory assignments*

solution number	volume of 1.00M unknown weak acid solution, mL	volume of 0.500M sodium hydroxide solution, mL
1	40.00	12.50
2	40.00	25.00
3	40.00	37.50
4	40.00	50.00

8. Add distilled or deionized water to the solution in the volumetric flask until the bottom of the meniscus of the solution is at the base of the flask neck. Stopper the flask. While firmly holding the stopper with your forefinger, invert the flask ten times to thoroughly mix the solution.

After the trapped air bubbles rise to the surface, add distilled water until the bottom of the meniscus coincides with the calibration mark on the flask neck. Stopper the flask. While firmly holding the stopper with your forefinger, invert the flask ten times to thoroughly mix the solution. Label the flask Solution 1. Record the total volume of the solution on Data Sheet 1.

> **Note:** Your laboratory instructor might give you additional directions for using your pH meter.

9. After your pH meter has warmed up, standardize it with a pH–7 buffer solution, using the following procedure. First, thoroughly rinse the electrode(s) with distilled water, using a wash bottle. Collect the rinse water in a small beaker. Then, obtain 50 mL of the pH–7 buffer solution in a clean, dry 100-mL beaker. Carefully immerse the electrode(s) in the buffer solution. **Remember that glass electrodes are fragile.**

Measure the temperature of the buffer solution. Set the temperature compensation knob to the solution temperature. Turn the function knob to pH. A pH value should appear on the display. Use the standardization knob to adjust the display until it agrees exactly with the pH of your buffer solution. Wait several seconds to be certain that the reading remains constant.

Turn the function knob to standby. Carefully remove the electrode(s) from the buffer solution. Return the buffer solution to your laboratory instructor. Thoroughly rinse the electrode(s) with distilled water, using a wash bottle. Collect the rinse water in the same small beaker.

10. Pour about 50 mL of Solution 1 into a clean, dry 100-mL beaker. Place the electrode(s) in this solution. Turn the function knob to pH. Read the pH to the nearest 0.01 pH unit and record it on Data Sheet 1. Turn the function knob to standby. Carefully remove the electrode(s) from the solution. Thoroughly rinse them with distilled water, collecting the rinse water in the same beaker. Discard the solution in the beaker and in the volumetric flask following the directions of your laboratory instructor. Thoroughly wash the beaker and flask with distilled water. Thoroughly rinse the beaker with three 10-mL portions of distilled water and the flask with three 20-mL portions of distilled water.

11. Use the amounts of reactants specified in Table 5 to prepare Solution 2. Measure the pH of Solution 2 to the nearest 0.01 pH unit following Steps 9 and 10. Record this pH on Data Sheet 1.

12. Repeat Step 11 two more times, once using the amounts of reactants specified in Table 5 to prepare Solution 3 and again to prepare Solution 4. Record these data on Data Sheet 1.

13. Before leaving the laboratory, thoroughly rinse the electrode(s) with distilled water, collecting the rinse water in the same small beaker. Leave the electrode(s) immersed in distilled water, following the directions of your laboratory instructor.

> *Caution:* Wash your hands thoroughly with soap or detergent before leaving the laboratory.

Calculations

Do the following calculations for each solution and record the results on Data Sheet 2.

1. Calculate the volume of unknown acid and of NaOH solution used.

2. Calculate the initial number of moles of HAn and of OH^- ion used to prepare Solutions 1, 2, 3, and 4, using Equation 8.

3. Record the number of moles of An^- ion in each solution at equilibrium. This value is equal to the number of moles of OH^- ion added.

4. Determine the number of moles of HAn in each solution at equilibrium, using Equation 10.

5. Calculate the molar equilibrium concentration of HAn for each solution, using Equation 10.

6. Calculate $[An^-]$, using Equation 10, and $1/[An^-]$ for each solution at equilibrium.

7. Calculate the molar concentration of H_3O^+ ion at equilibrium for each solution from your pH readings, using Equation 6.

8. Use the equilibrium concentrations of H_3O^+, An^-, and HAn determined for each of the four solutions, to calculate the K_a of your weak acid. Record the K_a for each solution on Data Sheet 2.

9. To calculate K_a using the graphical method, prepare a graph by plotting the reciprocal of the An^- ion concentrations on the ordinate and the corresponding H_3O^+ ion concentrations on the abscissa. Draw the best straight line through the plotted data points. Determine the slope of the line, using Equation 19, and calculate K_a, using Equation 5.

Post-Laboratory Questions

(Use the spaces provided for the answers and additional paper if necessary.)

1. Obtain the accepted dissociation constant for your unknown acid from your laboratory instructor. Calculate the percent error in your graphical determination of the K_a of this acid.

3. The equations used in this experiment would have to be modified slightly if an acid with a K_a of 1.0×10^{-3} or greater were used as an unknown. Assuming such a situation, explain the problems that would result from the use of the equations in their present form. What modifications would you make to give accurate results when using such an unknown?

2. (1) Compute the percent error for the dissociation constant determined from each individual solution.

(2) Compute the average K_a from the values for the four solutions. Compute the percent error for the average K_a.

Data Sheet 1

unknown number _____

molarity of unknown acid _____   of NaOH solution _____

solution

	1	2	3	4

unknown acid solution

	1	2	3	4
final buret reading, mL	_____	_____	_____	_____
initial buret reading, mL	_____	_____	_____	_____
volume of unknown acid, mL	_____	_____	_____	_____

NaOH solution

	1	2	3	4
final buret reading, mL	_____	_____	_____	_____
initial buret reading, mL	_____	_____	_____	_____
volume of NaOH solution, mL	_____	_____	_____	_____

	1	2	3	4
total volume of solution, mL	_____	_____	_____	_____
pH	_____	_____	_____	_____

34 • Dissociation Constant of a Weak Acid

Data Sheet 2

Calculating K_a of your unknown acid

solution

	1	2	3	4
initial number of moles				
HAn(aq)	_____	_____	_____	_____
OH⁻(aq)	_____	_____	_____	_____
number of moles at equilibrium				
HAn(aq)	_____	_____	_____	_____
An⁻(aq)	_____	_____	_____	_____
equilibrium concentration, mol L⁻¹				
HAn(aq)	_____	_____	_____	_____
An⁻(aq)	_____	_____	_____	_____
H₃O⁺(aq)	_____	_____	_____	_____
1/[An⁻]	_____	_____	_____	_____
K_a	_____	_____	_____	_____

slope of line _____

K_a, determined from slope _____

Pre-Laboratory Assignment

1. Review Experiment 4 for a discussion of buret use. Read an authoritative source for a discussion of pH meters.

2. A student doing this experiment spilled a small amount of 0.500M NaOH solution on the bench top. What procedure should be followed to clean up the spill?

(3) For each solution, calculate the number of moles of OH⁻ ion added.

(4) For each solution, determine the number of moles of HAn and of An⁻ ion at equilibrium.

3. A student following the procedure in this experiment prepared three solutions by adding 11.50, 23.00, and 34.50 mL of a $7.60 \times 10^{-2}M$ NaOH solution to 50.00 mL of a 0.112M solution of a weak acid. The solutions were labeled 1, 2, and 3, respectively. Each of the solutions was diluted to a total volume of 100.0 mL with distilled water. The pH readings of these solutions were:
(1) 6.42, (2) 6.81, and (3) 7.09.

(1) Convert the pH of each solution to an equivalent H_3O^+ ion concentration.

(5) For each solution, calculate the HAn and An⁻ ion molar equilibrium concentrations.

(2) For each solution, calculate the number of moles of acid added.

(6) Determine the reciprocal of the An⁻ ion concentration for each solution.

(7) On the graph paper provided on p. 13, plot the reciprocal of the An^- ion concentration on the ordinate against the H_3O^+ ion concentration on the abscissa. Draw the best straight line through the plotted points.

(8) Find the slope of the line.

_____
answer

(9) Determine the initial concentration of the acid, HAn. Use the concentration and the slope determined in (8) to calculate the K_a for the weak acid.

_____
answer

4. The student whose experiment is described in Pre-Laboratory Assignment 3 had not used a buret before and did not realize the importance of good buret technique. While filling the buret with $7.60 \times 10^{-2} M$ NaOH solution, the student forgot to open the stopcock, so the buret tip remained empty. Explain the effect this oversight has on the determination of the K_a of this acid, based on the data given in Pre-Laboratory Assignment 3.

A Study of pH, Dissociation, Hydrolysis, and Buffers

prepared by **Norma Robinson North**, State University of New York at Potsdam

---

Purpose of the Experiment

Compare the calculated pH and the measured pH of a series of solutions of hydrochloric acid and of acetic acid. From pH measurements, calculate the acid dissociation constant of acetic acid.

Measure the pH of various salt solutions. From pH measurements, calculate the hydrolysis constant of ammonium chloride.

Compare the measured pH and the calculated pH of a buffer solution; of the buffer solution and HCl; of the buffer solution and NaOH; of water with HCl; and of water with NaOH.

Background Information

Strong Acids

Strong acids in dilute solution are generally considered to be completely dissociated. In this case, the concentration of the hydronium ion, $[H_3O^+]$, is equal to the concentration of the acid solution. The brackets indicate the concentration of the enclosed species in $mol\ L^{-1}$. A strong base is also completely dissociated in dilute solution, and the $[OH^-]$ is the same as the concentration of the base solution.

The $[H_3O^+]$ and $[OH^-]$ are often very small. To alleviate some of the inconvenience of using these small numbers, Sørenson in 1909 proposed the adoption of an exponential system. The new term used to express the $[H_3O^+]$ is **pH**. Mathematically, pH can be defined as

$$pH = \log \frac{1}{[H^+]} = -\log [H^+] \qquad \text{(Eq. 1)}$$

$$[H_3O^+] = 10^{-pH} \qquad \text{(Eq. 2)}$$

Equations 1 and 2 are identical mathematically. By using the same type of definition, the $[OH^-]$ can be expressed in the logarithmic system as

$$pOH = \log \frac{1}{[OH^-]} = -\log [OH^-] \qquad \text{(Eq. 3)}$$

$$[OH^-] = 10^{-pOH} \qquad \text{(Eq. 4)}$$

Because the product $[H_3O^+] [OH^-]$ must always equal 10^{-14},

$$pH + pOH = 14.00 \qquad \text{(Eq. 5)}$$

Many dilute solutions have a pH between 1 and 14. This scale, which appears on a pH meter, is presented for both pH and pOH in Figure 1.

The following examples show some mathematical applications of Equations 1 and 3.

Copyright © 1983 by Chemical Education Resources, Inc., P.O. Box 357, 220 S. Railroad, Palmyra, Pennsylvania 17078
No part of this laboratory program may be reproduced or transmitted in any form or by any means, electronic or mechanical, including photocopying, recording, or any information storage and retrieval system, without permission in writing from the publisher. Printed in the United States of America

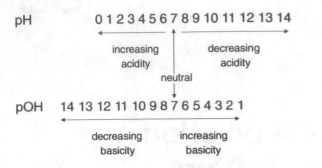

Figure 1   *pH and pOH scales*

Example 1.   The $[H_3O^+]$ of a solution is 3.4×10^{-3} mol L^{-1}. Calculate the pH of the solution.

$$pH = -\log [H_3O^+]$$
$$pH = -\log (3.4 \times 10^{-3})$$
$$pH = 2.47 \qquad \text{(Eq. 1)}$$

Example 2.   The pH of a solution is 4.3. Calculate the $[H_3O^+]$ of the solution.

$$pH = -\log [H_3O^+]$$
$$4.3 = -\log [H_3O^+]$$
$$-4.3 = \log [H_3O^+]$$
$$[H_3O^+] = 5 \times 10^{-5} \text{ mol } L^{-1}$$

Example 3.   The $[OH^-]$ of a solution is 1×10^{-2} mol L^{-1}. Calculate the pH of the solution.

$$pOH = -\log [OH^-] \qquad \text{(Eq. 3)}$$
$$pOH = -\log (1 \times 10^{-2})$$
$$pOH = -(0 - 2.0) = 2.0$$
$$pH + pOH = 14.0 \qquad \text{(Eq. 5)}$$
$$pH = 12.0$$

Dissociation

Weak acids are those that do not completely dissociate in aqueous solutions. The $[H_3O^+]$ is not equal to the concentration of the acid solution. In the case of a weak base, the $[OH^-]$ is less than the concentration of the base solution.

Aqueous solutions of acetic acid, CH_3CO_2H, which is commonly written as HOAc, dissociate according to the reversible reaction represented by Equation 6.

$$H_2O + HOAc(aq) \rightleftarrows H_3O^+(aq) + OAc^-(aq) \qquad \text{(Eq. 6)}$$

The acid dissociated constant, K_a, can be calculated from Equation 7.

$$K_a = \frac{[H_3O^+][OAc^-]}{[HOAc]} = 1.8 \times 10^{-5} \qquad \text{(Eq. 7)}$$

The general acid dissociation expression is shown in Equation 8.

$$K_a = \frac{[H_3O^+][A^-]}{[HA]} \qquad \text{(Eq. 8)}$$

Example 4.   The pH of a 0.10M HOAc solution is 2.87. Calculate the K_a of acetic acid.

$$pH = -\log [H_3O^+] \qquad \text{(Eq. 1)}$$
$$2.87 = -\log [H_3O^+]$$
$$-2.87 = \log [H_3O^+]$$
$$1.4 \times 10^{-3} \text{ mol } L^{-1} = [H_3O^+] = [OAc^-]$$
$$K_a = \frac{[H_3O^+][OAc^-]}{[HOAc]} \qquad \text{(Eq. 7)}$$
$$K_a = \frac{(1.4 \times 10^{-3})^2}{0.10}$$
$$K_a = 2.0 \times 10^{-5}$$

The concentration of undissociated acetic acid is considered to be the initial concentration of the acid. The relatively small dissociation to form $H_3O^+(aq)$ and $OAc^-(aq)$ may be safely neglected.

Hydrolysis

Hydrolysis is a reaction of a salt with water. Because of this reaction, aqueous solutions of salts are not necessarily neutral, that is, having a pH of 7. A salt solution is basic only if the anion undergoes hydrolysis, and it is acidic only if the cation undergoes hydrolysis. Salts may be classified generally into four types by considering them as neutralization products of acid–base reactions, as shown in Table 1.

Table 1   *Types of salts*

(1)	Strong base and strong acid (neutral, no hydrolysis)
(2)	Strong base and weak acid (basic, only anion hydrolyzes)
(3)	Weak base and strong acid (acidic, only cation hydrolyzes)
(4)	Weak base and weak acid (both ions hydrolyze, the pH depending on the extent of hydrolysis of each ion)

The ions that hydrolyze do so because a weak acid or a weak base is formed. The hydrolysis removes ions from the solution and serves as a driving force for the reaction. In the case of a Type (2) salt, NaOAc for example, Na^+ cannot undergo hydrolysis because no weak base is formed. However, the OAc⁻ can hydrolyze because the weak acid HOAc is formed, as shown in Equation 9.

$$OAc^-(aq) + H_2O\ (l) \rightleftarrows HOAc(aq) + OH^-(aq)$$

$$(Eq.\ 9)$$

The equilibrium expression for Equation 9 is

$$K_e = \frac{[HOAc\ [OH^-]}{[OAc^-]} \qquad (Eq.\ 10)$$

For the reaction shown in Equation 11,

$$2\ H_2O(l) \rightleftarrows H_3O^+(aq) + OH^-(aq) \qquad (Eq.\ 11)$$

the equilibrium expression is

$$K_w = [H_3O^+]\ [OH^-] = 1.0 \times 10^{-14}\ at\ 25\ ^\circ C$$

$$(Eq.\ 12)$$

By considering the concentration of water to be a constant, the hydrolysis constant, K_h, for sodium acetate can be expressed as in Equation 13.

$$K_h = \frac{[HOAc]\ [OH^-]}{[OAc^-]} \qquad (Eq.\ 13)$$

K_h can be calculated from the K_a for the weak electrolyte produced and K_w for water. This calculation can be made because the hydrolysis actually involves two equilibria, one for HOAc and one for water.

If [OH⁻] is replaced in Equation 13 by its equivalent, $K_w\ /\ [H_3O^+]$, K_h may be written as

$$K_{h_{HOAc}} = \frac{k_w}{K_{a_{HOAc}}} = \frac{1.0 \times 10^{-14}}{1.8 \times 10^{-5}} = 5.6 \times 10^{-10}$$

$$(Eq.\ 14)$$

For a Type (3) salt such as NH₄Cl, only the NH₄⁺ hydrolyzes, as shown in Equation 15.

$$H_2O + NH_4^+(aq) \rightarrow NH_3(aq) + H_3O^+(aq) \qquad (Eq.\ 15)$$

The equilibrium expression for Equation 15 can be written

$$K_{h_{NH_3}} = \frac{[NH_3]\ [H_3O^+]}{[NH_4^+]} \qquad (Eq.\ 16)$$

By replacing [H₃O⁺] with its equivalent, $K_w\ /\ [OH^-]$, Equation 17 is obtained.

$$K_{h_{NH_3}} = \frac{K_w}{K_{b_{NH_3}}} = \frac{1.0 \times 10^{-14}}{1.8 \times 10^{-5}} = 5.6 \times 10^{-10}$$

$$(Eq.\ 17)$$

Thus, NH₄⁺ hydrolyzes to the same extent as does OAc⁻.

The use of pH in calculating K_h is illustrated in Example 5.

Example 5. A 0.050M solution of NaC₂H₃O₂ (NaOAc) has a pH of 8.72. Calculate the K_h for NaOAc.

$$NaOAc(aq) \rightarrow Na^+(aq) + OAc^-(aq)$$

$$OAc^-(aq) + H_2O(l) \rightleftarrows HOAc(aq) + OH^-(aq)$$

$$K_h = \frac{[HOAc]\ [OH^-]}{[OAc^-]} \qquad (Eq.\ 13)$$

$$[HOAc] = [OH^-]$$
$$pH + pOH = 14.00$$
$$pOH = 14.00 - 8.72$$
$$pOH = 5.28$$
$$5.28 = -\log[OH^-]$$
$$-5.28 = \log[OH^-]$$
$$5.3 \times 10^{-6}\ mol\ L^{-1} = [OH^-] = [HOAc]$$

$$K_h = \frac{[HOAc]\ [OH^-]}{[OAc^-]}$$

$$K_h = \frac{(5.3 \times 10^{-6})^2}{0.050}$$

$$K_h = 5.6 \times 10^{-10}$$

The concentration of OAc⁻(aq) is assumed to be the same as the initial concentration of NaOAc.

Buffer Solutions

Solutions that undergo only a small change in pH when small quantities of acid or base are added to them are called **buffer solutions**. For example, human blood is an extremely complex buffered mixture with a normal pH range of 7.35 − 7.45. If, for any reason, the pH of the blood changes ±0.1 pH unit, serious damage to the body can result.

A buffer solution is prepared from a weak acid and a salt of this acid, or from a weak base and a salt of this base. These solutions, as a result of the common ion in each case, tend to resist a change in pH. The dependence of [H₃O⁺] on the ratio of the concentration of the undissociated acid to the concentration of the anion of the acid can be seen in the following equations:

$$HA(aq) \rightleftarrows H_3O^+(aq) + A^-(aq) \qquad (Eq.\ 18)$$

$$K_a = \frac{[H_3O^+]\ [A^-]}{[HA]} \qquad (Eq.\ 8)$$

$$[H_3O^+] = \frac{K_a\ [HA]}{[A^-]} \qquad (Eq.\ 19)$$

Equation 19 shows that the pH of the solution is governed by the ratio [HA] / [A⁻].

The dependence of [OH⁻] on the ratio of the concentration of the undissociated base to the concentration of the anion of the base can be seen in the following equations:

$$BOH(aq) \rightleftarrows B^+(aq) + OH^-(aq) \quad (Eq.\ 20)$$

$$K_h = \frac{[B^+][OH^-]}{[BOH]} \quad (Eq.\ 21)$$

$$[OH^-] = \frac{K_h[BOH]}{[B^+]} \quad (Eq.\ 22)$$

Equation 22 shows that the pOH of the solution is governed by the ratio [BOH] / [B⁺].

If a small quantity of base is added to a buffered acid solution, the OH⁻ will react with the H_3O^+ to form water. This reaction will tend to lower the $[H_3O^+]$. However, the weak acid in the buffer solution will then dissociate to produce additional H_3O^+ and maintain the equilibrium. The net result is that the [HA] will decrease while the [A⁻] will increase. The ratio of [HA] / [A⁻], which controls the pH, changes only slightly and relatively constant pH will be maintained.

Solutions of weak bases and their salts behave in the same manner to maintain a relatively constant pH.

Calculations involving buffer solutions and buffering action are illustrated by the following examples.

Example 6. Calculate the pH of a buffer solution that contains 1.0 mol L⁻¹ each of sodium acetate, NaOAc, and acetic acid, HOAc.

$$[H_3O^+] = \frac{K_a[HOAc]}{[OAc^-]}$$

$$[H_3O^+] = \frac{(1.8 \times 10^{-5})(1.0\ mol\ L^{-1})}{(1.0\ mol\ L^{-1})}$$

$$[H_3O^+] = 1.8 \times 10^{-5}$$
$$pH = -\log[H_3O^+]$$
$$pH = -\log(1.8 \times 10^{-5})$$
$$pH = 4.74$$

Example 7. If sufficient HCl is added to the buffer solution in Example 6 to give an HCl concentration of 0.2 mol L⁻¹, what will be the resulting pH?

The completely dissociated HCl gives a concentration of 0.2 mol L⁻¹ of H_3O^+. The H_3O^+ will combine with the OAc⁻ to form HOAc at a concentration of 0.2 mol L⁻¹. Thus, the concentration of HOAc will be increased to 1.2 mol L⁻¹ and the concentration of OAc⁻ will be reduced to 0.8 mol L⁻¹.

$$[H_3O^+] = \frac{K_a[HOAc]}{[OAc^-]}$$

$$[H_3O^+] = (1.8 \times 10^{-5})\left(\frac{1.0\ mol\ L^{-1} + 0.2\ mol\ L^{-1}}{1.0\ mol\ L^{-1} - 0.2\ mol\ L^{-1}}\right)$$

$$[H_3O^+] = (1.8 \times 10^{-5})\left(\frac{1.2}{0.8}\right)$$

$$[H_3O^+] = 2.7 \times 10^{-5}\ mol\ L^{-1}$$
$$pH = -\log[H_3O^+]$$
$$pH = -\log(2.7 \times 10^{-5})$$
$$pH = 4.57$$

Example 8. If sufficient NaOH were added to the buffer solution in Example 6 to give a NaOH concentration of 0.2 mol L⁻¹, what would be the resulting pH?

The completely dissociated NaOH will given an OH⁻ concentration of 0.2 mol L⁻¹. This OH⁻ will combine with H_3O^+ to form water. The H_3O^+ is produced by the dissociation of HOAc. Thus, the HOAc concentration is reduced to 0.8 mol L⁻¹ and the OAc⁻ concentration will be increased to 1.2 mol L⁻¹.

$$[H_3O^+] = \frac{K_a[HOAc]}{[OAc^-]}$$

$$[H_3O^+] = (1.8 \times 10^{-5})\left(\frac{1.0\ mol\ L^{-1} + 0.2\ mol\ L^{-1}}{1.0\ mol\ L^{-1} - 0.2\ mol\ L^{-1}}\right)$$

$$[H_3O^+] = (1.8 \times 10^{-5})\left(\frac{0.8}{1.2}\right)$$

$$[H_3O^+] = 1.2 \times 10^{-5}\ mol\ L^{-1}$$
$$pH = -\log[H_3O^+]$$
$$pH = -\log(1.2 \times 10^{-5})$$
$$pH = 4.92$$

Example 9. If sufficient HCl is added to 1.0 L of distilled water to make the solution 0.2M in HCl, what will be the pH?

$$[H_3O^+] = 0.2\ mol\ L^{-1}$$
$$pH = -\log[H_3O^+]$$
$$pH = -\log(2 \times 10^{-1})$$
$$pH = 0.7$$

In this experiment, the pH of a number of different solutions will be calculated from the stated concentrations of the solutions. The measured pH and the calculated pH will be compared. In addition, the acid dissociation constant of acetic acid and the hydrolysis constant of ammonium chloride will be calculated from pH measurements.

Procedure

> **Caution:** Wear departmentally approved eye-protection while doing this experiment.

I. The pH of Hydrochloric Acid Solutions

1. Obtain 100 mL of a 0.1M hydrochloric acid solution from the laboratory instructor.

2. Using a graduated cylinder, pour 50 mL of this solution into a 150-mL beaker. Label the beaker with the concentration of acid.

3. Measure 5 mL of the remaining acid solution into a graduated cylinder and dilute to 50 mL with distilled water.

4. Transfer the solution prepared in Step 3 to a second 150-mL beaker and mix thoroughly. Label this beaker with the new concentration of the acid solution.

5. Repeat Steps 3 and 4 with 5 mL of the solution prepared in Step 4.

6. Repeat Steps 3 and 4 with 5 mL of the solution prepared in Step 5.

7. Using a pH meter, measure the pH of each of the four solutions just prepared. Contact the laboratory instructor for the proper use of the pH meter and for directions on cleaning its electrodes.

8. Record each pH on Data Sheet 1 under "Measured pH."

9. Discard all solutions in accordance with the laboratory instructor's directions.

II. The pH of Acetic Acid Solutions

1. Obtain 100 mL of a 0.1M acetic acid solution from the laboratory instructor.

2. Transfer 50 mL of this solution to a 150-mL beaker by using a graduated cylinder. Label the beaker with the acid concentration.

3. Add 5 mL of the remaining acid solution to a graduated cylinder and dilute to 50 mL with distilled water.

4. Transfer the solution prepared in Step 3 to a second 150-mL beaker. Label the beaker with the new concentration of the acid solution.

5. Repeat Steps 3 and 4 with 5 mL of the solution prepared in Step 4.

6. Repeat Steps 3 and 4 with 5 mL of the solution prepared in Step 5.

7. Measure the pH of each of the four solutions with a pH meter. Follow the directions of the laboratory instructor for cleaning the electrodes of the pH meter.

8. Record each pH on Data Sheet 1 under "Measured pH."

9. Discard all solution in accordance with the laboratory instructor's directions.

III. The pH of Various Salt Solutions

1. Obtain samples of the following aqueous solutions: 0.1M NH_4OAc, 0.1M NaCl, 0.1M NaOAc, 0.1M NH_4Cl, 0.1M $NaHCO_3$, and 0.1M Na_2CO_3.

2. Place 50 mL of NH_4OAc solution in a clean 150-mL beaker.

3. Measure the pH of the solution with a pH meter.

4. Record the pH on Data Sheet 1 under "Measured pH."

5. Discard the solution in accordance with the laboratory instructor's directions.

6. Rinse the beaker twice with distilled water. Follow the directions of the laboratory instructor for cleaning the electrodes of the pH meter.

7. Repeat Steps 1 through 6 with each of the remaining solutions.

8. Discard all solutions in accordance with the laboratory instructor's directions.

IV. The pH of Buffer Solutions

1. Weigh a clean 250-mL beaker on a centigram balance to the nearest 0.1 g.

2. Add 4.1 g of solid sodium acetate to the beaker.

3. Remove the beaker from the balance and add 8.5 mL of 6M acetic acid from a graduated cylinder to the beaker and sodium acetate.

4. Add 91.5 mL of distilled water from a graduated cylinder to the solution prepared in Step 3. Stir the solution until all sodium acetate is dissolved.

5. Label four clean 150-mL beakers 1, 2, 3, and 4.

6. Using a graduated cylinder, pour 40 mL of the buffer solution into beakers 1 and 3.

7. Add 40 mL of distilled water from a graduated cylinder to beakers 2 and 4.

8. Record the pH of each solution on Data Sheet 2 under "Measured pH."

9. Remove the electrodes of the pH meter from the solution and rinse them with distilled water into an empty beaker.

10. Pipet 1.0 mL of 6M hydrochloric acid solution to beakers 1 and 2. Measure the pH for each solution and record on Data Sheet 2 under "Measured pH."

11. Pipet 1.0 mL of 6M sodium hydroxide solution to beakers 3 and 4. Measure the pH for each solution and record on Data Sheet 2 under "Measured pH."

12. Follow the directions of the laboratory instructor for preparing the pH meter for use by the next student.

13. Dispose of all solutions in accordance with the laboratory instructor's directions.

Calculations

I. The pH of Hydrochloric Acid Solutions

Calculate the theoretical pH for each concentration of the hydrochloric acid solutions, using Equation 1.

$$pH = -\log [H_3O^+] \qquad \text{(Eq. 1)}$$

The hydrogen ion concentration is the same as the acid concentration. Record each calculated pH on Data Sheet 1, Part I., under "Calculated pH."

II. The pH of Acetic Acid Solutions

1. Assuming that the hydrogen ion concentration is the same as the acetate ion concentration, calculate the pH of each acetic acid solution from Equation 7, using $K_a = 1.8 \times 10^{-5}$. Record each calculated pH on Data Sheet 1, Part II., Under "Calculated pH."

$$K_a = \frac{[H_3O^+][OAc^-]}{[HOAc]} \qquad \text{(Eq. 7)}$$

2. Calculate the dissociation constant, K_a, for acetic acid for each solution, using the pH obtained from the pH meter. Assume the concentration of undissociated acetic acid is the same as the initial concentration.

$$K_a = \frac{[H_3O^+][OAc^-]}{[HOAc]} \qquad \text{(Eq. 7)}$$

Record each calculated K_a on Data Sheet 1, Part II., under "Experimental."

III. The pH of Various Salt Solutions

Calculate the hydrolysis constant, K_h, for the ammonium chloride solution, using Equation 16.

$$K_h = \frac{[NH_3][H_3O^+]}{[NH_4^+]} \qquad \text{(Eq. 16)}$$

Assume that $[NH_4^+]$ is the same as the initial concentration of ammonium chloride. Enter the K_h on Data Sheet 2.

IV. The pH of Buffer Solutions

1. Calculate the concentrations of the solutions in moles L^{-1} from the measured pH. Enter on Data Sheet 2, Part IV., under "Concentration, mol L^{-1}."

2. Calculate the theoretical pH of the original buffer solution from Equations 19 and 1.

$$[H_3O^+] = \frac{K_a[HA]}{[A^-]} \qquad \text{(Eq. 19)}$$

$$pH = -\log [H_3O^+] \qquad \text{(Eq. 1)}$$

Record the calculated pH on Data Sheet 2, Part IV., under "Calculated pH."

3. Calculate the theoretical pH for each of the following solutions. Enter all data on Data Sheet 2, Part IV., under "Calculated pH."

(a) Distilled water.

$$pH = -\log [H_3O^+] \qquad \text{(Eq. 1)}$$

(b) Buffer solution with added HCl.

$$[H_3O^+] = \frac{K_a[HA]}{[A^-]} \qquad \text{(Eq. 19)}$$

$$pH = -\log [H_3O^+] \qquad \text{(Eq. 1)}$$

(c) Buffer solution with added NaOH.

$$[H_3O^+] = \frac{K_a[HA]}{[A^-]} \qquad \text{(Eq. 19)}$$

$$pH = -\log [H_3O^+] \qquad \text{(Eq. 1)}$$

(d) Distilled water with added HCl.

$$pH = -\log [H_3O^+] \qquad \text{(Eq. 1)}$$

(e) Distilled water with added NaOH.

$$pOH = -\log [OH^-] \qquad \text{(Eq. 3)}$$

$$pH + pOH = 14.00 \qquad \text{(Eq. 5)}$$

Post-Laboratory Questions

(Use the spaces provided for the answers and additional paper if necessary.)

1. A buffer solution is made from acetic acid and sodium acetate. Write a chemical equation that describes what happens when hydrochloric acid is added to this solution. Write chemical equations that describe what happens when sodium hydroxide is added to the original buffer solution.

(2)   If K_a for the dissociation of HSO_4^- is 1.2×10^{-2}, what is the pH of a solution containing equal quantities of SO_4^{2-} and HSO_4^-?

---
answer

2. In aqueous solution, the hydrogen sulfate ion, HSO_4^-, dissociates according to

$$H_2O + HSO_4^-(aq) \rightleftarrows H_3O^+(aq) + SO_4^{2-}(aq)$$

(1)   How might sodium hydrogen sulfate be used to prepare a buffer solution?

3. The dissociation constant for formic acid, HCO_2H, is 1.8×10^{-4}. Find the hydrolysis constant, K_h, for formate.

---
answer

Data Sheet 1

I. The pH of Hydrochloric Acid Solutions

concentration of HCl, M	calculated pH	measured pH
1.0×10^{-1}	_____	_____
1.0×10^{-2}	_____	_____
1.0×10^{-3}	_____	_____
1.0×10^{-4}	_____	_____

II. The pH of Acetic Acid Solutions

concentration of HOAc, M	calculated pH	measured pH
1.0×10^{-1}	_____	_____
1.0×10^{-2}	_____	_____
1.0×10^{-3}	_____	_____
1.0×10^{-4}	_____	_____

Calculation of K_a for Acetic Acid

	acid dissociation constant for HOAc	
concentration of HOAc, M	experimental	literature
1.0×10^{-1}	_____	_____
1.0×10^{-2}	_____	
1.0×10^{-3}	_____	
1.0×10^{-4}	_____	

III. The pH of Various Salt Solutions

salt	measured pH	explanation of pH difference from 7.00
NH_4OAc	_____	_____
		_____
$NaCl$	_____	_____
		_____
		_____
$NaOAc$	_____	_____
		_____

Data Sheet 2

salt	measured pH	explanation of pH difference from 7.00
NH_4Cl	_____	_____
		_____
		_____
$NaHCO_3$	_____	_____
		_____
		_____
Na_2CO_3	_____	_____
		_____
		_____
		_____

Calculation of K_h for 0.1M NH_4Cl from the experimental pH

Ans._____

IV. The pH of Buffer Solutions

solution	measured pH	hydrogen ion concentration mol L^{-1}	calculated pH
original buffer solution	_____	_____	_____
	_____	_____	_____
distilled water	_____	_____	_____
	_____	_____	_____
buffer with HCl	_____	_____	_____
buffer with NaOH	_____	_____	_____
distilled water with HCl	_____	_____	_____
distilled water with NaOH	_____	_____	_____

Pre-Laboratory Assignment

1. A solution of hydrochloric acid was found to have a pH of 2.60. What is the $[H_3O^+]$ of the solution?

(2) Calculate the pH of a buffer solution that contains 1.5M sodium propionate and 1.0M propionic acid.

answer

2. The $[OH^-]$ of a solution is 8.5×10^{-8} mol L^{-1}. What is the pH of the solution?

answer

(c) If 1.0×10^{-1} mol of HCl is added to a liter of a buffer solution that contains 1.5M sodium propionate and 1.0M propionic acid, what is the pH of the resulting solution?

answer

3. (1) The pH of a $2.0 \times 10^{-1}M$ propionic acid solution is 2.79. Calculate the K_a of propionic acid.

$$H_2O + CH_3CH_2CO_2H(aq) \rightleftarrows$$
$$CH_3CH_2COO^-(aq) + H_3O^+(aq)$$

answer

answer

The Chemistry and Qualitative Analysis of Anions

prepared by **James G. Boyles**, Bates College,
Judith C. Foster and **David S. Page**, Bowdoin College

Purpose of the Experiment

Develop the chemistry of selected anions, including Cl^-, Br^-, I^-, SO_4^{2-}, CO_3^{2-}, SO_3^{2-}, PO_4^{3-}, NO_2^-, and NO_3^- ions. Use this chemistry to develop an analysis procedure for solutions containing these anions.

Background Information

I. Classes of Reactions and Chemical Processes

In this experiment, you will be performing a number of chemical reactions and observing the results. Some of the important types of chemical reactions and processes you will be conducting are summarized below.

Acid–base reactions occur when an ion reacts with either hydronium ion (H_3O^+) or hydroxide ion (OH^-) in aqueous solution. No oxidation or reduction is involved. For example

$$SO_4^{2-}(aq) + H_3O^+(aq) \rightleftarrows HSO_4^-(aq) + H_2O(l) \quad \text{(Eq. 1)}$$

Hydrolysis reactions take place when the elements of one or more moles of water react with an ion or compound. For example

$$Fe^{3+}(aq) + 6\ H_2O(l) \rightleftarrows$$
$$Fe(OH_3)(s, \text{rusty}) + 3\ H_3O^+(aq) \quad \text{(Eq. 2)}$$

Oxidation–reduction reactions occur when one reagent containing an element with a high electron affinity reacts with a second reagent containing an element of lower electron affinity. The oxidizing agent (high electron affinity) gains electrons from the reducing agent (lower electron affinity). The oxidizing agent is reduced and the reducing agent is oxidized. For example

$$2\ I^-(aq) + Cl_2(aq) \rightleftarrows I_2(aq) + 2\ Cl^-(aq) \quad \text{(Eq. 3)}$$

Disproportionation reactions are a special form of oxidation–reduction reactions. In a disproportionation reaction, the same species serves as an electron donor (reducing agent) and electron acceptor (oxidizing agent). Some of the species are oxidized and some are reduced, the exact proportions depending on the oxidation number of the oxidized and reduced species relative to the original reactant. This is also called **autoredox** (self-oxidation and self-reduction). For example

$$3\ HNO_2(aq) \rightleftarrows$$
$$2\ NO(aq) + NO_3^-(aq) + H_3O^+(aq) \quad \text{(Eq. 4)}$$

Copyright © 1989 by Chemical Education Resources, Inc., P.O. Box 357, 220 S. Railroad, Palmyra, Pennsylvania 17078
No part of this laboratory program may be reproduced or transmitted in any form or by any means, electronic or mechanical, including photocopying, recording, or any information storage and retrieval system, without permission in writing from the publisher. Printed in the United States of America

Decomposition occurs when an unstable compound or ion spontaneously decomposes into one or more new products. For example

$$H_2CO_3(aq) \rightleftarrows H_2O(l) + CO_2(aq) \quad \text{(Eq. 5)}$$

Oxidation–reduction may also be involved in the decomposition (as shown in Equation 4).

Precipitation occurs when the ion product of ions in solution exceeds the solubility product constant, K_{sp}, of the compound formed by a combination of these ions. This results in the formation of a solid, or precipitate. For example

$$Ag^+(aq) + Cl^-(aq) \rightleftarrows AgCl(s, white) \quad \text{(Eq. 6)}$$

The **ion product** is the value obtained when the actual concentrations of ions in solution are substituted into the K_{sp} expression.

Dissolution occurs when the ion product of the dissolved species is less than the K_{sp} of the solid, resulting in the solid dissolving. Often, an added reagent reacts with one of the ions formed in the dissolution, lowering the ion product and shifting the equilibrium in favor of the dissolution. For example

$$Ag_2CO_3(s, white) + 2\,H_3O^+(aq) \rightleftarrows$$
$$2\,Ag^+(aq) + 3\,H_2O(l) + CO_2(aq) \quad \text{(Eq. 7)}$$

Extraction is not, strictly speaking, a chemical reaction, but it is a chemical process. It involves the migration of chemical species from one layer to another of a multi-phase system of immiscible solvents. Often, one solvent is aqueous and another is organic, as is the case in this experiment. The migration from one phase to another occurs because the species is more soluble in one solvent than in the other. For example

$$I_2(aq, yellow) \rightleftarrows I_2(organic, violet) \quad \text{(Eq. 8)}$$

In this experiment, extraction is used to identify certain compounds that have similar colors in aqueous solution but very different colors when extracted into an organic solvent, methylene chloride (CH_2Cl_2). This color change is due to differences in solvent–solute interaction.

II. Chemistry of Selected Anions

A. Sulfate Ion (SO_4^{2-})

In aqueous solution, the following equilibria involving SO_4^{2-} ion are established.

$$SO_4^{2-}(aq) + H_2O(l) \rightleftarrows HSO_4^-(aq) + OH^-(aq)$$
$$K_b = 1.0 \times 10^{-12} \quad \text{(Eq. 9)}$$

$$HSO_4^-(aq) + H_2O(l) \rightleftarrows H_2SO_4(aq) + OH^-(aq)$$
$$K_b = 1.0 \times 10^{-15} \quad \text{(Eq. 10)}$$

K_b is the **base dissociation constant** for the reactions shown in Equations 9 and 10. As the constants for these equilibria indicate, SO_4^{2-} ion is a very weak Brønsted base, and hydrogen sulfate ion (HSO_4^-, bisulfate ion) is a relatively strong acid. Sulfate ion undergoes negligible hydrolysis. Therefore, SO_4^{2-} ion is stable in acidic and basic solutions.

Most sulfates are water soluble, but two notable exceptions are lead(II) sulfate ($PbSO_4$) and barium sulfate ($BaSO_4$), as shown in Equation 11 and 12.

$$PbSO_4(s, white) \rightleftarrows Pb^{2+}(aq) + SO_4^{2-}(aq)$$
$$K_{sp} = 1.4 \times 10^{-8} \quad \text{(Eq. 11)}$$

$$BaSO_4(s, white) \rightleftarrows Ba^{2+}(aq) + SO_4^{2-}(aq)$$
$$K_{sp} = 1.1 \times 10^{-10} \quad \text{(Eq. 12)}$$

The affinity of SO_4^{2-} ion for protons in aqueous solution is so small, as seen in Equation 9, and the solubility product constants of $PbSO_4$ and $BaSO_4$ are so small, that these salts are classified as insoluble sulfates even in moderately strong acid solutions. The formation of the white, acid-insoluble precipitate of $BaSO_4$ upon addition of a source of barium ion (Ba^{2+}) to a solution is a reliable classification reaction for the presence of SO_4^{2-} ion.

Sulfate ion is neither a strong oxidizing nor a strong reducing agent and therefore is not involved in significant redox reactions.

B. Sulfite Ion (SO_3^{2-})

In aqueous solution, SO_3^{2-} ion is involved in the following equilibria.

$$SO_3^{2-}(aq) + H_2O(l) \rightleftarrows HSO_3^-(aq) + OH^-(aq)$$
$$K_b = 1.8 \times 10^{-7} \quad \text{(Eq. 13)}$$

$$HSO_3^-(aq) + H_2O(l) \rightleftarrows H_2SO_3(aq) + OH^-(aq)$$
$$K_b = 1.0 \times 10^{-12} \quad \text{(Eq. 14)}$$

$$H_2SO_3(aq) \rightleftarrows H_2O(l) + SO_2(aq) \quad \text{(Eq. 15)}$$

Sulfite ion is a weak Brønsted base, but hydrolysis (see Equations 13 and 14) does give basic solutions. These equilibria are shifted strongly to the right in highly acidic solutions (low pH). Sulfurous acid (H_2SO_3), as such, has never been isolated. If an acidified solution containing SO_3^{2-} ion is heated, sulfur(IV) oxide (SO_2, sulfur dioxide) gas is evolved, because the solubility of SO_2 decreases with increasing temperature.

Sulfur(IV) oxide is a colorless gas with characteristic choking fumes with a sharp odor. Consequently its presence, even in small amounts, is easily detected. An acceptable procedure for determining the presence of SO_3^{2-} ion in a solution is to acidify and heat the solution, and carefully check for the odor of SO_2.

The sulfites of most cations are insoluble in water, with the notable exceptions being the sulfites of ammonium ion (NH_4^+) and the alkali metal cations. The insolubility of barium sulfite ($BaSO_3$) will play an important role in this experiment. Sulfite ion has a much stronger affinity for a proton than does SO_4^{2-} ion, as you can see from the magnitude of the equilibrium constants of the reactions in Equations 9 and 13. Consequently, sulfites, which are insoluble in water, are at least moderately soluble in acidic solutions. Their dissolution is accompanied by the evolution of SO_2 gas.

Sulfite ion is easily oxidized to SO_4^{2-} ion by atmospheric oxygen. When a solution is left standing for a long time, SO_3^{2-} ion in the solution may be converted almost entirely to SO_4^{2-} ion, as shown in Equation 16.

$$2\ SO_3^{2-}(aq) + O_2(g) \rightleftarrows 2\ SO_4^{2-}(aq) \qquad (Eq.\ 16)$$

Many reducing agents will reduce SO_3^{2-} ion to elemental sulfur, $S(s)$, or even to hydrogen sulfide (H_2S), where sulfur is in the –2 oxidation state. Thus, SO_3^{2-} ion can act as either an oxidizing or a reducing agent.

C.   Nitrate Ion (NO_3^-)

Nitrate ion is the anion of a strong Brønsted acid and undergoes negligible hydrolysis in aqueous solution. Nitrate ion can exist in significant concentration in both acidic and basic solutions.

No insoluble inorganic salts are formed with NO_3^- ion.

The nitrogen in NO_3^- ion is in its highest possible oxidation state (+5), hence NO_3^- ion can react only as an oxidizing agent, resulting in the reduction of the nitrogen. In redox reactions involving NO_3^- ion, there can be a variety of products, but for a given system only one product predominates. The concentration of NO_3^- ion and the acidity of the solution largely determine the identity of the product. For example, only the following redox reaction occurs in strongly acidic solutions.

$$NO_3^-(aq) + 4\ H_3O^+(aq) + 3\ Fe^{2+}(aq) \rightleftarrows$$
$$NO(aq) + 3\ Fe^{3+}(aq) + 6\ H_2O(l) \qquad (Eq.\ 17)$$

This reaction can form the basis for a useful confirmatory reaction. The product, nitrogen(II) oxide (NO, nitric oxide), which is dissolved in the solution, can be released from the reaction mixture as NO gas

by heating. This release shifts the equilibrium of Equation 17 to the right. Gaseous NO in the presence of oxygen (O_2) from the air is oxidized rapidly to gaseous nitrogen(IV) oxide (NO_2, nitrogen dioxide), which is readily detected as a red-brown gas with an antiseptic odor.

$$2\ NO(g,\ colorless) + O_2(g) \rightleftarrows$$
$$2\ NO_2(g,\ red\text{-}brown) \qquad (Eq.\ 18)$$

Thus, when a strongly acidic solution containing NO_3^- ion is heated in the presence of a reducing agent such as iron(II) ion (Fe^{2+}, ferrous ion), a rapid reaction occurs, accompanied by the evolution of a red-brown gas, which serves as a confirmatory reaction for NO_3^- ion. An intermediate product is usually seen. As NO is produced, the unstable complex ion, pentaaquonitrosyliron(II), $[Fe(H_2O)_5(NO)]^{2+}$, briefly colors the solution dark brown. This ion decomposes as the NO(aq) is converted to NO(g), which evolves from the solution on heating.

D.   Nitrite Ion (NO_2^-)

In aqueous solution, NO_2^- ion is involved in the following equilibrium:

$$NO_2^-(aq) + H_2O(l) \rightleftarrows HNO_2(aq) + OH^-(aq)$$

$$K_b = 2.2 \times 10^{-11} \qquad (Eq.\ 19)$$

As the equilibrium constant for this hydrolysis reaction indicates, NO_2^- ion is a very weak Brønsted base and undergoes negligible hydrolysis. Upon addition of H_3O^+ ion, the equilibrium in Equation 19 shifts far to the right. This shift greatly increases the concentration of HNO_2. Pure nitrous acid has never been isolated, and in warm aqueous solutions it decomposes, via disproportionation, as shown in Equation 20.

$$3\ HNO_2(aq) \rightleftarrows$$
$$2\ NO(aq) + NO_3^-(aq) + H_3O^+(aq) \qquad (Eq.\ 20)$$

When an acidified solution containing NO_2^- ion is heated, NO gas, formed from the disproportionation, is evolved, because its solubility decreases with increasing temperature. Nitrogen(II) oxide is a colorless gas that reacts rapidly with atmospheric oxygen to produce red-brown NO_2 gas.

$$2\ NO(g,\ colorless) + O_2(g) \rightleftarrows$$
$$2\ NO_2(g,\ red\text{-}brown) \qquad (Eq.\ 18)$$

Thus, a convenient procedure for determining the presence of NO_2^- ion in a solution would be to acidify and heat the solution and to look for red-brown NO_2 gas with an antiseptic odor.

The nitrites of most cations are soluble in water, with the notable exception of silver nitrite ($AgNO_2$). With the exception of the alkali-metal nitrites, all solid nitrites decompose on heating.

Nitrite ion is a weak reducing agent and can be oxidized to NO_3^- ion only by strong oxidizing agents such as permanganate ion.

(MnO_4^-) and basic hydrogen peroxide (H_2O_2). Nitrite ion is a reasonably strong oxidizing agent. It is reduced to NO by such reducing agents as SO_3^{2-}, H_2S, Fe^{2+}, and I^- (iodide) ions.

Nitrite ion reacts similarly to NO_3^- ion in strongly acidic media. A differentiation between NO_2^- ion and NO_3^- ion can be made, however, because an acidic solution containing NO_2^- ion requires no reducing agent to produce the red-brown NO_2 gas.

E. Carbonate Ion (CO_3^{2-})

In aqueous solution, CO_3^{2-} ion is involved in the following equilibria:

$$CO_3^{2-}(aq) + H_2O(l) \rightleftarrows HCO_3^-(aq) + OH^-(aq)$$
$$K_b = 2.0 \times 10^{-4} \qquad \text{(Eq. 21)}$$

$$HCO_3^-(aq) + H_2O(l) \rightleftarrows H_2CO_3(aq) + OH^-(aq)$$
$$K_b = 2.5 \times 10^{-8} \qquad \text{(Eq. 22)}$$

$$H_2CO_3(aq) \rightleftarrows H_2O(l) + CO_2(aq) \qquad \text{(Eq. 23)}$$

Carbonate ion is a moderately strong Brønsted base and undergoes considerable hydrolysis in aqueous solution. The predominant species present is determined by the pH of the solution. At pH 11, CO_3^{2-} ion is the predominant species; at pH 8, hydrogen carbonate ion (HCO_3^-, bicarbonate) is most abundant; at pH 4 and lower, carbonic acid (H_2CO_3) and its decomposition products, carbon(IV) oxide (CO_2, carbon dioxide) and H_2O, are dominant. Therefore, if acid is added to a carbonate-containing solid or solution, CO_2, a colorless, odorless gas, will be evolved, because the equilibria in Equations 21, 22, and 23 are shifted to the right.

The barium and calcium salts of carbonates are insoluble in distilled water. Because the addition of H_3O^+ ion will shift the equilibrium shown in Equations 21, 22, and 23 strongly to the right, both substances are quite soluble in acid solutions. Their dissolution is accompanied by the evolution of CO_2 gas.

$$BaCO_3(s, white) \rightleftarrows Ba^{2+}(aq) + CO_3^{2-}(aq)$$
$$K_{sp} = 5.0 \times 10^{-9} \qquad \text{(Eq. 24)}$$

$$CaCO_3(s, white) \rightleftarrows Ca^{2+}(aq) + CO_3^{2-}(aq)$$
$$K_{sp} = 7.5 \times 10^{-9} \qquad \text{(Eq. 25)}$$

Thus, a convenient reaction for the presence of CO_3^{2-} ion in a solution is to acidify the solution and to observe CO_2 evolution, noting the lack of odor. Many common cations form insoluble carbonates, notable exceptions being those of NH_4^+ ion, sodium ion (Na^+), and potassium ion (K^+).

Neither CO_3^{2-} ion nor HCO_3^- ion is a strong oxidizing or reducing agent. These ions are not involved in significant redox reactions.

F. Phosphate Ion (PO_4^{3-})

The following equilibria are established in aqueous solution containing PO_4^{3-} ion:

$$PO_4^{3-}(aq) + H_2O(l) \rightleftarrows HPO_4^{2-}(aq) + OH^-(aq)$$
$$K_b = 1.0 \times 10^{-2} \qquad \text{(Eq. 26)}$$

$$HPO_4^{2-}(aq) + H_2O(l) \rightleftarrows H_2PO_4^-(aq) + OH^-(aq)$$
$$K_b = 1.6 \times 10^{-7} \qquad \text{(Eq. 27)}$$

$$H_2PO_4^-(aq) + H_2O(l) \rightleftarrows H_3PO_4(aq) + OH^-(aq)$$
$$K_b = 1.3 \times 10^{-12} \qquad \text{(Eq. 28)}$$

Phosphate ion is a reasonably strong Brønsted base and hydrolyzes appreciably to form a basic solution containing PO_4^{3-} ion, hydrogen phosphate ion (HPO_4^{2-}), dihydrogen phosphate ion ($H_2PO_4^-$), and orthophosphoric acid (H_3PO_4). The ratio of these species in aqueous solution is determined by the pH of the solution.

Most phosphates, except those of the alkali metals, are sparingly soluble in neutral solution. Examples of these sparingly soluble phosphates are barium phosphate, $Ba_3(PO_4)_2$, calcium phosphate, $Ca_3(PO_4)_2$, silver phosphate, Ag_3PO_4, lead(II) phosphate, [$Pb_3(PO_4)_2$, plumbous phosphate], bismuth phosphate, $BiPO_4$, and iron(II) phosphate [$Fe_3(PO_4)_2$, ferrous phosphate]. These substances are soluble in strongly acidic solutions, because the PO_4^{3-} ion produced can become involved in the hydrolysis reactions shown in Equations 26, 27, and 28.

Phosphorus is in its highest possible oxidation state (+5) in PO_4^{3-} ion, hence PO_4^{3-} ion can react only as an oxidizing agent, because the phosphorus can only be reduced. In aqueous solution, PO_4^{3-} ion is a very weak oxidizing agent and can be reduced only under rather strong reducing conditions. Redox reactions are not an important part of aqueous PO_4^{3-} ion chemistry.

Phosphate ion forms an insoluble bright yellow precipitate when treated with ammonium molybdate, $(NH_4)_2MoO_4$, in an acidic solution.

$$PO_4^{3-}(aq) + 3\ NH_4^+(aq)$$
$$+\ 12\ MoO_4^{2-}(aq) + 24\ H_3O^+(aq) \rightleftarrows$$
$$(NH_4)_3PMo_{12}O_{40}(s,\ bright\ yellow) + 36\ H_2O(l)$$
$$\text{(Eq. 29)}$$

The precipitate, ammonium molybdophosphate, $(NH_4)_3PMo_{12}O_{40}$, or ammonium phosphomolybdate, varies in composition, but the P to Mo ratio remains constant at 1 to 12. This reaction is quite sensitive and can be used to confirm the presence of PO_4^{3-} ion. Arsenate ion (AsO_4^{3-}) gives a similar precipitate, but only if the solution is heated strongly.

G. Chloride (Cl⁻), Bromide (Br⁻), and Iodide (I⁻) Ions

As anions of strong Brønsted acids, Cl^-, Br^-, and I^- undergo negligible hydrolysis in aqueous solution. The halide ions are stable in aqueous solution over a broad pH range, encompassing both acidic and basic solutions.

Most halide salts are soluble, notable exceptions being those of silver, lead(II), and mercury(I). Indeed, precipitation of Cl^-, Br^-, or I^- ion as a silver salt is a common first step in the identification of halide ions. Because of the negligible hydrolysis of the halides, the silver halides are insoluble even in strongly acidic solution, except HCl.

$$AgCl(s,\ white) \rightleftarrows Ag^+(aq) + Cl^-(aq)$$
$$K_{sp} = 1.2 \times 10^{-10} \qquad \text{(Eq. 30)}$$

$$AgBr(s,\ cream) \rightleftarrows Ag^+(aq) + Br^-(aq)$$
$$K_{sp} = 4.8 \times 10^{-13} \qquad \text{(Eq. 31)}$$

$$AgI(s,\ light\ yellow) \rightleftarrows Ag^+(aq) + I^-(aq)$$
$$K_{sp} = 1.4 \times 10^{-16} \qquad \text{(Eq. 32)}$$

The colors of these three substances are so similar that a confirmation reaction is needed to distinguish them from one another. Because these three halides have only one negative oxidation state, −1, they are capable of acting only as reducing agents. The decreasing relative strengths of the halide ions as reducing agents is: $I^- > Br^- > Cl^-$. Hence, the decreasing relative strengths of their oxidized forms, as the neutral diatomic halogens, as oxidizing agents is: $Cl_2 > Br_2 > I_2$. Therefore, chlorine (Cl_2) will oxidize Br^- ion to elemental bromine (Br_2) and will oxidize I^- ion to elemental iodine (I_2). Bromine will only oxidize I^- ion to I_2.

$$Cl_2(aq,\ colorless) + 2\ Br^-(aq) \rightleftarrows$$
$$Br_2(aq,\ yellow\text{-}brown) + 2\ Cl^-(aq) \qquad \text{(Eq. 33)}$$

$$Cl_2(aq,\ colorless) + 2\ I^-(aq) \rightleftarrows$$
$$I_2(aq,\ yellow\text{-}brown) + 2\ Cl^-(aq) \qquad \text{(Eq. 34)}$$

Because the colors of $Br_2(aq)$ and $I_2(aq)$ are very similar, we need a further reaction to distinguish them. We will take advantage of the fact that bromine and iodine have very different colors when dissolved in the organic solvent methylene chloride (CH_2Cl_2). Methylene chloride is much less polar than water, so the nonpolar halogens dissolve preferentially in the less polar solvent. Methylene chloride and water are not mutually soluble, and CH_2Cl_2, being denser than water (1.33 g mL^{-1}), will form the lower layer. [If hexane (C_6H_{14}), density 0.66 g mL^{-1}, is used in place of CH_2Cl_2, it will form the upper layer.] The halogen reactions occur in the aqueous layer. When the two layers are vigorously mixed by shaking, the halogen product will migrate into the CH_2Cl_2 layer (or C_6H_{14} layer). Upon separation, the color of the organic layer will be distinctive for each halogen.

Bromine dissolves in CH_2Cl_2 to give a solution of red-amber color, and iodine dissolves in CH_2Cl_2 to give a solution of violet color. This color difference can be used in a confirmatory reaction. Chlorine does not impart any color to CH_2Cl_2.

The formation of an acid-insoluble precipitate following the addition of aqueous silver nitrate ($AgNO_3$) to an unknown solution indicates the presence of one or more of the halide ions. Positive identification of the anion can be achieved by reacting the precipitate with the oxidizing agent, Cl_2 (as chlorine water), and CH_2Cl_2, followed by thorough shaking of the solution. Observation of the color imparted to the CH_2Cl_2 layer will confirm the presence of a given halide ion in the original solution. If the CH_2Cl_2 layer is colorless, Cl^- ion was present initially. If the CH_2Cl_2 layer is red-amber, Br^- ion was present. If the CH_2Cl_2 layer is violet, I^- ion was present.

Be careful when interpreting the observations, because the reaction for Cl^- ion results in two colorless layers. This reaction is only a negative confirmation reaction, and is only to be used if the classification reaction indicates a halide. Almost all other anions except Br^- and I^- ions will also yield two colorless layers with this reaction. If both steps of the $AgNO_3$ reaction indicate the presence of a halide, then do the CH_2Cl_2 confirmatory reaction.

In this experiment, you will observe the behavior of known anions with a number of reagents. Then you will determine which anion is present in each of the unknown samples assigned to you. Initially, you will use three classification reactions to determine to which group of anions your anion belongs. These reactions are:

(1) formation of insoluble barium salts, such as $BaCO_3$, $BaSO_4$, $BaSO_3$, and $Ba_3(PO_4)_2$;

(2) formation of volatile products when H_3O^+ ion is added to such anions as CO_3^{2-}, NO_2^-, and SO_3^{2-} ions; and

(3) formation of insoluble silver salts, such as AgBr, AgCl, and AgI.

Then, you will do the confirmatory reactions suggested by the results of your elimination reactions.

Procedure

Chemical Alert

0.5M ammonium molybdate—toxic and irritant
0.1M barium chloride—highly toxic
chlorine water—toxic, irritant, and oxidant
dichloromethane—toxic and irritant
hexane—flammable and irritant
1M iron(II) sulfate—toxic
3M nitric acid—toxic, corrosive, and strong oxidant
0.1M silver nitrate—toxic, corrosive, and oxidant
sodium carbonate—irritant
sodium chloride—irritant
sodium hydrogen phosphate—irritant
sodium iodide—irritant
sodium nitrate—toxic, irritant, and oxidant
sodium nitrite—toxic and oxidant
sodium sulfate—irritant
sodium sulfite—irritant
sulfuric acid—toxic and corrosive

Caution: Wear departmentally approved eye protection while doing this experiment.

In this investigation, you will use solid samples as sources of the anion for analysis. The use of a large amount of salt in any given reaction is neither necessary nor desirable. In the following procedure, references to "an amount" of a given salt should be taken to mean a portion of solid approximately equal in bulk to a grain of rice. You should first perform the experiments using known samples, as instructed. After completing the procedure with the known samples of anions, use the reactions given in the **Procedure** to identify the anion in each of several unknown salts. To avoid contamination, rinse your spatula and stirring rod with distilled or deionized water after every use.

Note: In all parts of this experiment, follow the directions of your laboratory instructor for discarding reaction mixtures and unused reagents.

I. Classification Reactions

It is convenient to categorize the anions being studied in terms of several types of chemical reactions.

A. Barium Salts Insoluble in Distilled Water (SO_4^{2-}, SO_3^{2-}, CO_3^{2-}, and PO_4^{3-} ions)

Dissolve an amount of solid sodium sulfate (Na_2SO_4) in 10 drops of distilled water in a 75-mm test tube. Add 5 drops of 0.1M barium chloride ($BaCl_2$) solution.

Caution: 3M nitric acid is a corrosive, toxic solution that can cause severe burns and discolor your skin. Prevent contact with your eyes, skin, and clothing. Avoid inhaling vapors and ingesting the solution.

Stir the solution, and then add 5 drops of 3M HNO_3. Stir thoroughly. Note the color and odor of any gas evolved. Record your observations on Data Sheet 1. Repeat this procedure using solid sodium sulfite (Na_2SO_3), sodium carbonate (Na_2CO_3), and sodium phosphate (Na_3PO_4) as anion samples. In each case, record your observations on Data Sheet 1.

B. Anions Forming Volatile Products (SO_3^{2-}, CO_3^{2-}, and NO_2^- ions)

Dissolve an amount of solid Na_2SO_3 in 5 drops of distilled water in a 75-mm test tube.

Caution: 3M sulfuric acid is a corrosive, toxic solution that can cause severe burns. Prevent contact with your eyes, skin, and clothing. Avoid ingesting the solution.

Add 5 drops of 3M H_2SO_4 to the tube. Carefully note the odor and color of any gas evolved. The color can be noted by holding the tube against a white background. If necessary, gently heat the tube and its contents in a warm water bath. Record your observations on Data Sheet 1.

Repeat the procedure with a solid sample of Na_2CO_3 and with a sample of sodium nitrite

(NaNO$_2$). In each case, record your observations on Data Sheet 1.

C. Silver Salts Insoluble in Distilled Water and in Acid Solution (Cl$^-$, Br$^-$, and I$^-$ ions)

Dissolve an amount of sodium chloride (NaCl) in 10 drops of distilled water in a 75-mm test tube.

> **Caution:** 0.1 M silver nitrate is a toxic solution that can cause skin burns or discolor your skin. Prevent contact with your eyes, skin, and clothing. If any of the solution comes in contact with your skin, wash immediately with copious amounts of water, and notify your laboratory instructor.

Add to this solution 5 drops of 0.1 M AgNO$_3$ solution. Stir thoroughly. Next, add 5 drops of 3 M HNO$_3$, and stir thoroughly. Record your observations on Data Sheet 1.

Repeat this procedure with solid samples of sodium bromide (NaBr) and of sodium iodide (NaI). In each case, record your observations on Data Sheet 1.

II. Confirmatory Reactions

Each of the following reactions is used to confirm the presence of a given anion.

A. Sulfate Ion (SO$_4^{2-}$)

Dissolve an amount of Na$_2$SO$_4$ in 10 drops of distilled water in a 75-mm test tube. Add 5 drops of 0.1 M BaCl$_2$ solution. Stir the contents of the tube. Add 5 drops of 3 M HNO$_3$ to the mixture and stir thoroughly. Record your observations on Data Sheet 2.

B. Sulfite Ion (SO$_3^{2-}$)

Dissolve an amount of Na$_2$SO$_3$ in 5 drops of distilled water in a 75-mm test tube. Add 5 drops of 3 M H$_2$SO$_4$ and stir thoroughly. Carefully note the odor and color of any gas evolved. If necessary, gently heat the tube and its contents in a warm water bath. Record your observations on Data Sheet 2.

C. Nitrate Ion (NO$_3^-$)

Dissolve an amount of NaNO$_3$ in 5 drops of a 1 M solution of iron(II) sulfate (FeSO$_4$, ferrous sulfate), available as 1 M FeSO$_4$ in 1 M H$_2$SO$_4$, in a 75-mm test tube. Add 3–5 drops of concentrated H$_2$SO$_4$. The tube and its contents will become warm. This reaction should occur without stirring.

Be sure that the mouth of the test tube is not pointing toward anyone, because a gas may be evolved rapidly. If necessary, gently heat the tube and its contents in a warm water bath for 1 min or until a reaction has taken place. Record your observations on Data Sheet 2.

D. Nitrite Ion (NO$_2^-$)

Dissolve an amount of NaNO$_2$ in 5 drops of distilled water in a 75-mm test tube. Carefully add 5 drops of 3 M H$_2$SO$_4$. If necessary, gently heat the tube and its contents in a warm water bath. Record your observations on Data Sheet 2.

E. Carbonate Ion (CO$_3^{2-}$)

Place an amount of Na$_2$CO$_3$ in 5 drops of distilled water in a 75-mm test tube. Add 5 drops of 3 M H$_2$SO$_4$. Note the odor and color of any gas evolved. Record your observations on Data Sheet 2.

F. Phosphate Ion (PO$_4^{3-}$)

Dissolve an amount of Na$_3$PO$_4$ in 8 drops of 3 M nitric acid in a 75-mm test tube. Add 4 drops of 0.5 M (NH$_4$)$_2$MoO$_4$ solution. Stir thoroughly. Record your observations on Data Sheet 2.

G. Halide Ions (Cl$^-$, Br$^-$, and I$^-$)

Dissolve an amount of NaCl in 5 drops of distilled water in a 75-mm test tube. Add 10 drops of chlorine water, and then 10 drops of CH$_2$Cl$_2$. Stopper the tube, and shake the tube and its contents thoroughly. Note the color of the CH$_2$Cl$_2$ (lower) layer. Record your observations on Data Sheet 2.

Repeat this procedure using NaBr and using NaI. In each case, record your observations on Data Sheet 2.

III. Analyzing Unknowns

Obtain from your laboratory instructor several unknown samples to identify. Record the number of your samples on Data Sheet 3. Each sample will contain only one anion. For each sample, perform **all three** classification reactions, even if the first or second one appears positive. Then, perform those confirmatory reactions suggested by the results of the classification reactions. Record your observations on Data Sheet 3, and report the identity of each unknown anion.

> **Caution:** Wash your hands thoroughly with soap or detergent before leaving the laboratory.

Post-Laboratory Questions

(Use the spaces provided for the answers and additional paper if necessary.)

1. Suppose that a solution is tested for NO_2^- ion by adding $3M$ H_2SO_4 and heating, repeating this process until no further reaction occurs. This solution is then analyzed for NO_3^- ion by adding $FeSO_4$ solution and more H_2SO_4, heating the test tube and its contents. The NO_3^- ion reaction is positive. In this case, is it accurate to state that NO_3^- ion was present in the original solution? Briefly explain. Write appropriate equations to support your answer.

2. A solution contains a mixture of Cl^- and Br^- ions. Can both be positively identified? Briefly explain. Write appropriate equations to support your answer.

3. The solubility product constants of AgCl, AgBr, and AgI are, respectively, 1.7×10^{-10}, 4.1×10^{-13}, and 1.5×10^{-16}. The K_{sp} expression for each silver halide, AgX, is:

$$K_{sp} = [Ag^+][X^-]$$

When the ion product, $[Ag^+][X^-]$, exceeds the K_{sp}, AgX(s) will precipitate until the ion product once again equals K_{sp}. If the Ag^+ ion concentration in the mixture is $0.067M$ (diluted from $0.1M$), calculate the detection limit of this reaction for each halide, the value below which there will be no observed reaction.

4. A solution is known to contain Cl^-, NO_3^-, NO_2^-, and SO_4^{2-} ions. Which of these ions can be positively identified? Describe the reactions used and the results.

5. Why wouldn't HCl be a good acid to use to determine the solubility of a silver precipitate in acid solution?

Data Sheet 1

(Record below your observations made while performing Section I of this experiment.)

I. Classification Reactions

A. Barium salts insoluble in distilled water	B. Anions forming volatile products	C. Silver salts insoluble in distilled water and in acid solution
sulfate	sulfite	chloride
sulfite	carbonate	bromide
carbonate	nitrite	iodide
phosphate		

Data Sheet 2

(Record below your observations made while performing Section II of this experiment.)

II. Confirmatory Reactions

bromide	carbonate	chloride
iodide	nitrate	nitrite
phosphate	sulfate	sulfite

Data Sheet 3

III. Analyzing Unknowns

unknown identification number _____

1. Classification reactions performed and observed results

2. Confirmatory reactions performed and observed results

3. Equations for the reactions performed that led to the identification of the unknown

4. Identity _____

_____

unknown identification number _____

1. Classification reactions performed and observed results

2. Confirmatory reactions performed and observed results

3. Equations for the reactions performed that led to the identification of the unknown

4. Identity _____

_____

unknown identification number _____

1. Classification reactions performed and observed results

2. Confirmatory reactions performed and observed results

3. Equations for the reactions performed that led to the identification of the unknown

4. Identity _____

_____

Pre-Laboratory Assignment

1. Of the following anions, Br^-, CO_3^{2-}, Cl^-, I^-, NO_3^-, NO_2^-, PO_4^{3-}, SO_4^{2-}, and SO_3^{2-}, list those ions that meet the following criteria.

    (1)   Gives a precipitate with Ag^+ ion that is insoluble in HNO_3.

    (2)   Is involved in an oxidation–reduction reaction as part of a confirmation reaction.

    (3)   Gives a precipitate with Ba^{2+} ion that is soluble in HNO_3.

    (4)   Evolves a gas when acid is added.

    (5)   Releases an odorous gas as a product.

    (6)   Does not produce a precipitate in any classification or confirmatory reaction described in this experiment.

2. What types of reactions are involved when acid is added to a solution containing SO_3^{2-} ion, resulting in the release of gaseous SO_2?

3. In analyzing an unknown salt containing one of the anions discussed in this module, a student obtains negative results for all three classification reactions. Briefly explain.

4. A solution is known to contain at least one of the anions discussed in this module.

    (1)   The yellowish precipitate that forms when $AgNO_3$ solution is added to the solution dissolves when HNO_3 is added. Which anion(s) may be present?

    (2)   A $BaCl_2$ solution is added to this solution. A white precipitate forms that dissolves when HNO_3 is added. No color, odor, or evolved gas is observed. Which anion is present? Describe a confirmatory reaction for this anion.

5. In some of the classification reactions, it is important when there is *no* reaction upon addition of a second reagent. For which anions is this true?

Studying Electrochemistry and Establishing the Relative Reactivities of a Series of Metals

prepared by **J. N. Spencer**, Franklin and Marshall College,
and **H. A. Neidig**, Lebanon Valley College

Purpose of the Experiment

Establish the relative electromotive forces, or reactivities, of a series of metals by observing the direction of electron exchange when pairs of the metals, each suspended in a solution of its ions, are connected to form a closed electrical system, or electrochemical cell.

Background Information

Chemical reactions that proceed by electron transfer from one reactant to another are called **oxidation–reduction reactions,** or **redox reactions.** The reactant that gives up electrons is said to be **oxidized**, and the reactant that gains electrons is **reduced.**

For example, when we place a strip of zinc (Zn) metal in an aqueous copper(II) sulfate solution ($CuSO_4$), we immediately begin to see evidence that a chemical reaction is occurring. The Zn begins to dissolve, and a deposit of copper (Cu) metal appears on the Zn strip. We can write an overall equation for this reaction once we understand each of the two component processes separately.

In this reaction, Zn atoms lose electrons to become zinc ions (Zn^{2+}) as the Zn strip dissolves. We can express the dissolution of the Zn as shown in Equation 1, where "e^-" represents an electron.

$$Zn(s) \rightarrow Zn^{2+}(aq) + 2\ e^- \qquad \text{(Eq. 1)}$$

This equation also demonstrates that Zn is the reactant being oxidized. At the same time, copper(II) ions (Cu^{2+}) in the solution gain electrons to become Cu atoms. We can express the formation of Cu, as shown in Equation 2.

$$Cu^{2+}(aq) + 2\ e^- \rightarrow Cu(s) \qquad \text{(Eq. 2)}$$

The Cu^{2+} ion is the reactant being reduced.

We can write the overall equation for this oxidation–reduction reaction by combining Equations 1 and 2.

$$Zn(s) + Cu^{2+}(aq) \rightarrow Zn^{2+}(aq) + Cu(s) \quad \text{(Eq. 3)}$$

The electrons appearing in Equations 1 and 2 cancel each other when we add the two equations. In order for a redox reaction to occur, the number of electrons provided by the oxidized reactant must be equal to the number of electrons accepted by the reduced reactant. There can be no net gain or loss of electrons in the overall equation.

Copyright © 1993 by Chemical Education Resources, Inc., P.O. Box 357, 220 S. Railroad, Palmyra, Pennsylvania 17078
No part of this laboratory program may be reproduced or transmitted in any form or by any means, electronic or mechanical, including photocopying, recording, or any information storage and retrieval system, without permission in writing from the publisher. Printed in the United States of America

We refer to the reactions in Equations 1 and 2 as **half-reactions**, because each reaction constitutes half of the overall chemical reaction. We refer to the combination of the oxidized and reduced substances in a half reaction as a **redox couple**. Thus, as shown in Equations 1 and 2, Cu/Cu^{2+} ion and Zn/Zn^{2+} ion are examples of redox couples in their respective half-reactions.

If we place a Cu strip in an aqueous solution of a zinc(II) salt such as zinc(II) sulfate ($ZnSO_4$), we see no visible evidence of a reaction. This experiment shows that indeed the reverse reaction of Equation 3 does not occur. Thus, we conclude that it is easier for Zn atoms to give up electrons to Cu^{2+} ions than it is for Cu atoms to give up electrons to Zn^{2+} ions.

A remarkable feature of oxidation–reduction reactions is their ability to proceed spontaneously even if we confine the redox couples to separate containers, called **half-cells**. In order to do so, we must connect the half-cells with a wire, and, to maintain a charge balance in each half-cell, we must also connect the two half-cells with a device that permits ionic flow into and out of the cells. We call such a device a **salt bridge**.

In our example, we place a Zn strip in a container with an aqueous Zn^{2+} ion solution, and a Cu strip in a separate container with an aqueous Cu^{2+} ion solution. When we connect the two metal strips with a wire and add a salt bridge between the two solutions as shown in Figure 1, electrons will pass through the wire from the Zn strip to the Cu strip. We call these metal strips **electrodes**. These are the components of the respective half-cells on the surface of which the actual half-reactions occur. They are also the portions of the half-cells that directly connect to the external circuit or wire. We call the electrode in the half-cell that receives the electrons the **cathode**, and the electrode in the half-cell that supplies the electrons the **anode**. In our example, Cu is the cathode, and Zn is the anode. The passage of electrons from the anode to the cathode through the wire is called an electrical **current**. We measure current in amperes (A).

Electrons are driven from the Zn/Zn^{2+} ion half-cell to the Cu/Cu^{2+} ion half-cell by the difference in electrical potential, called the **electromotive force** or **EMF**, between the two half-cells. We measure this potential difference in volts (V). The assembly of two half-cells connected to form a circuit such as that shown in Figure 1 is called an **electrochemical cell**, or a **galvanic** or **voltaic cell**. A spontaneous redox reaction results in an electrochemical cell with a positive voltage.

We can measure the net EMF for a pair of half-reactions in an electrochemical cell by connecting a voltmeter between the half-cells. To do so, we attach the anode to the negative voltmeter terminal and the cathode to the positive voltmeter terminal. Thus, for the cell shown in Figure 1, we attach the wire from the Zn/Zn^{2+} ion half-cell to the negative voltmeter terminal and the wire from the Cu/Cu^{2+} ion half-cell to the positive voltmeter terminal.

The magnitude of the cell EMF and the direction of electron flow through the external wire depend on the relative abilities of the metal/metal ion couples to give up or accept electrons. These abilities depend on the inherent chemical natures of the specific metal/metal ion couples we use to construct the cell. For instance, suppose that we replace the Zn/Zn^{2+} ion half-cell in Figure 1 with a strip of silver metal (Ag) suspended in an aqueous silver ion solution (Ag^+). When we connect the Ag/Ag^+ ion half-cell to the positive voltmeter termi-

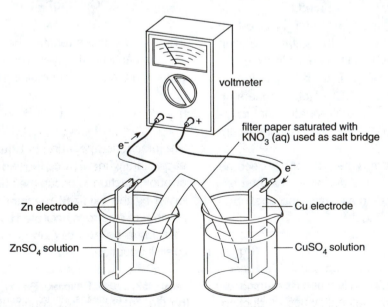

Figure 1 *An electrochemical cell*

nal and the Cu/Cu^{2+} ion half-cell to the negative terminal we observe a positive voltage reading. In practice, without knowing beforehand which electrode will be the anode and which the cathode in the spontaneous redox reaction, the wires are connected to the voltmeter by trial-and-error so as to yield a positive voltage. This reading tells us that the Cu/Cu^{2+} ion half-cell gives up electrons more readily than the Ag/Ag$^+$ ion half-cell does. Hence, the electron flow is from the Cu anode to the Ag cathode in the external circuit. This observation also indicates that if we place a Cu strip in an aqueous Ag$^+$ ion solution, the Cu will dissolve, and Ag will deposit on the Cu strip.

By testing different pairs of metals in solutions of their salts in electrochemical cells, we can rank the various metal/metal ion couples in order of their relative abilities to accept electrons. This type of listing is called an **electromotive force series**. Table 1 shows the reduction half-reactions for a series of common metal/metal ion half-cells, arranged in order of increasing EMF. As the half-cell EMF becomes increasingly positive, the ability for the half-cell to accept electrons, or to be reduced, increases. Accordingly, these half-cell EMFs are sometimes called the reduction potentials for the half-reactions. From Table 1, we can predict the direction of electron flow for a cell composed of any two metal/metal ion combinations listed in the table. We can also predict the chemical reactions that might occur when various redox couples are mixed in aqueous solution.

Table 1 *Reduction half-cell reactions arranged in order of increasing ability to accept electrons.**

reduction half-cell reaction	EMF, V
$Mg^{2+}(aq) + 2\ e^- \rightarrow Mg(s)$	−2.37
$Al^{3+}(aq) + 3\ e^- \rightarrow Al(s)$	−1.66
$Cr^{2+}(aq) + 2\ e^- \rightarrow Cr(s)$	−0.91
$Fe^{2+}(aq) + 2\ e^- \rightarrow Fe(s)$	−0.44
$Ni^{2+}(aq) + 2\ e^- \rightarrow Ni(s)$	−0.26
$Sn^{2+}(aq) + 2\ e^- \rightarrow Sn(s)$	−0.14
$Cu^{2+}(aq) + 2\ e^- \rightarrow Cu(s)$	+0.34
$Ag^+(aq)\ + e^- \rightarrow Ag(s)$	+0.80

* The EMF can also be defined as the standard reduction potential for that half-reaction.

As an illustration of how we can use Table 1, consider the Ag/Ag$^+$ ion–Cu/Cu^{2+} ion cell just described. The EMF of the Ag/Ag$^+$ ion half-cell, whose reduction half-reaction is shown in Equation 4,

$$Ag^+(aq) + e^- \rightarrow Ag(s) \qquad \text{(Eq. 4)}$$

is +0.80 V. The EMF of the Cu/Cu^{2+} ion half-cell, whose reduction half-reaction is shown in Equation 5,

$$Cu^{2+}(aq) + 2\ e^- \rightarrow Cu(s) \qquad \text{(Eq. 5)}$$

is +0.34 V. The more positive EMF of the Ag/Ag$^+$ ion half-cell indicates that this half-cell has a greater ability to accept electrons than does the Cu/Cu^{2+} ion half-cell. Thus, the Cu/Cu^{2+} ion half-cell will give up electrons to the Ag/Ag$^+$ ion half-cell. Accordingly, we reverse Equation 5 to obtain Equation 6, which shows the oxidation half-reaction for the Cu/Cu^{2+} ion half-cell. The EMF we associate with the oxidation reaction has the same magnitude but the opposite sign as the EMF of the reduction reaction.

$$Cu(s) \rightarrow Cu^{2+}(aq) + 2\ e^- \qquad \text{(Eq. 6)}$$

Equation 6 would then have an EMF of −0.34 V.

Remember that, in an oxidation–reduction reaction, we must have the same number of electrons transferred in each half-reaction. Thus, before we combine half-reactions, we must multiply Equation 4 by 2.

$2\ Ag^+(aq) + 2\ e^- \rightarrow 2\ Ag(s)$	+0.80 V
$Cu(s) \rightarrow Cu^{2+}(aq) + 2\ e^-$	−0.34 V
$2\ Ag^+(aq) + Cu(s) \rightarrow 2\ Ag(s) + Cu^{2+}(aq)$	?

Adding the EMFs of the two half-cells gives us a net EMF of 0.46 V for the cell. Note that, even though we multiply the Ag/Ag$^+$ ion half-cell reaction by 2 to obtain an equivalent transfer of electrons, we do not double the EMF for that half-cell. This is because EMF is related to inherent chemical nature, not to the stoichiometric relationships in a particular redox reaction. In other words, the EMF of a particular half-cell is independent of the number of electrons accepted or given up, since the EMF only represents the potential for that half-reaction to occur.

The electromotive force series, shown in Table 1, is tabulated using $1\,M$ ionic solutions at 25 °C, which are defined as standard conditions. Thus, if the Ag$^+$ ion and the Cu^{2+} ion concentrations are each $1\,M$ in the Ag/Ag$^+$ ion–Cu/Cu^{2+} ion cell, the EMF generated by the cell would be +0.46 V. A similar cell in which the ionic solution concentrations are other than $1\,M$ will produce a somewhat different EMF.

In this experiment, you will study electrochemical cells composed of various pairs of half-cells containing the Cu/Cu^{2+} ion redox couple, the Zn/Zn^{2+} ion redox couple, and the lead (Pb)/lead ion (Pb^{2+}) redox couple.

These half-cells will be ranked in order of increasing EMF. You will use a piece of filter paper soaked in potassium nitrate solution (KNO_3) as a salt bridge. You will use a $CuSO_4$ solution as a Cu^{2+} ion source, a $ZnSO_4$ solution as a Zn^{2+} ion source, and a lead nitrate solution, $Pb(NO_3)_2$, as a Pb^{2+} ion source. Even though the concentrations of these solutions will be less than that of standard solutions, this will not affect the measured EMFs significantly.

Procedure

Chemical Alert

0.1 M copper(II) sulfate solution—toxic and irritant
0.1 M lead(II) nitrate solution—toxic, irritant, and oxidant
0.5 M potassium nitrate solution—irritant and oxidant
0.1 M zinc(II) sulfate solution—toxic and irritant

Caution: Wear departmentally approved eye protection while doing this experiment.

I. Studying Half-Cell Potentials

A. The Zn/Zn²⁺ Ion–Cu/Cu²⁺ Ion Cell

Caution: 0.1 M $ZnSO_4$ and 0.1 M $CuSO_4$ solutions are toxic irritants. Prevent contact with your eyes, skin, and clothing.

 If you spill any solution on your skin, thoroughly wash your skin.

Measure 20 mL of 0.1 M $ZnSO_4$ solution in a clean graduated cylinder, and pour it into a clean 50-mL beaker. Label the beaker.

Rinse the cylinder three times using 20 mL of tap water each time and two times using 20 mL of distilled water each time. Pour the rinses into the drain, diluting with a large amount of running water. Allow the cylinder to drain.

Measure 20 mL of 0.1 M $CuSO_4$ solution in a clean graduated cylinder, and pour it into a clean 50-mL beaker. Label the beaker. Rinse the cylinder and discard the rinses as you did previously.

Note: All metal strips used in this experiment should be completely free of any surface oxide

---

coating. This oxide coating would give a dull, instead of glossy, appearance to the metal. Use sandpaper to remove any coating on the metal surface.

If necessary, clean the surface of a Zn strip with sandpaper. Place the Zn strip in the beaker containing $ZnSO_4$ solution.

If necessary, clean the surface of a Cu strip with sandpaper. Place the Cu strip in the beaker containing $CuSO_4$ solution.

Pour about 50 mL of 0.5 M KNO_3 solution into a clean 150-mL beaker. Fold a piece of 11.0-cm filter paper lengthwise until it makes a strip about 1.5 cm wide. Place the filter paper in the KNO_3 solution, and allow the filter paper to become thoroughly soaked. Use the saturated filter paper as the salt bridge for the cell. *Do not allow this filter paper to dry at any time during the experiment.* If the paper does become dry, resoak it in the KNO_3 solution.

Place the filter paper so that it contacts the solutions in both beakers, as shown in Figure 1. Do not use your hands to position the filter paper as the KNO_3 solution is a toxic irritant. Position the filter paper with your tweezers or forceps.

Connect a Cu wire to the Zn strip, using an alligator clip. Connect another Cu wire to the Cu strip, using another clip. Make sure the clips are not exposed to the solutions, but are connected to the portions of the strips outside the solutions. Connect the other end of the wire from the Zn strip to the negative terminal of the voltmeter. Connect the other end of the wire from the Cu strip to the positive voltmeter terminal. Read the voltage on the voltmeter. Record this voltage, which is the net or cell EMF, on Data Sheet 1. Also record on your Data Sheet which strip you connected to each terminal.

Remove the filter paper salt bridge, and place it on a clean watch glass. Unclip the Cu wire from the Zn strip. Save the $ZnSO_4$ solution and the Zn strip for use in Part C.

B. The Pb/Pb²⁺ Ion–Cu/Cu²⁺ Ion Cell

Caution: 0.1 M $Pb(NO_3)_2$ solution is a toxic irritant and an oxidant. Prevent contact with your eyes, skin, and clothing.

 If you spill any solution on your skin, thoroughly wash your skin.

Repeat the steps in Part *A*, using a 0.1 M $Pb(NO_3)_2$ solution and a strip of Pb in place of the $ZnSO_4$ solution and Zn strip. Connect the wires from the strips to the voltmeter terminals so as to obtain a positive reading, and observe the cell EMF. Record this voltage on Data Sheet 1. Also record which strip you attached to each terminal.

Remove the salt bridge from the cell, and place it on a clean watch glass. Unclip the wire from the Cu strip. Save the $CuSO_4$ solution and Cu strip and the $Pb(NO_3)_2$ solution and Pb strip for later use.

C. The Pb/Pb²⁺ Ion–Zn/Zn²⁺ Ion Cell

Repeat the steps in Part B, using $0.1 M$ $ZnSO_4$ solution and a Zn strip in place of the $CuSO_4$ solution and Cu strip. Connect the wires from the strips to the voltmeter terminals so as to obtain a positive reading, and observe the cell EMF. Record this voltage on Data Sheet 1. Also record which strip you attached to each terminal.

At this time your laboratory instructor may ask you to answer Questions 1–6 on Data Sheet 1.

II. Studying Cell Reactions

If necessary, clean your Zn strip with sandpaper again. Place the Zn strip in the $CuSO_4$ solution retained from Part I. Observe the solution for 5 min, then remove the

Zn strip from the solution. Record all observations on Data Sheet 2. Retain the $CuSO_4$ solution for later use.

If necessary, clean your Pb strip with sandpaper again. Place the Pb strip in the same $CuSO_4$ solution. Observe the solution for 5 min, then remove the Pb strip from the solution. Record all observations on Data Sheet 2. Pour the used $CuSO_4$ solution into the container provided by your laboratory instructor and labeled "Used $CuSO_4$ Solution."

If necessary, clean your Zn strip with sandpaper again. Place the Zn strip in the $Pb(NO_3)_2$ solution saved from Part I. Observe the solution for 5 min, then remove the Zn strip from the solution. Record all observations on Data Sheet 2.

Pour the used $Pb(NO_3)_2$ solution into the container provided by your laboratory instructor and labeled "Used $Pb(NO_3)_2$ Solution." Place the used Cu, Pb, and Zn strips in the appropriately labeled containers provided by your laboratory instructor.

Write chemical equations that describe the three reactions in Part II.

> **Caution:** Wash your hands thoroughly with soap or detergent before leaving the laboratory.

Post-Laboratory Questions

(Use the spaces provided for the answers and additional paper if necessary.)

1. Acidic solutions, such as hydrochloric acid solutions (HCl), contain relatively high hydronium ion (H_3O^+) concentrations. The chemical equation and EMF of the H_2/H_3O^+ ion half-cell is:

$$2 H_3O^+(aq) + 2 e^- \rightarrow H_2(g) + 2 H_2O(l) \qquad E = 0.000 \text{ V}$$

(1)   Based on your laboratory results from this experiment, would a piece of Zn dissolve in a $1 M$ HCl solution? Briefly explain.

(2)   Write the equation for the reaction that occurs when Zn is placed in a $1 M$ HCl solution, and determine the net EMF for the reaction. If no reaction occurs, write N.R.

(3)   What would you observe if you placed a piece of Zn in a $1 M$ HCl solution?

(4)   What would you observe if you placed a piece of Cu in a $1 M$ HCl solution? Briefly Explain.

2. (1)   It is not a good idea to cook acidic foods such as tomatoes in aluminum (Al) cookware. Briefly explain.

(2)   Write a chemical equation describing your answer to (2).

3. How would your experimental results from Part I of this Procedure have been different if you had reversed the connections to the voltmeter? Briefly explain.

4. Silver-plated tableware is prepared by electrochemically depositing Ag over the surface of a utensil fabricated from a less expensive material. Use the data in Table 1 to determine whether or not it would be possible to silver-plate an Fe fork by suspending the fork in a $1 M$ $AgNO_3$ solution. Briefly explain, including an equation for the reaction and a calculation of the net EMF.

name _____

section _____

date _____

Data Sheet 1

Experimental Data

I. Studying Half-Cell Potentials

section	cell	strip attached to positive terminal	strip attached to negative terminal	net or cell EMF, V
A	Zn/Zn^{2+} ion–Cu/Cu^{2+} ion	_____	_____	_____
B	Pb/Pb^{2+} ion–Cu/Cu^{2+} ion	_____	_____	_____
C	Pb/Pb^{2+} ion–Zn/Zn^{2+} ion	_____	_____	_____

Treating the Data

1. Arrange the three half-cells in order of their ability to accept electrons.

   first _____   second _____   third _____

2. Justify your answer to 1 using observations made during the experiment.

3. For each cell listed below, write the half-equations for each of the component half-cells.

   Zn/Zn^{2+} ion–Cu/Cu^{2+} ion cell

Zn/Zn^{2+} ion half-cell

Cu/Cu^{2+} ion half-cell

Pb/Pb²⁺ ion–Cu/Cu²⁺ ion cell

Pb/Pb²⁺ ion half-cell

Cu/Cu²⁺ ion half-cell

Pb/Pb²⁺ ion–Zn/Zn²⁺ ion cell

Pb/Pb²⁺ ion half-cell

Zn/Zn²⁺ ion half-cell

4.   Use the half-cell EMF for Cu/Cu²⁺ ion given in Table 1 and the net EMF observed for the Zn/Zn²⁺ ion–Cu/Cu²⁺ ion cell to calculate the Zn/Zn²⁺ ion half-cell EMF. Show all calculations and chemical equations.

_____
answer

5.   Use the Zn/Zn²⁺ ion half-cell EMF and the net EMF observed for the Pb/Pb²⁺ ion–Zn/Zn²⁺ ion cell to calculate the Pb/Pb²⁺ ion half-cell EMF. Show all calculations and chemical equations.

_____
answer

6.   Arrange the three half-cell reactions equations and their corresponding EMFs in order of increasing ability to accept electrons.

Data Sheet 2

Experimental Data

II. Studying Cell Reactions

metal	solution of metallic salt	observations	chemical equation for reaction
Zn	$CuSO_4$		
Pb	$CuSO_4$		
Zn	$Pb(NO_3)_2$		

Treating the Data

Are the chemical equations you wrote to describe the reactions consistent with your answer to Question 6 on Data Sheet 1? Briefly comment.

Pre-Laboratory Assignment

1. Briefly explain why it is necessary to be careful when disposing of the filter paper salt bridges used in this experiment.

2. Based on the EMFs for the magnesium/magnesium ion (Mg/Mg^{2+}) and the Ag/Ag^+ ion half-cells listed in Table 1, answer the following questions:

(1) Which half-cell has the greater ability to give up electrons, Ag/Ag^+ ion or Mg/Mg^{2+} ion? Briefly explain.

(2) Sketch a diagram of the cell that would be formed by connecting the two half-cells in Question 2(1).

(3) Which half-cell should be connected to the positive terminal of the voltmeter, and which to the negative terminal? Briefly explain.

(4) In which direction would the electrons flow through the external wire? Briefly explain.

(5) Calculate what the net EMF of this cell would be if the concentrations of the Ag^+ ion and Mg^{2+} ion solutions were each 1 M.

(6) Write the equation for the overall reaction for this cell.

(7)   What would you observe if you placed an Ag strip in an aqueous $MgCl_2$ solution? Briefly explain.

3.   When you multiply the equation for a half-reaction by a factor before combining it with another half-equation, why don't you multiply the EMF for the half-reaction by the same factors? Briefly explain.

Studying Electrochemical Cells and Reduction Potentials

prepared by **R. L. Marks,** University of Arizona

Purpose of the Experiment

Determine the reduction potentials for the three half-reactions, $Fe^{2+}(aq) + 2 e^- \rightarrow Fe(s)$, $Cu^{2+}(aq) + 2 e^- \rightarrow Cu(s)$, and $Pb^{2+}(aq) + 2 e^- \rightarrow Pb(s)$, from appropriate cell potentials and a selected standard reduction half-reaction. Compare predicted and measured potentials of cells constructed from combinations of these three half-reactions.

Background Information

Much of the damage that necessitated the refurbishment of the Statue of Liberty in 1986 resulted from oxidation-reduction reactions. In the years since the statue was erected, the copper (Cu) and iron (Fe) framework has undergone **oxidation reactions** in which the uncombined elements have lost electrons to form ions. These reactions are shown in Equations 1 and 2.

$$Cu \rightarrow Cu^{2+} + 2 e^- \qquad \text{(Eq. 1)}$$

$$Fe \rightarrow Fe^{3+} + 3 e^- \qquad \text{(Eq. 2)}$$

Coupled with every oxidation reaction is a **reduction reaction**, wherein a chemical species gains the electrons lost by another chemical species in the oxidation process. Because the oxidation and reduction processes are coupled, we refer to them together as **redox reactions**. Often it is important to focus on only an oxidation or a reduction process. We do this when we write an equation for only one part of a redox reaction. We call such an equation a **half-reaction**. Equations 1 and 2 are examples of oxidation half-reactions.

We say a redox reaction is **spontaneous** if the electron exchange occurs without outside influence. For example, a spontaneous redox reaction occurs when solid zinc (Zn) is placed in an aqueous solution of copper(II) ion (Cu^{2+}, or cupric ion), produced by the dissolution of copper(II) sulfate ($CuSO_4$). Equation 3 represents this reaction.

$$Cu^{2+}(aq, \text{blue}) + Zn(s, \text{gray}) \rightarrow$$
$$Cu(s, \text{brown}) + Zn^{2+}(aq, \text{colorless}) \qquad \text{(Eq. 3)}$$

We see several indications that this reaction proceeds spontaneously: the color intensity of the initially blue Cu^{2+} ion solution decreases, the size of the Zn sample decreases, and solid brown Cu forms. We can divide Equation 3 into two half-reactions, shown in Equations 4 and 5.

$$Zn(s, \text{gray}) \rightarrow Zn^{2+}(aq, \text{colorless}) + 2 e^- \qquad \text{(Eq. 4)}$$

$$Cu^{2+}(aq, \text{blue}) + 2 e^- \rightarrow Cu(s, \text{brown}) \qquad \text{(Eq. 5)}$$

Equation 4 is the oxidation half-reaction, and Equation 5 is the reduction half-reaction.

Copyright © 1992 by Chemical Education Resources, Inc., P.O. Box 357, 220 S. Railroad, Palmyra, Pennsylvania 17078
No part of this laboratory program may be reproduced or transmitted in any form or by any means, electronic or mechanical, including photocopying, recording, or any information storage and retrieval system, without permission in writing from the publisher. Printed in the United States of America

Energy is released in a spontaneous chemical reaction. If properly managed, this energy can be used to do work. In the experimental setup described above, the energy produced by the reaction of Cu^{2+} ion solution with Zn cannot be used. However, if we set up the experiment differently, we can both use this energy and evaluate its magnitude as well.

Galvanic Cells

If we carry out the reaction in Equation 3 but physically separate the half-reactions, we can cause electrons to flow through an external circuit from Zn to Cu^{2+} ion. We call such a system a **galvanic** or **voltaic cell**. A diagram of a galvanic cell utilizing the reaction in Equation 3 is shown in Figure 1.

The Zn electrode, immersed in a zinc sulfate solution ($ZnSO_4$), is called the **anode**. The Cu electrode, immersed in a $CuSO_4$ solution, is called the **cathode**. We refer to the $ZnSO_4$ and $CuSO_4$ solutions as **electrolytes**, because they are composed of solvated ions and can conduct an electric current. As Zn^{2+} ions are released into solution from the anode, electrons flow to the cathode through the external circuit. Once at the cathode, these electrons combine with Cu^{2+} ions in the $CuSO_4$ solution, depositing additional Cu on the cathode.

Formation of Zn^{2+} ion at the anode causes the $ZnSO_4$ solution to become positively charged. In contrast, the $CuSO_4$ solution becomes negatively charged

as Cu^{2+} ion is converted into Cu at the cathode. Without a means of neutralizing these solution charges, the reaction would quickly stop. To overcome this problem, a **salt bridge** containing a non-interfering electrolyte such as potassium nitrate (KNO_3) is used to join the two reaction cells. Potassium ions (K^+) flow from the salt bridge into the $CuSO_4$ solution to balance the excess negative charge, and nitrate ions (NO_3^-) flow from the salt bridge into the $ZnSO_4$ solution to balance the excess positive charge. Thus, the circuit of current flow through the cell is closed even though the half-reactions occur in separate containers.

The type of cell in Figure 1 can be conveniently described using the expression, $A \mid A^+ \parallel B^+ \mid B$. In this expression, $A \mid A^+$ represents the oxidation half-reaction at the anode, $\parallel$ represents the salt bridge, and $B^+ \mid B$ represents the reduction half-reaction at the cathode. Thus, the specific electrochemical cell in Figure 1 would be represented by $Zn \mid Zn^{2+} \parallel Cu^{2+} \mid Cu$. We make slight modifications to this expression when reactants and/or products are gases.

Studying Cell Potentials and Electrode Potentials

We measure the force of the electron flow from anode to cathode with a voltmeter. We call this force the **cell potential** (E_{cell}) or **electromotive force** (**emf**) of the cell. The cell potential, expressed in volts (V), is a measure of the work that can be done by the energy

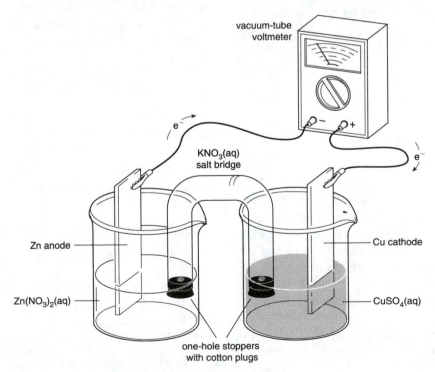

Figure 1   *A galvanic cell for the reaction of Cu^{2+} ion with Zn*

produced by the cell. Mathematically speaking, the cell potential represents the sum of the power of ions at the cathode to attract electrons ($E_{cathode}$) and the resistance of atoms at the anode to releasing electrons (E_{anode}). This relationship is shown in Equation 6.

$$E_{cell} = E_{anode} + E_{cathode} \qquad \text{(Eq. 6)}$$

Thus characteristic potentials are associated with each cathode and anode reaction, so E_{cell} depends upon the specific chemical system chosen. Cell potentials can be positive or negative. Reactions with positive cell potentials proceed spontaneously. Reactions with larger positive cell potentials can produce more energy to do work than reactions with smaller positive cell potentials. If a cell potential is negative, work must be done to cause the reaction to occur. We expect the cell potential for the reaction in Equation 3 to be positive, because we know that this reaction proceeds spontaneously.

Redox reactions are reversible. For example, we could place a piece of Cu into a $ZnSO_4$ solution in an effort to observe the reverse of the reaction in Equation 3. The result is shown as Equation 7.

$$Cu(s, \text{ brown}) + Zn^{2+}(aq, \text{ colorless}) \rightarrow$$
$$\text{no reaction} \qquad \text{(Eq. 7)}$$

The cell potential for the reaction in Equation 7 is as large as the one for the reaction in Equation 3, but it is negative instead of positive. Because the reaction in Equation 3 is spontaneous ($E_{cell} < 0$), the reverse reaction is not spontaneous ($E_{cell} > 0$). Half-reactions can also be written in reverse. The potential of the reversed half-reaction is the negative of the potential of the original half-reaction. Thus, if the electrode potential (E) for the Zn | Zn^{2+} half-reaction is -0.76 V, then E for the Zn^{2+} | Zn half-reaction is $+0.76$ V.

Calculating Electrode Potentials from Cell Potentials

We find it useful to quantify the relative abilities of ions and molecules to attract electrons. To do this, we measure the reduction potential of each ion or molecule. A listing of such reduction potentials is called the **electromotive series**.

However, we cannot directly measure an isolated reduction potential, because the necessary electrons must come from a simultaneous oxidation. Therefore we measure the cell potential of cells using a series of reduction reactions, each coupled with a common oxidation process. If we assign a voltage to the oxidation

reaction, we can calculate the voltage of the reduction reaction.

Suppose we want to determine the reduction potential of the half-reaction of tin (Sn), shown in Equation 8.

$$Sn^{2+}(aq) + 2 \ e^- \rightarrow Sn(s) \qquad \text{(Eq. 8)}$$

We begin with an electrode reaction for which E has been assigned or is known. We refer to such an electrode as a **reference electrode**. For example, we can assign a voltage to the oxidation of nickel(Ni) to nickel(II) ion (Ni^{2+}, nickelous ion), shown in Equation 9.

$$Ni(s) \rightarrow Ni^{2+}(aq) + 2 \ e^- \quad E = 0.25 \text{ V} \quad \text{(Eq. 9)}$$

Next we construct and determine the potential of an electrochemical cell using the reference electrode reaction in Equation 9 with the reaction in Equation 8. We find E_{cell} of this Ni | Ni^{2+} || Sn^{2+} | Sn cell to be 0.11 V. The cell reaction is shown in Equation 10.

$$Ni(s) + Sn^{2+}(aq) \rightarrow Ni^{2+}(aq) + Sn(s) \qquad \text{(Eq. 10)}$$

We use Equation 6 to calculate the reduction potential of the Sn^{2+} | Sn half-reaction.

$$0.11 \text{ V} = 0.25 \text{ V} + E_{cathode}$$

$$E_{cathode} = 0.11 \text{ V} - 0.25 \text{ V}$$
$$= -0.14 \text{ V}$$

Once we have established the reduction potential of the Sn^{2+} | Sn half-reaction, we can couple the Sn electrode with an electrode whose reaction potential is unknown. The measured cell voltage, combined with the reduction potential of the Sn^{2+} | Sn half-reaction, can be used in Equation 6 to determine the unknown potential.

In this experiment you will construct and determine the potential of several galvanic cells. You will use a vacuum-tube voltmeter or other appropriate instrument offering a high resistance to current flow to measure each cell potential. Initially you will use a reference electrode to determine the potentials of three reduction half-reactions. The reference electrode is Zn in $Zn(NO_3)_2$ solution, with an assigned reduction potential of -0.76 V, as shown in Equation 11.

$$Zn^{2+}(aq) + 2 \ e^- \rightarrow Zn(s) \quad E = -0.76 \text{ V} \quad \text{(Eq. 11)}$$

You will couple the Zn^{2+} | Zn system with an Fe electrode immersed in iron(II) sulfate solution ($FeSO_4$), a Cu electrode immersed in $CuSO_4$ solution, and a lead (Pb) electrode immersed in lead(II) nitrate solution, $Pb(NO_3)_2$. Based on the measured cell potentials, you will determine the reduction potentials for the reactions in Equations 12, 5, and 13, respectively.

$$Fe^{2+}(aq) + 2\ e^- \rightarrow Fe(s) \qquad (Eq.\ 12)$$

$$Pb^{2+}(aq) + 2\ e^- \rightarrow Pb(s) \qquad (Eq.\ 13)$$

You will then consult your chemistry text or other chemistry reference, in order to compare the experimental reduction potentials with the corresponding reference potentials. The calculated reduction potentials may not exactly match the reference potentials, because the reference potentials are determined under standard conditions, which are not used in this experiment. Potentials determined under standard conditions are called **standard potentials** and are designated E^o.

Next you will construct Fe | Fe^{2+} || Cu^{2+} | Cu, Fe | Fe^{2+} || Pb^{2+} | Pb, and Pb| Pb^{2+}|| Cu^{2+}| Cu cells and measure their cell potentials. You will compare these measured cell potentials with the cell potentials calculated from data obtained in Part I of the experiment. Finally, from your collected data, you will list the four reduction half-reactions in order of increasing reduction potential.

Procedure

Chemical Alert

0.1M copper(II) sulfate—toxic and irritant
0.1M iron(II) sulfate—toxic
0.1M potassium nitrate—irritant and oxidant
0.1M lead(II) nitrate—toxic, irritant, and oxidant
0.1M zinc(II) nitrate—irritant and oxidant

Caution: Wear departmentally approved eye protection while doing this experiment.

Note: The apparatus you use will be similar to that shown in Figure 1. Your laboratory instructor may substitute an ordinary voltmeter for the vacuum-tube voltmeter (VTVM).

Your laboratory instructor will describe the salt bridge(s) you will use in your cells.

I. Determining Cell Potentials Using a Zn^{2+} | Zn Reference Electrode

A. The Zn | Zn^{2+} || Fe^{2+} | Fe Cell

Use a 50-mL graduated cylinder to transfer 30 mL of 0.1M $Zn(NO_3)_2$ solution to a clean, 100-mL beaker. Use a clean graduated cylinder to transfer 30 mL of 0.1M $FeSO_4$ solution into a second clean, 100-mL beaker. Appropriately label these and all other beakers used in this experiment.

Use sandpaper or steel wool to clean the surfaces of a Zn electrode and an Fe electrode to ensure good contact of the metals with the electrolytes and the wire. Place the Zn electrode in the $Zn(NO_3)_2$ solution and the Fe electrode in the $FeSO_4$ solution. Position the salt bridge so it makes good contact with both solutions.

Set up the apparatus as shown in Figure 1 by connecting a Cu wire from the Zn electrode to the negative VTVM terminal (ground). In the same manner, connect the Fe electrode to the positive VTVM terminal. Clamp the electrodes in place so that they do not touch the beaker walls.

Read E_{cell} to the nearest 0.01 V on the voltmeter. Record this reading on Data Sheet 1.

Disconnect the wire from the Fe electrode and remove the salt bridge from the $FeSO_4$ solution. Retain the beaker of $FeSO_4$ solution and the Fe electrode for Parts II D and E.

If you are using a U-tube or an unglazed porcelain cup salt bridge, do **not** remove it from the beaker containing the $Zn(NO_3)_2$ solution. If you are using a filter paper salt bridge, discard the paper in the container provided by your laboratory instructor and labeled, "Discarded Filter Paper."

B. The Zn | Zn^{2+} || Cu^{2+} | Cu Cell

Add 30 mL 0.1M $CuSO_4$ solution to another clean, 100-mL beaker. Place a polished Cu electrode in the $CuSO_4$ solution.

If you are using a U-tube or an unglazed porcelain cup salt bridge, rinse the end not suspended in $Zn(NO_3)_2$ solution with distilled or deionized water, collecting the rinses in a beaker labeled, "Discarded Salt Bridge Rinses." Position the rinsed end of the salt bridge so that it makes good contact with the $CuSO_4$ solution. If you are using a filter paper salt bridge, prepare a new salt bridge as you did for Part I A.

Connect the Cu electrode to the positive VTVM terminal. Read the cell potential to the nearest 0.01 V. Record this potential on Data Sheet 1. Retain the $CuSO_4$ solution and electrode for Parts IIE and IIG.

Repeat the salt bridge treatment procedure described in Part I A.

C. The Zn | Zn²⁺ || Pb²⁺ | Pb Cell

Transfer 30 mL 0.1M Pb(NO$_3$)$_2$ to another clean, 100-mL beaker. Follow the salt bridge rinsing procedure described in Part IA.

Place a cleaned Pb electrode in the Pb(NO$_3$)$_2$ solution and connect the electrode to the positive VTVM terminal. Connect the Pb electrode to the positive VTVM terminal. Read the cell potential to the nearest 0.01 V. Record this potential on Data Sheet 1. Retain both solutions and electrodes.

Using the solutions you saved, repeat procedures A through C and record your observations on Data Sheet 1. If your results for determinations 1 and 2 for a particular cell vary by more than 0.05 V, do a third determination for that cell.

II. Determining Cell Potentials Without Using a Reference Electrode

D. The Fe | Fe²⁺ || Cu²⁺ | Cu Cell

Connect the FeSO$_4$ solution to the CuSO$_4$ solution using a freshly rinsed salt bridge. Connect the Fe electrode to the negative VTVM terminal. Connect the Cu electrode to the positive VTVM terminal. Read the cell potential to the nearest 0.01 V. Record this potential on Data Sheet 1. Retain both solutions for further studies.

E. The Fe | Fe²⁺ || Pb²⁺ | Pb Cell

Repeat Part D using Pb(NO$_3$)$_2$ solution and electrode in place of the CuSO$_4$ solution. Connect the Pb electrode to the positive VTVM terminal. Read the cell potential to the nearest 0.01 V. Record this potential on Data Sheet 1. Retain both solutions.

F. The Pb | Pb²⁺ || Cu²⁺ | Cu Cell

Repeat Part E, using CuSO$_4$ solution and electrode in place of FeSO$_4$ solution. Connect the Pb electrode to the negative VTVM terminal. Connect the Cu electrode to the positive VTVM terminal. Read the cell potential to the nearest 0.01 V. Record this potential on Data Sheet 1.

Using the solutions you saved, repeat procedures D through F and record your observations on Data Sheet 1. If your results for determinations 1 and 2 for a particular cell vary by more than 0.05 V, do a third determination for that cell.

Transfer all solutions into appropriately marked waste containers provided by your laboratory instructor and labeled, "Discarded CuSO$_4$ Solutions," "Discarded FeSO$_4$ Solutions," "Discarded Pb(NO$_3$)$_2$ Solutions," and "Discarded Zn(NO$_3$)$_2$ Solutions."

> **Caution:** Wash your hands thoroughly with soap or detergent before leaving the laboratory.

Calculations

Using the experimental data you recorded on Data Sheet 1, do the following calculations and record the results on Data Sheet 2.

I. If two of your determinations for a specific cell are within 0.05 V of each other, use the average of those two determinations in the following calculations. Otherwise, use data from the first determination only.

  A. Calculate the reduction potential of the Fe²⁺ | Fe half-reaction.

  B. Calculate the reduction potential of the Cu²⁺ | Cu half-reaction.

  C. Calculate the reduction potential of the Pb²⁺ | Pb half-reaction.

II. On Data Sheet 2, Part II, list the reduction half-reactions and calculated potentials in order of increasing potential. Include the Zn²⁺ | Zn half-reaction in this list.

III. Using your table of reduction potentials, calculate an **anticipated** cell potential for the following.

  A. Fe | Fe²⁺ || Cu²⁺ | Cu cell

  B. Fe | Fe²⁺ || Pb²⁺ | Pb cell

  C. Pb | Pb²⁺ || Cu²⁺ | Cu cell

Using Equation 14, compare the **anticipated** cell potentials with the **observed** cell potentials as recorded on Data Sheet 1. Record the results on Data Sheet 2, Part III.

$$\text{percent error, \%} = \left(\frac{\text{anticipated } E_{cell} - \text{observed } E_{cell}}{\text{anticipated } E_{cell}}\right)(100\%) \quad \text{(Eq. 14)}$$

IV. Consult your chemistry text or other appropriate reference to determine the standard reduction potential for each of the half-reactions studied in this experiment. Record the information on Data Sheet 2, Part IVA.

Using Equation 14, calculate the percent error in each of your **observed** cell potentials relative to the cell potentials calculated using the standard reduction potentials, and record your results on Data Sheet 2, Part IVB.

Post-Laboratory Questions

(Use the spaces provided for the answers and additional paper if necessary.)

1. Briefly comment on the effect of solution concentration on cell potential. Note that E^o is measured using $1M$ solutions, while your measurements were made using $0.1M$ solutions.

3. If the cells you constructed in this experiment were doubled in size, would the measured cell potentials double? Briefly explain.

2. A student doing this experiment using a U-tube salt bridge is having difficulty obtaining useful voltage readings. Examining the experimental setup, the student finds an air pocket between the stopper at one end of the U-tube and the electrolyte solution inside the tube. Briefly explain how the air pocket could be the source of the erroneous voltage readings.

_____

name

Data Sheet 1
Experimental Observations

I. Determining Cell Potentials Using a Zn^{2+} | Zn Reference Electrode

			determination	
		1	2	3
A.	**The Zn	Zn^{2+} || Fe^{2+}	Fe Cell**			
	cell potential, V	_____	_____	_____
B.	**The Zn	Zn^{2+} || Cu^{2+}	Cu Cell**			
	cell potential, V	_____	_____	_____
C.	**The Zn	Zn^{2+} || Pb^{2+}	Pb Cell**			
	cell potential, V	_____	_____	_____

II. Determining Cell Potentials Without Using a Reference Electrode

			determination	
D.	**The Fe	Fe^{2+} || Cu^{2+}	Cu Cell**			
	cell potential, V	_____	_____	_____
E.	**The Fe	Fe^{2+} || Pb^{2+}	Pb Cell**			
	cell potential, V	_____	_____	_____
F.	**The Pb	Pb^{2+} || Cu^{2+}	Cu Cell**			
	cell potential, V	_____	_____	_____

Data Sheet 2
Treatment of Data

I. Experimentally Determined Reduction Potentials

half-reaction	*calculated reduction potential*
A. Fe^{2+}	Fe	_____
B. Cu^{2+}	Cu	_____
C. Pb^{2+}	Pb	_____

II. Table of Reduction Potentials

half-reaction	potential
_____	_____
_____	_____
_____	_____
_____	_____

III. Cell Potentials

cell	observed cell potential from Data Sheet 1, V	anticipated cell potential, V	percent error, %
A. Fe | Fe^{2+} || Cu^{2+} | Cu	_____	_____	_____
B. Fe | Fe^{2+} || Pb^{2+} | Pb	_____	_____	_____
C. Pb | Pb^{2+} || Cu^{2+} | Cu	_____	_____	_____

IV. A. Standard Reduction Potentials for Half-Reactions Studied

Reference consulted: _____

half-reaction	$E°$, V
Fe | Fe^{2+}	_____
Cu | Cu^{2+}	_____
Pb | Pb^{2+}	_____

B. A Comparison of Observed Cell Potentials with Cell Potentials Calculated from Standard Reduction Potentials

cell	observed cell potential from Data Sheet 1, V	cell potential derived from standard reduction potentials, V	percent error, %
Fe | Fe^{2+} || Cu^{2+} | Cu	_____	_____	_____
Fe | Fe^{2+} || Pb^{2+} | Pb	_____	_____	_____
Pb | Pb^{2+} || Cu^{2+} | Cu	_____	_____	_____

Pre-Laboratory Assignment

1. Briefly explain why it is important to wash your hands after you complete this experiment, and before you leave the laboratory.

2. Briefly define the following terms as they relate to this experiment.

   (1)   reference electrode

   (2)   oxidation half-reaction

   (3)   salt bridge

   (4)   cell potential

3. Briefly distinguish between the reaction that occurs at the anode and the reaction that occurs at the cathode of a galvanic cell.

4. (1) Sketch an apparatus that could be used to study the reaction shown in Equation 15.

$$Ti(s) + Cd^{2+}(aq) \rightarrow Ti^{2+}(aq) + Cd(s) \qquad (Eq.15)$$

   (2)   Use the A I A$^+$ II B$^+$ I B expression to represent the cell you have drawn in (1).

5. A student constructed a Ti I Ti^{2+} II Zn^{2+} I Zn cell and measured a cell potential of 0.87 V. Next, the student constructed a Mg I Mg^{2+} II Ti^{2+} I Ti cell and measured a cell potential of 0.74 V. Given that the assigned reduction potential of Zn is −0.76 V (Equation 11):

   (1)   Calculate the **reduction** potential of the Ti^{2+} I Ti half-reaction.

   (2)   Write the cell reaction for the Ti I Ti^{2+} II Zn^{2+} I Zn cell.

(3) Calculate the *reduction* potential of the Mg^{2+} | Mg half-reaction.

(4) Write the cell reaction for the Mg | Mg^{2+} || Ti^{2+} | Ti cell.

Preparing Tris(2,4-pentanedionato)iron(III), an Iron Coordination Complex

prepared by **George S. Patterson,** Suffolk University, MA

Purpose of the Experiment

Prepare tris(2,4-pentanedionato)iron(III). Recrystallize the compound from a methanol–water solution. Calculate the percent yield of the product.

Background Information

Transition metals form cations that stabilize in solution by bonding with molecules or ions with unshared electrons. We call such molecules, which are all Lewis bases, **ligands**. Some ligands, such as water (H_2O) and the chloride ion (Cl^-), have one site where they can attach to a metal cation. Under proper pH conditions, other ligands, such as 2,4-pentanedione ($C_5H_8O_2$), have two sites for attachment to a metal ion. The structure of 2,4-pentanedione, also known as acetylacetone, is shown in Figure 1.

Figure 1 *The structure of 2,4-pentanedione*

Still other ligands, such as ethylenediaminetetraacetate ion, $EDTA^{4-}$, have six sites for attachment to a metal ion. Depending on the charge on the metal cation, the charge, if any, on the ligands, and the number of ligands that associate with the cation, the resulting species may or may not have a charge. We call the stabilized metal ion coupled with the ligands a **coordination complex**, or, if the complex has a charge, a **complex ion**. Using appropriate solvent control, we can isolate these complexes as insoluble compounds.

In this experiment, you will prepare the coordination complex tris(2,4-pentanedionato)iron(III), $Fe(C_5H_7O_2)_3$, by mixing an aqueous solution of iron(III) chloride hexahydrate ($FeCl_3 \cdot 6\, H_2O$), containing sodium acetate trihydrate ($CH_3CO_2Na \cdot 3\, H_2O$), with a solution of 2,4-pentanedione in methanol (MeOH). In solution, 2,4-pentanedione [I] is in equilibrium with its tautomeric enol form, 4-hydroxy-3-penten-2-one [II], as shown in Equation 1.

Acetate ion, CH_3COO^-, acts as a base and removes a proton (H^+) from 4-hydroxy-3-penten-2-one, as shown in Equation 2 at the top of p. 2.

The 2,4-pentanedionate ion, $C_5H_7O_2^-$, formed by the reaction in Equation 2, complexes with iron(III) ion (Fe^{3+}) to form $Fe(C_5H_7O_2)_3$ as shown in Equation 3.

Copyright © 1993 by Chemical Education Resources, Inc., P.O. Box 357, 220 S. Railroad, Palmyra, Pennsylvania 17078
No part of this laboratory program may be reproduced or transmitted in any form or by any means, electronic or mechanical, including photocopying, recording, or any information storage and retrieval system, without permission in writing from the publisher. Printed in the United States of America

$$CH_3 - \overset{\overset{O}{\|}}{C} - O^-(aq) + CH_3\overset{\overset{OH}{|}}{C} = CH - \overset{\overset{O}{\|}}{C} - CH_3(MeOH) \rightarrow$$

$$CH_3 - \overset{\overset{O}{\|}}{C} - OH(MeOH/aq) + CH_3\overset{\overset{O^-}{|}}{C} = CH - \overset{\overset{O}{\|}}{C} - CH_3(MeOH/aq) \quad \text{(Eq. 2)}$$

Both $FeCl_3 \cdot 6\ H_2O$ and $CH_3CO_2Na \cdot 3\ H_2O$ are soluble in water. 2,4-Pentanedione and $Fe(C_5H_7O_2)_3$ are only slightly soluble in water, but each dissolves in methanol. Therefore, you will prepare an aqueous solution containing $FeCl_3 \cdot 6\ H_2O$ and CH_3CO_2Na. To this you will add a solution of 2,4-pentanedione dissolved in methanol. As the reaction proceeds, red crystals of $Fe(C_5H_7O_2)_3$ will precipitate from solution. You will collect the precipitate and then remove some of its impurities by washing it with water. To further purify the solid $Fe(C_5H_7O_2)_3$, you will dissolve it in methanol and reprecipitate it by adding water to the solution. Then you will dry and weigh the crystals to determine the actual yield of the preparation. You will also calculate the theoretical yield, based on the reaction stoichiometry and the mass of limiting reactant present. Based on the actual and theoretical yields, you will use Equation 4 to determine the percent yield.

percent yield, % =

$$\left(\frac{\text{actual yield of } Fe(C_5H_7O_2)_3, \text{ g}}{\text{theoretical yield of } Fe(C_5H_7O_2)_3, \text{ g}} \right)(100\%) \quad \text{(Eq. 4)}$$

Procedure

Chemical Alert
iron(III) chloride hexahydrate—corrosive
methanol—flammable and toxic
2,4-pentanedione (acetylacetone)—flammable,
toxic, and mild irritant

Caution: Wear departmentally approved eye protection while doing this experiment.

I. Preparing $Fe(C_5H_7O_2)_3$

Note: The compounds used in this preparation are corrosive and can stain. If possible, wear protective gloves while performing the experiment.

Note: Your laboratory instructor will demonstrate an acceptable technique for weighing a solid and transferring the solid to a beaker.

1. Weigh 2.7 g of $FeCl_3 \cdot 6\ H_2O$ on a weighing paper and transfer it into a 100-mL beaker. Record on your Data Sheet the exact mass used. Measure 6 mL of distilled or deionized water in a 10-mL graduated cylinder. Add the water to the solid in the beaker. Stir the mixture with a glass stirring rod to dissolve the solid. Set the beaker aside.

2. Weigh 5.0 g of $CH_3CO_2Na \cdot 3\ H_2O$ on another piece of weighing paper. Record on your Data Sheet the exact mass used. Transfer the $CH_3CO_2Na \cdot 3\ H_2O$ to a 150-mL beaker. Dissolve the solid by adding 10 mL distilled water and stirring.

Drain and dry the graduated cylinder for use in Step 4.

3. Slowly add the $CH_3CO_2Na \cdot 3\ H_2O$ solution to the $FeCl_3 \cdot 6\ H_2O$ solution, with constant stirring. After you add all the $CH_3CO_2Na \cdot 3\ H_2O$ solution, set the mixture aside.

Rinse the glass stirring rod with distilled water from a wash bottle and dry it. Collect the rinse water in the container provided by your laboratory instructor and labeled "Discarded Filtrates."

Caution: Methanol is flammable and toxic. Be certain that there are no open flames in your work area when you work with this solvent. Avoid inhaling any vapors or ingesting any of the compound. Use a fume hood if your laboratory instructor directs you to do so.

4. Measure 10 mL of methanol in a clean, dry 10-mL graduated cylinder. Transfer the methanol to a clean, dry 50-mL beaker. Carefully measure 4 mL of 2,4-pentanedione in the 10-mL graduated cylinder. Record the exact volume of 2,4-pentanedione on your Data Sheet. Add the 2,4-pentanedione to the methanol. Stir the mixture with a clean glass stirring rod.

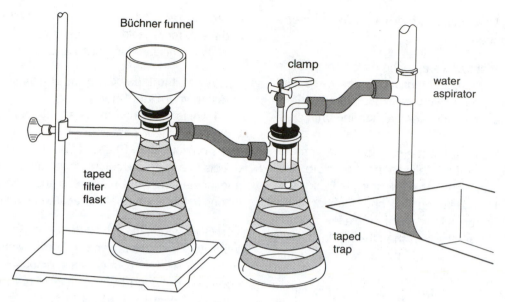

Figure 2   *Filtering apparatus*

5. Slowly add the 2,4-pentanedione solution to the FeCl$_3$/CH$_3$CO$_2$Na solution with stirring. Continue to stir the mixture for about 5 min to be sure the reaction is complete.

> **Note:** Your laboratory instructor will inform you whether to use a rubber filtering ring or a rubber stopper to fit your Büchner funnel in the filter flask in Step 6.

6. Set up a filtering apparatus as shown in Figure 2. Begin by clamping a 250-mL filtering flask to a ring stand. Fit a Büchner funnel and a rubber filtering ring in the mouth of the filter flask. Insert a piece of filter paper into the Büchner funnel. Moisten the paper with distilled water.

Attach one end of a piece of pressure tubing to the arm of the filter flask. Attach the other end to a trap, as shown in Figure 2. Attach the trap to the water aspirator with another piece of pressure tubing. If a trap is not available, attach the pressure tubing on the filter flask directly to the aspirator.

Turn on the aspirator, and gently press down on the top of the Büchner funnel with your palm. As you remove your hand from the funnel, you should be able to feel the vacuum suction. If you do not feel any suction, re-check your setup.

7. With the aspirator on, pour the reaction mixture into the Büchner funnel. Be careful not to overfill the funnel. Use a stream of distilled water from a wash

bottle to rinse the precipitate from the beaker into the Büchner funnel.

8. To wash the precipitate, begin by opening the clamp on the top of the trap and turning off the aspirator. Cover the precipitate with distilled water from the wash bottle. Gently stir the water-precipitate mixture with the glass stirring rod, being careful to avoid tearing the wet filter paper. Close the clamp, turn on the aspirator, and draw the water through the funnel. Repeat this washing process two more times.

Use a spatula to loosen the edge of the filter paper from the funnel. Carefully transfer the precipitate *and* filter paper to a clean watch glass. Transfer the water in your filter flask to the "Discarded Filtrates" container.

Rinse the flask and funnel with distilled water, transferring the rinse water to the "Discarded Filtrates" container. Allow the flask to drain and dry.

II.   Recrystallizing and Drying the Product

9. Use the spatula to carefully scrape the precipitate from the filter paper into a clean, dry 250-mL Erlenmeyer flask. Avoid transferring any wet filter paper with the precipitate.

Discard the filter paper in the container provided by your laboratory instructor and labeled "Discarded Filter Paper." Wash and dry the watch glass for use in Step 13.

10. Dissolve the precipitate in the Erlenmeyer flask in methanol. To do this, add 2–3 mL methanol to the precipitate, and stir with a stirring rod. Repeat this pattern of adding 2–3 mL methanol and stirring until the precipitate is completely dissolved. About 30 mL of methanol will be required.

11. Add 25 mL distilled water to the methanol solution in the Erlenmeyer flask and thoroughly stir. Keep adding water and stirring until crystallization begins. Continue adding the distilled water with stirring until crystallization appears to stop.

To test the methanol–water solution in order to determine whether or not crystallization is complete, filter 5–10 mL of the solution in the Erlenmeyer flask using the cleaned and dried filtering flask and Büchner funnel fitted with a new piece of filter paper. Add a few drops of distilled water to the filtrate in the filtering flask. If no crystals appear when you add the water, the crystallization process is complete.

If crystallization is not complete, pour the filtrate into the methanol solution in the Erlenmeyer flask. Continue to add water to the methanol solution in the Erlenmeyer flask as before until crystallization appears to stop, and retest 5–10 mL of the filtered solution.

12. When you are sure that crystallization is complete, filter the crystals using the cleaned and dried Büchner funnel and a new piece of filter paper. Wash the crystals once with distilled water. Leave the aspirator on for a few minutes in order to partially dry the crystals. Pour the filtrate into the "Discarded Filtrates" container.

13. Use the spatula to transfer the crystals and filter paper to a watch glass. Leave the container and its contents uncovered until the next laboratory period or until they are dry.

Discard the filtrate in the filter flask into the container labeled "Discarded Filtrates."

14. Prepare a label for the vial in which you will store your crystals. Include the following information on the label: your name, the date, your laboratory section, the chemical name (or formula) of the product, and leave

spaces to record the mass of the crystals produced and the percent yield of the synthesis.

Attach this label to your vial or container.

15. Weigh the labeled vial or container. Record this mass on your Data Sheet.

Transfer your dry crystals to the vial or container. Weigh the vial or container plus crystals. Record this mass on your Data Sheet. If your laboratory instructor directs you to do so, calculate the mass of the crystals and the percent yield at this time and record your calculations in the appropriate spaces on the vial.

16. Give your labeled container of crystals to your laboratory instructor. If your laboratory instructor does not wish to collect your crystals, dispose of them in the container provided by your laboratory instructor and labeled "Discarded $Fe(C_5H_7O_2)_3$."

17. Wash and store your glassware as directed by your laboratory instructor.

> **Caution:** Wash your hands thoroughly with soap or detergent before leaving the laboratory.

Calculations

(Do the following calculations and record the results on your Data Sheet.)

1. Calculate the actual yield of product. To do so, subtract the mass of the vial from the mass of the vial plus crystals.

2. Calculate the percent yield of product. Use the following molar masses for your calculations: $FeCl_3 \cdot 6 H_2O = 270.3$ g mol^{-1}; $CH_3CO_2Na \cdot 3 H_2O = 136.1$ g mol^{-1}; and, 2,4-pentanedione $= 100.1$ g mol^{-1}.

Post-Laboratory Questions

(Use the spaces provided for the answers and additional paper if necessary.)

1. (1)   Briefly explain the role of CH_3CO_2Na in the preparation of $Fe(C_5H_7O_2)_3$.

(2)   Briefly explain why the addition of water to the filtrate causes the crystals to precipitate from the solution.

(3)   Briefly explain why, in Step 11, it is more efficient to examine a small portion of the filtrate rather than the entire solution, when checking to see if crystalization is complete.

(2)   How would using 0.42 g $CH_3CO_2Na \cdot 3\,H_2O$ instead of the amount called for in the procedure change the theoretical yield of $Fe(C_5H_7O_2)_3$? Briefly explain.

(4)   Briefly explain why, in Step 12, you are instructed to wash the collected crystals with distilled water instead of a water–methanol mixture.

2. A student reached Step 11 of the Procedure and saw crystals form after she added a few drops of water to the filtrate.

(1)   Write the chemical formula of the crystals.

Data Sheet

I. Preparing Fe(C$_5$H$_7$O$_2$)$_3$

mass of FeCl$_3$ $\cdot$ 6 H$_2$O used, g _____

mass of CH$_3$CO$_2$Na $\cdot$ 3 H$_2$O used, g _____

volume of 2,4-pentanedione used, mL _____

number of moles of FeCl$_3$ $\cdot$ 6 H$_2$O used, mol _____

number of moles of CH$_3$CO$_2$Na $\cdot$ 3 H$_2$O used, mol _____

number of moles of 2,4-pentanedione used, mol _____

theoretical yield of Fe(C$_5$H$_7$O$_2$)$_3$, mol _____

theoretical yield of Fe(C$_5$H$_7$O$_2$)$_3$, g _____

II. Recrystallizing and Drying the Product

mass of Fe(C$_5$H$_7$O$_2$)$_3$ plus vial, g _____

mass of vial, g _____

mass of Fe(C$_5$H$_7$O$_2$)$_3$, g _____

percent yield of Fe(C$_5$H$_7$O$_2$)$_3$, % _____

Pre-Laboratory Assignment

1. Briefly describe the hazards associated with using methanol.

2. (1) Calculate the number of moles of $FeCl_3 \cdot 6 H_2O$ you will use in this experiment. Record this number on your Data Sheet.

(2) Calculate the number of moles of $CH_3CO_2Na \cdot 3 H_2O$ you will use in this experiment. Record this number on your Data Sheet.

(3) Calculate the number of moles of 2,4-pentanedione you will use in this experiment (density of 2,4-pentanedione = 0.976 g mL^{-1}, GMM: 100.11 g mol^{-1}). Record this number on your Data Sheet.

(4) Calculate the theoretical yield, in moles of $Fe(C_5H_7O_2)_3$, for this experiment. Record this number on your Data Sheet.

(5) For this experiment, calculate the theoretical yield in g $Fe(C_5H_7O_2)_3$ (GMM: 353.18 g mol^{-1}). Record the theoretical yield on your Data Sheet.

3. A chemist needed some potassium nitrate, KNO_3, for an experiment. Because there was none in her laboratory, she decided to make it from potassium iodide, KI, and sodium nitrate, $NaNO_3$, by means of the following reaction:

$$KI(aq) + NaNO_3(aq) \rightarrow KNO_3(s) + NaI(aq)$$

The success of this synthesis depends on the fact that KNO_3 is much less soluble in water than KI, $NaNO_3$, or NaI. When she dissolved 0.50 mol KI and 0.75 mol $NaNO_3$ in 100 mL of water and cooled the solution to 0 °C, 37 g of pure KNO_3 crystallized from solution.

(1) What is the theoretical yield, in grams, of KNO_3 for this synthesis?

(2) What is her percent yield for this synthesis?

Synthesizing and Analyzing a Coordination Compound of Nickel(II) Ion, Ammonia, and Chloride Ion

prepared by **George S. Patterson**, Suffolk University, MA

Purpose of the Experiment

Synthesize a coordination compound of nickel(II) ion, ammonia, and chloride ion. Analyze the compound for mass percent nickel and mass percent ammonia. From these data, determine the empirical formula of the compound, and calculate the percent yield of the synthesis.

Background Information

Two important tasks many chemists perform are the synthesis and analysis of compounds. Synthesis involves not only preparing the compound, but also maximizing the yield of pure product. After isolating the product, the chemist must analyze it to ascertain its chemical composition or formula. Both tasks require good technique and close attention to what might seem minor procedural details. Therefore, a technically skilled chemist with a good understanding of the purposes of each step in both the synthesis and analysis procedures will get the most accurate results.

In this experiment you will prepare a coordination compound containing nickel(II) ion (Ni^{2+}), ammonia (NH_3), and chloride ion (Cl^-). Then you will determine the empirical formula of the compound. Until you determine the exact formula, we will represent it as $Ni(NH_3)_nCl_2$, with n representing a small whole number.

Synthesizing $Ni(NH_3)_nCl_2$

You will synthesize $Ni(NH_3)_nCl_2$ by reacting nickel chloride hexahydrate ($NiCl_2 \cdot 6\,H_2O$) and NH_3. This reaction is shown in Equation 1.

$$Ni^{2+}(aq, green) + 2\,Cl^-(aq) + n\,NH_3(aq) \rightarrow$$
$$Ni(NH_3)_nCl_2(s, bluish\ purple) \quad (Eq.\ 1)$$

A complication arises because NH_3 in aqueous solution is involved in the equilibrium reaction shown in Equation 2.

$$NH_3(aq) + H_2O(l) \rightleftarrows NH_4^+(aq) + OH^-(aq) \quad (Eq.\ 2)$$

Although the equilibrium constant for the reaction in Equation 2 is small, 1.75×10^{-5}, some of the Ni^{2+} ion can react with hydroxide ion (OH^-) to form nickel(II) hydroxide, $Ni(OH)_2$, as shown in Equation 3.

$$Ni^{2+}(aq, green) + 2\,OH^-(aq) \rightarrow$$
$$Ni(OH)_2(s, green) \quad (Eq.\ 3)$$

Copyright © 1994 by Chemical Education Resources, Inc., P.O. Box 357, 220 S. Railroad, Palmyra, Pennsylvania 17078
No part of this laboratory program may be reproduced or transmitted in any form or by any means, electronic or mechanical, including photocopying, recording, or any information storage and retrieval system, without permission in writing from the publisher. Printed in the United States of America

To the extent that the reaction in Equation 3 occurs, the product formed in Equation 1 will be impure, and the synthesis reaction yield will therefore decrease.

Water is a convenient solvent for the synthesis reaction because the reactants are water soluble. However, because $Ni(NH_3)_nCl_2$ is also somewhat soluble in water, you must keep the volume of water you use in the synthesis to an absolute minimum. Nickel(II) chloride hexahydrate is more soluble in hot water than in cold water, so heating the reactants will enable you to dissolve more of this compound in a smaller volume of water. Unfortunately, the water solubility of NH_3 is greatly decreased with increasing temperature. In this case, at temperatures approaching 100 °C, NH_3 volatilizes before it can react with the $NiCl_2$. By maintaining the reaction temperature at 60 °C, you will maximize the $Ni(NH_3)_nCl_2$ yield.

Once the reaction is complete, you will cool the reaction mixture to 0 °C in an ice–water bath. Because $Ni(NH_3)_nCl_2$ is less soluble in cold water than in hot water, this step decreases the solubility of the product. You will add cold ethanol to the cold reaction mixture to further reduce the product solubility, because $Ni(NH_3)_nCl_2$ is insoluble in ethanol.

You will filter the $Ni(NH_3)_nCl_2$ crystals from the cold ethanolic solution and wash them with cold concentrated NH_3. This treatment will convert any $Ni(OH)_2$ on the crystals to $Ni(NH_3)_nCl_2$, as shown in Equation 4.

$$Ni(OH)_2(s, green) + n\,NH_3(aq) + 2\,Cl^-(aq) \rightarrow$$
$$Ni(NH_3)_nCl_2(s, bluish\ purple) + 2\,OH^-(aq) \quad (Eq.\ 4)$$

Finally, you will dry and weigh the crystals to determine the actual yield of your synthesis.

Analyzing $Ni(NH_3)_nCl_2$

Determining the Mass Percent NH_3

Ammonia is a base. Theoretically, you could determine the NH_3 content of $Ni(NH_3)_nCl_2$ by titrating a known mass of the compound with an acid solution of known concentration, called a **standardized solution**. However, in this case, the procedure gives inacurate results for two reasons. First, aqueous $Ni(NH_3)_nCl_2$ solutions *slowly* evolve NH_3, so some NH_3 may evaporate before the sample is completely titrated. Second, the procedure takes longer than normal because the bonds between NH_3 and Ni^{2+} ion must be broken before NH_3 can react with the titrant. This bond breaking is slow and the titration reaction itself is fast, so some of the indicator may react with the acid before all the NH_3 is released, causing a premature end point.

To minimize NH_3 evolution and ensure complete disruption of all NH_3–Ni^{2+} bonds, you will use an indirect titration method. You will add a known volume of standard hydrochloric acid solution (HCl) that contains more moles of HCl than there are moles of NH_3 in the $Ni(NH_3)_nCl_2$ sample you are analyzing. The HCl will react with the $Ni(NH_3)_nCl_2$ sample as shown in Equation 5.

$$Ni(NH_3)_nCl_2(s) + n\,H_3O^+(aq) \rightarrow n\,NH_4^+(aq) +$$
$$Ni^{2+}(aq, green) + (n\,H_2O + 2)Cl^-(aq) \quad (Eq.\ 5)$$

The excess acid and a favorable equilibrium constant drive the reaction in Equation 5 to completion. Then you will titrate the excess HCl with standard sodium hydroxide solution (NaOH) in a process called **back titration**. The back titration reaction is shown in Equation 6.

$$H_3O^+(aq) + OH^-(aq) \rightarrow 2\,H_2O(l) \quad (Eq.\ 6)$$

You will detect the **equivalence point** of the back titration, the point at which the number of moles of NaOH added is stoichiometrically equivalent to the number of moles of HCl present, by observing an indicator color change. The point at which the indicator changes color is the **end point** of the titration. In order to select the proper indicator for a particular titration, one that will produce an end point that is close to the equivalence point, you must consider the chemical behavior of the species present in solution at the equivalence point. In this case, if the titration mixture contained only the products shown in Equation 6, the equivalence point would be at pH 7; hence, phenolphthalein, which changes color at pH 8, would be a satisfactory indicator. However, the reaction mixture also contains ammonium ion (NH_4^+), which hydrolyzes as shown in Equation 7, causing the titration mixture to be acidic at the equivalence point.

$$NH_4^+(aq) + H_2O(l) \rightleftharpoons NH_3(aq) + H_3O^+(aq) \quad (Eq.\ 7)$$

Therefore, phenolphthalein is not a good indicator for this titration, because the end point would occur after more than an equivalent amount of NaOH had been added. Instead, you will use a mixed indicator solution, composed of bromcresol green and methyl red. Bromcresol green changes from yellow to blue over a pH range of 3.8 to 5.4, and methyl red changes from red to yellow over a pH range of 4.2 to 6.2. The mixed indicator changes from rose to green at pH 5.1 which is the pH at the equivalence point of the titration. Due to the presence of green Ni^{2+} ion in the mixture, the color change you will see is orange-yellow to green-blue.

From the volume and concentration of HCl solution added, you will calculate the number of moles of HCl added, using Equation 8.

$$\text{number of moles of HCl added, mol} = \left(\begin{array}{c}\text{volume of HCl}\\\text{solution added, mL}\end{array}\right) \times$$

$$\left(\frac{1\ L}{1000\ mL}\right)\left(\begin{array}{c}\text{concentration of}\\\text{HCl solution, mol L}^{-1}\end{array}\right) \quad \text{(Eq. 8)}$$

You will calculate the number of moles of HCl remaining in solution after Ni^{2+}–NH_3 separation by substituting into Equation 9, which is a variation of Equation 8, the volume and concentration of NaOH solution required for the back titration.

$$\begin{array}{c}\text{number of moles of}\\\text{HCl remaining after}\\\text{neutralization of NH}_3\text{, mol}\end{array} = \left(\begin{array}{c}\text{volume of}\\\text{NaOH solution}\\\text{added, mL}\end{array}\right) \times$$

$$\left(\frac{1\ L}{1000\ mL}\right)\left(\begin{array}{c}\text{concentration of}\\\text{NaOH solution,}\\\text{mol L}^{-1}\end{array}\right)\left(\frac{1\ mol\ HCl}{1\ mol\ NaOH}\right) \quad \text{(Eq. 9)}$$

The number of moles of NH_3 in the $Ni(NH_3)_nCl_2$ sample is the difference between the number of moles of HCl added and the number of moles of NaOH added, as shown in Equation 10.

$$\begin{array}{c}\text{number of moles of NH}_3\\\text{in Ni(NH}_3\text{)}_n\text{Cl}_2\text{ sample, mol}\end{array} =$$

$$\left(\begin{array}{c}\text{number of}\\\text{moles of HCl}\\\text{added, mol}\end{array}\right) - \left(\begin{array}{c}\text{number of}\\\text{moles of NaOH}\\\text{added, mol}\end{array}\right) \quad \text{(Eq. 10)}$$

Finally, you will use Equations 11 and 12 to calculate the mass of NH_3 in the sample and the mass percent NH_3 in $Ni(NH_3)_nCl_2$.

$$\text{mass of NH}_3\text{ in Ni(NH}_3\text{)}_n\text{Cl}_2\text{ sample, g} =$$

$$\left(\begin{array}{c}\text{number of moles of}\\\text{NH}_3\text{ in Ni(NH}_3\text{)}_n\text{Cl}_2\\\text{sample, mol}\end{array}\right)\left(\frac{17.03\ g\ NH_3}{1\ mol\ NH_3}\right) \quad \text{(Eq. 11)}$$

$$\text{mass percent NH}_3\text{ in Ni(NH}_3\text{)}_n\text{Cl}_2\text{, \% } =$$

$$\left(\frac{\text{mass of NH}_3\text{ in Ni(NH}_3\text{)}_n\text{Cl}_2\text{ sample, g}}{\text{mass of Ni(NH}_3\text{)}_n\text{Cl}_2\text{ sample, g}}\right)(100\%)$$

$$\text{(Eq. 12)}$$

Determining the Mass Percent Ni^{2+} Ion

The $[Ni(NH_3)_n]^{2+}$ ion absorbs light in the visible region of the spectrum. You will take advantage of this light-absorbing property in order to determine the mass percent Ni^{2+} ion in $Ni(NH_3)_nCl_2$. Solutions containing $[Ni(NH_3)_n]^{2+}$ ion are colored. The color is produced by those visible wavelengths that are not absorbed. You can determine which wavelengths are absorbed by using a spectrophotometer to measure the absorbance of the solution throughout the visible region of the spectrum. The wavelength at which the species absorbs the most light is called the **analytical wavelength** (λ_{max}) for that species.

Absorbance is directly proportional to the concentration of the absorbing species in solution. This relationship, known as **Beer's law**, is represented by Equation 13. A is absorbance, ε is molar absorptivity, b is the length of the light path through the solution, and c is the molar concentration of the absorbing species.

$$A = \varepsilon bc \quad \text{(Eq. 13)}$$

Molar absorptivity is a proportionality constant relating absorbance and molar concentration of the absorbing species at the wavelength being measured. The value of ε varies with wavelength, reaching a maximum at the analytical wavelength.

Practically speaking, when using a spectrophotometer, we can often read percent transmittance (%T) more accurately than we can absorbance. This is because the %T scale is linear, while the absorbance scale is logarithmic. Therefore, unless you use a spectrophotometer with a digital absorbance readout, you should record %T readings and use Equation 14 to convert these readings to their equivalent absorbances.

$$A = 2.000 - \log \%T \quad \text{(Eq. 14)}$$

Note that a solution that absorbs none of the light transmits 100% of the light. Therefore, a small absorbance translates to a large %T, and vice versa.

You will prepare a standard $[Ni(NH_3)_n]^{2+}$ ion solution by dissolving a known mass of nickel(II) sulfate hexahydrate ($NiSO_4 \cdot 6\ H_2O$) in water and adding excess concentrated NH_3. Then you will dilute the mixture with water to a known volume. The Ni^{2+} ion in the sample converts to $[Ni(NH_3)_n]^{2+}$ ion, as shown in Equation 15.

$$Ni^{2+}(aq,\ green) \xrightarrow{\text{excess NH}_3} [Ni(NH_3)_n]^{2+}(aq,\ bluish\ purple) \quad \text{(Eq. 15)}$$

You will use your standard $[Ni(NH_3)_n]^{2+}$ ion solution to establish the analytical wavelength (λ_{max}) for $[Ni(NH_3)_n]^{2+}$ ion. Then, you will compare the absorbance of your standard $[Ni(NH_3)_n]^{2+}$ ion solution with that of a solution you will prepare from a known mass of the $Ni(NH_3)_nCl_2$ you synthesized. Because the absorbing species, the $[Ni(NH_3)_n]^{2+}$ ion, is identical in both solutions, ε is the same for both solutions. Also, if you use identical cuvettes, the light path length through each sample will be the same. Therefore, using the subscript s to indicate the standard $[Ni(NH_3)_n]^{2+}$ ion solution and the subscript x to indicate the synthesized $Ni(NH_3)_nCl_2$ solution, you can write Equations 16 and 17.

$$\frac{A_s}{c_s,\ mol\ L^{-1}} = \varepsilon b = \frac{A_x}{c_x,\ mol\ L^{-1}} \qquad (Eq.\ 16)$$

$$c_x,\ mol\ L^{-1} = \frac{(A_x)(c_x,\ mol\ L^{-1})}{A_s} \qquad (Eq.\ 17)$$

Because one mole of $Ni(NH_3)_nCl_2$ contains one mole of Ni^{2+} ion, the concentration of Ni^{2+} ion is equal to the concentration of $[Ni(NH_3)_n]^{2+}$ ion and of $Ni(NH_3)_nCl_2$, or c_x. Hence, the relationship of c_x to the number of moles of Ni^{2+} ion dissolved is shown in Equation 18.

$$\begin{pmatrix} \text{number of moles} \\ \text{of } Ni^{2+} \text{ ion} \\ \text{dissolved, mol} \end{pmatrix} = \begin{pmatrix} c_x, \text{ concentration} \\ \text{of } Ni(NH_3)_nCl_2 \\ \text{solution, mol } L^{-1} \end{pmatrix} \times$$

$$\begin{pmatrix} \text{total volume} \\ \text{of } Ni(NH_3)_nCl_2 \\ \text{solution, mL} \end{pmatrix} \begin{pmatrix} \dfrac{1\ L}{1000\ mL} \end{pmatrix} \quad (Eq.\ 18)$$

From the number of moles of Ni^{2+} ion dissolved and the mass of compound dissolved, you can calculate the mass percent Ni^{2+} ion in $Ni(NH_3)_nCl_2$, using Equations 19 and 20.

mass of Ni^{2+} ion in $Ni(NH_3)_nCl_2$ analyzed, g =

$$\begin{pmatrix} \text{number of moles of} \\ Ni^{2+} \text{ ion in } Ni(NH_3)_nCl_2 \\ \text{analyzed, mol} \end{pmatrix} \begin{pmatrix} \dfrac{58.69\ g\ Ni^{2+}\ ion}{1\ mol\ Ni^{2+}\ ion} \end{pmatrix} \quad (Eq.\ 19)$$

mass percent Ni^{2+} ion in $Ni(NH_3)_nCl_2$, % =

$$\left(\frac{\text{mass of } Ni^{2+} \text{ ion in } Ni(NH_3)_nCl_2 \text{ analyzed, g}}{\text{mass of } Ni(NH_3)_nCl_2 \text{ analyzed, g}} \right) (100\%)$$
$$(Eq.\ 20)$$

Note that before you can use your calculated absorbances in Equation 17, you must establish whether or not the cuvettes you use are truly identical, or matched. **Matched** cuvettes respond identically at λ_{max}, allowing you to attribute any absorbance differences between solutions in the cuvettes solely to differences in solution concentrations. To determine if your two cuvettes are matched, you will fill them with distilled or deionized water. With the water-filled reference cuvette in the cuvette compartment, you will adjust the spectrophotometer to read 100%T. Then you will measure the %T of the water-filled sample cuvette. If the reading for the water-filled sample cuvette is 100%T,

the two cuvettes are matched. If the %T is less than 100%T, you must subtract the calculated absorbance for the sample cuvette from the absorbance of both the standard and unknown $[Ni(NH_3)_n]^{2+}$ ion solutions at λ_{max}.

For example, suppose a sample cuvette filled with distilled water has a relative absorbance of 0.005, and the same cuvette filled with standard $[Ni(NH_3)_n]^{2+}$ ion solution has an absorbance of 0.315, both at λ_{max}. The corrected absorbance of the standard $[Ni(NH_3)_n]^{2+}$ ion solution is $0.315 - 0.005 = 0.310$.

If the water-filled sample cuvette has a negative relative absorbance, it means that the sample cuvette is more transparent than the reference cuvette. To compensate for this, you should replace the water-filled sample cuvette in the spectrophotometer compartment and adjust the spectrophotometer to obtain a meter reading of 100%T. Then you should insert the water-filled reference cuvette and determine its relative absorbance. You must add the absorbance of the water-filled reference cuvette to the absorbance of the standard and unknown $[Ni(NH_3)_n]^{2+}$ ion solutions at λ_{max}.

Determining the Empirical Formula of the Coordination Compound

The coordination compound you will synthesize is composed of Ni^{2+} ion, NH_3, and Cl^- ion. Therefore, you can calculate the mass percent Cl^- ion in the compound by subtracting the mass percents of Ni^{2+} ion and NH_3 from 100%, as shown in Equation 21.

mass percent Cl^- ion in $Ni(NH_3)_nCl_2$, % = (100%) −

$$\begin{pmatrix} \text{mass percent} \\ Ni^{2+} \text{ ion in} \\ Ni(NH_3)_nCl_2, \% \end{pmatrix} + \begin{pmatrix} \text{mass percent} \\ NH_3 \text{ in} \\ Ni(NH_3)_nCl_2, \% \end{pmatrix} \quad (Eq.\ 21)$$

Assume you have 100 g of the compound. Then, the mass percent of each component would be equivalent to the number of grams of each component in the 100 g sample. Thus, based on the mass percents of the components, you can calculate the empirical formula of the compound, using Equations 22–24.

number of moles of Ni^{2+} ion, mol =

$$\begin{pmatrix} \text{mass of} \\ Ni^{2+} \text{ ion, g} \end{pmatrix} \begin{pmatrix} \dfrac{1\ mol\ Ni^{2+}\ ion}{58.69\ g\ Ni^{2+}\ ion} \end{pmatrix} \quad (Eq.\ 22)$$

number of moles of NH_3, mol =

$$\begin{pmatrix} \text{mass of} \\ NH_3,\ g \end{pmatrix} \begin{pmatrix} \dfrac{1\ mol\ NH_3}{17.04\ g\ NH_3} \end{pmatrix} \quad (Eq.\ 23)$$

number of moles of Cl^- ion, mol =

$$\left(\begin{array}{c} \text{mass of} \\ Cl^- \text{ ion, g} \end{array}\right)\left(\frac{1 \text{ mol } Cl^- \text{ ion}}{35.45 \text{ g } Cl^- \text{ ion}}\right) \quad \text{(Eq. 24)}$$

The **empirical formula** of a compound is the simplest whole-number ratio among the numbers of moles of the components, that is $Ni_x(NH_3)_nCl_y$ where x, n, and y are whole numbers. Often, we can find the values of x, n, and y by dividing each number of moles obtained using Equations 22–24 by the smallest number of moles. The results of these divisions, rounded to the nearest whole numbers, will give the empirical formula of the product you synthesized.

From the number of moles of $NiCl_2 \cdot 6 H_2O$ used in your synthesis, you will calculate the theoretical yield of product in moles, using Equation 25.

$$\begin{array}{c} \text{number of moles of} \\ NiCl_2 \cdot 6 H_2O \text{ used, mol} \end{array} = \begin{array}{c} \text{number of moles} \\ \text{of product, mol} \end{array} \quad \text{(Eq. 25)}$$

From the mass of product obtained and the molar mass of the product, you will calculate the actual yield of product in moles, using Equation 26.

actual yield of product, mol =

$$\frac{\text{mass of product obtained, g}}{\text{molar mass of product, g mol}^{-1}} \quad \text{(Eq. 26)}$$

Finally, you will calculate the percent yield of your synthesis, using Equation 27.

percent yield of product, % =

$$\left(\frac{\text{actual yield of product, mol}}{\text{theoretical yield of product, mol}}\right)(100\%) \quad \text{(Eq. 27)}$$

Procedure

> **Chemical Alert**
> acetone—flammable and irritant
> concentrated ammonia—toxic, corrosive, and lachrymator
> 95% ethanol—toxic and flammable
> 0.25M hydrochloric acid solution—toxic and corrosive
> nickel(II) chloride hexahydrate—toxic and suspected carcinogen
> nickel(II) sulfate hexahydrate—toxic, irritant, oxidant, and suspected carcinogen
> 0.10M sodium hydroxide solution—toxic and corrosive

Caution: Wear departmentally approved eye protection while doing this experiment.

You are strongly urged to wear latex or vinyl gloves while performing all parts of this experiment.

I. Synthesizing $Ni(NH_3)_nCl_2$

> **Caution:** Nickel(II) chloride hexahydrate is toxic and a suspected carcinogen. Prevent eye, skin, and clothing contact. Avoid inhaling dust and ingesting the compound.
> If you should spill any of the solid, immediately notify your laboratory instructor.
> After handling the solid, immediately wash your hands and face with soap or detergent before proceeding with the experiment.

> **Note:** Your laboratory instructor will describe and demonstrate a satisfactory method for weighing and transferring a solid sample and will inform you as to the number of significant digits to the right of the decimal point to record masses.

Prepare a warm-water bath. Half fill a 600-mL or larger beaker with tap water. Attach a large ring support to a ring stand. Place the beaker through the ring. Adjust the ring so that the beaker is stabilized while sitting on a hot plate, as shown in Figure 1.

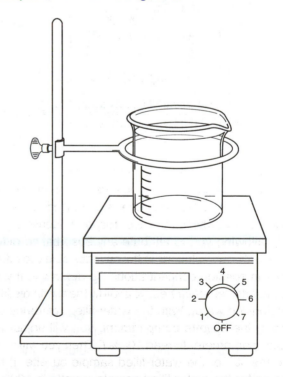

Figure 1 *Stabilizing a beaker on a hot plate*

Monitor the water temperature with a thermometer that you have carefully inserted through a one-hole stopper. Clamp the stopper with the inserted thermometer to the ring stand, making sure that the thermometer extends into the water but does not touch the side or bottom of the beaker. Heat the beaker and its contents until the water temperature reaches 50 °C. Adjust the hot plate setting so that the water temperature remains between 50 and 60 °C.

Weigh a piece of weighing paper and record the mass on Data Sheet 1. Weigh 8.0 g of $NiCl_2 \cdot 6\,H_2O$ on the weighed piece of weighing paper. Record the mass of the solid and paper on Data Sheet 1. Transfer the solid to a clean 125-mL Erlenmeyer flask.

Add 10 mL of distilled or deionized water to the $NiCl_2 \cdot 6\,H_2O$ in the flask. Place the flask in the 60 °C water bath and clamp the flask in position, as shown in Figure 2. Stir the mixture in the flask with a clean 125-mm glass stirring rod until the $NiCl_2 \cdot 6\,H_2O$ has dissolved. Loosen the clamp on the ring stand, and while holding the end of the clamp, remove the flask from the bath. Attach the clamp to another ring stand, and let the flask and contents cool in air for 1–2 min.

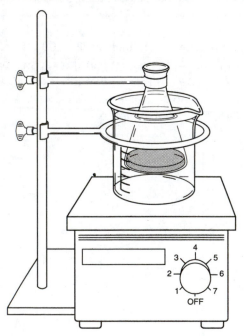

Figure 2 *Suspending the flask in the warm-water bath*

Caution: Concentrated NH_3 solution is toxic, corrosive, and a lachrymator. Prevent eye, skin, and clothing contact. Permanent fogging of soft contact lenses may result from NH_3 vapors. Avoid inhaling vapors and ingesting the solution. Unless your laboratory instructor tells you otherwise, restrict all work with this reagent to a fume hood.

If you spill any NH_3 solution on yourself, immediately rinse with a large amount of running water. If you spill any NH_3 solution on the laboratory bench, add water and wipe it up immediately with a damp paper towel. Immediately notify your laboratory instructor of any NH_3 solution spills. Dispose of NH_3-soaked paper towels as directed by your laboratory instructor.

You should cover the top of the reaction flask and any other containers of concentrated NH_3 with a wet paper towel, in order to limit your and others' exposure to NH_3 vapors. The moisture in the towel will absorb NH_3 fumes, forming aqueous NH_3.

Slowly, with stirring, add 25 mL of concentrated NH_3 solution to the $NiCl_2 \cdot 6\,H_2O$ solution in the flask. Cover the top of the flask with a wet paper towel. After adding the concentrated NH_3 solution, suspend the flask in the warm-water bath by clamping it to the ring stand. Make sure the water temperature is between 50 and 60 °C. Leave the flask in the bath for 15 min. During this time, periodically swirl the mixture.

Note: For the filtering apparatus described below, your laboratory instructor will inform you whether you are to place a trap between your filter flask and the water aspirator and whether you are to place a Büchner funnel in a rubber FilterVac, a bored rubber stopper, or a rubber filtering adapter.

Do not turn off the aspirator before removing the tubing from the side arm of the filter flask.

While the above reaction proceeds, assemble your filtering apparatus, as shown in Figure 3. Clamp a 250-mL taped filtering flask to your other ring stand, and place a Büchner funnel in the flask.

Place a flat circle of filter paper in the funnel. Wet the paper with 1–2 mL of distilled water. Attach a piece of pressure tubing to the flask and to the water aspirator. Turn on the aspirator to draw the water in the funnel into the flask and to snugly seal the filter paper to the funnel. Disconnect the tubing from the filter flask and turn off the aspirator. Remove the funnel, and pour the water from the filtering flask into the drain. Reassemble the vacuum filtration assembly.

Caution: 95% Ethyl alcohol is flammable and should not be exposed to an open flame in the laboratory.

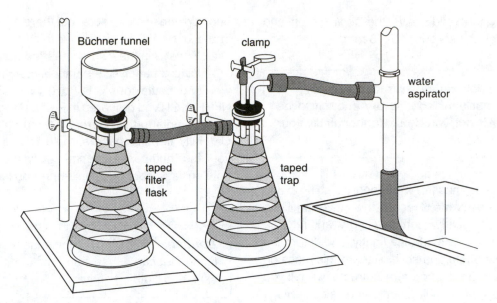

Figure 3   *A vacuum filtration apparatus*

Prepare an ice–water bath in another 600-mL beaker by adding 150 mL of water and several pieces of ice to the beaker. Transfer 20 mL of concentrated NH_3 solution into a labeled, 18 x 150-mm test tube. Stopper the test tube with a No. 2 solid rubber stopper. Place the test tube in the ice–water bath. Obtain 60 mL of 95% ethanol in a labeled 100-mL beaker. Place the beaker and its contents in the ice–water bath.

Assemble a second ice–water bath in a 600-mL beaker. After the reaction in the warm-water bath has proceeded for 15 min, unclamp the flask from the ring stand. Carefully clamp the reaction flask on another ring stand so that the flask is suspended in the second ice–water bath. Remove the damp paper towel covering the mouth of the flask. While holding the flask, loosen the clamp and swirl the reaction mixture for 5 min while it is cooling. Add 10 mL of ice-cold 95% ethanol to the flask and stir. Remove the reaction flask from the ice–water bath. Wipe any water off the bottom of the flask using a paper towel.

After reassembling the vacuum filtration apparatus, turn on the aspirator. Slowly pour the liquid–solid mixture from the flask into the Büchner funnel, as follows. Decant as much supernatant liquid as possible into the funnel. Use a stirring rod to guide the liquid from the flask onto the filter paper, in order to prevent splashing and product loss. Then use a rubber policeman attached to another glass stirring rod to help transfer the solid from the flask into the funnel. When all the liquid has been drawn through the funnel, disconnect the tubing from the filter flask arm, and turn off the aspirator.

Rinse any remaining solid down the inside wall of the reaction flask using 5 mL of ice-cold, concentrated

NH_3 solution from your test tube. Swirl the solid and rinse solution mixture, and quickly pour the mixture into the Büchner funnel. Reattach the tubing to the filter flask, and turn on the aspirator to draw the rinse solution through the funnel. Then disconnect the tubing and turn off the aspirator.

In the same manner, rinse any remaining solid from the flask using two additional 5-mL portions of the cold, concentrated NH_3 solution. After pouring any remaining concentrated NH_3 rinse solution over the solid in the funnel, reattach the tubing, and turn on the aspirator.

Dry the solid by drawing air through the solid for 3–5 min. Disconnect the aspirator, and turn off the aspirator. Break up the solid with a spatula, being careful not to tear the filter paper. Pour 15 mL of cold 95% ethanol over the solid, reattach the tubing, and turn on the aspirator. Repeat the ethanol washing two more times, using 15 mL of 95% ethanol each time. Make sure to disconnect the tubing, turn off the aspirator, and carefully break up the solid before each ethanol wash. When you have finished the third wash and have dried the solid, disconnect the tubing and turn off the aspirator.

> **Caution:** Acetone is flammable and should not be exposed to an open flame in the laboratory.

Pour 15 mL of acetone over the solid in the funnel. Break up the solid with a spatula to expose all its surfaces to the acetone. Reattach the tubing and turn on the aspirator to draw the acetone through the funnel. To completely dry the solid, leave the aspirator on and draw air through the solid for 10–15 min.

> **Note:** Your laboratory instructor will tell you whether or not your solid is dry enough to be weighed. If it is not, you should further air dry your solid.

Determine the mass of a capped weighing bottle labeled with your name. Record this mass on Data Sheet 1. After the solid is completely dry, add it to the weighing bottle. Close the bottle tightly, and determine the mass of the bottle and solid. Record this mass on your Data Sheet. Your laboratory instructor will tell you where to store your weighing bottle and its contents.

Transfer the solution in the filter flask into the container provided by your laboratory instructor and labeled "Discarded $NiCl_2$/Ethanol/Acetone Solution Mixture." Rinse the filter flask and Büchner funnel once with 20 mL of tap water. Transfer the rinse to the same discard container.

II. Determining the Mass Percent NH_3 in $Ni(NH_3)_nCl_2$ by Titration

> **Note:** Your laboratory instructor will demonstrate and describe proper techniques for the quantitative use of a buret and pipet.
>
> Record all buret readings to the nearest 0.02 mL.

> **Caution:** $0.1M$ NaOH solution is corrosive and should be handled with care. Prevent eye, skin, and clothing contact. Avoid ingesting the solution.

Obtain 200 mL of standardized NaOH solution in a clean, dry, stoppered 250-mL Erlenmeyer flask. Record the exact NaOH solution concentration on Data Sheet 2. Number four clean 125-mL Erlenmeyer flasks #1–4.

Clean your 50-mL buret if necessary. Rinse the clean buret with distilled water and then with 5 mL of the NaOH solution. Transfer the rinses into the container provided by your laboratory instructor and labeled "Discarded $Ni(NH_3)_nCl_2$/HCl/NaOH Solution Mixture."

Fill the buret to or just below the 0-mL mark with NaOH solution. Drain 1–2 mL of the NaOH solution from the buret into the discard beaker, both to fill the

buret tip and to make sure that there are no air bubbles in the buret tip. Record your initial buret reading on Data Sheet 2 in the column headed "determination 1."

Using an analytical balance, weigh out from your weighing bottle four 0.12 to 0.14 g samples of the $Ni(NH_3)_nCl_2$ you prepared in Part I. Record the mass of each sample to the nearest 0.1 mg on Data Sheet 2. Carefully transfer each sample to a different one of your four numbered Erlenmeyer flasks. Make sure to note on flask labels which sample you transfer to which flask.

> **Caution:** $0.25M$ HCl solution is a corrosive solution that can cause skin irritation. Prevent eye, skin, and clothing contact.

Obtain 125 mL HCl solution. Record the exact concentration of this solution on Data Sheet 2. Using a *clean* 25-mL volumetric pipet, transfer 25.00 mL of approximately $0.25M$ HCl solution to each numbered flask. Using your 100-mL graduated cylinder, add 10 mL of distilled water to each flask. Then add 2 or 3 drops of the bromcresol green–methyl red mixed indicator solution to each flask.

Titrate the sample in flask #1 to determine the approximate volume of NaOH solution required and to observe the end point. While swirling the flask with one hand, control the buret stopcock with your other hand. When the orange-yellow color of the titration mixture first begins to change, continue to add NaOH solution dropwise, just until the mixture turns green-blue. Record the final buret volume in the column for determination 1 on Data Sheet 2.

> **Note:** Remember that a larger mass of $Ni(NH_3)_nCl_2$ will liberate more NH_3 and consume more HCl than will a smaller mass of $Ni(NH_3)_nCl_2$. Thus, the amount of excess HCl will be greater for a smaller $Ni(NH_3)_nCl_2$ sample than for a larger one.

Using the mass of $Ni(NH_3)_nCl_2$ in flask #1 and volume of NaOH solution used for flask #1 as a guide, titrate the solutions in flasks #2, #3, and #4. Record initial and final buret readings in the appropriate columns on Data Sheet 2.

Transfer any NaOH solution remaining in your buret as well as the titration solutions to the "Discarded $Ni(NH_3)_nCl_2$/HCl/NaOH Solution Mixture" container.

Rinse the buret, pipet, graduated cylinder, and Erlenmeyer flasks twice each, first with tap water, and then with distilled water. Transfer all rinses to the discard container.

III. Determining the Mass Percent Ni²⁺ Ion in $Ni(NH_3)_nCl_2$ by Spectrophotometry

> **Caution:** Nickel(II) sulfate hexahydrate is toxic and a suspected carcinogen. Prevent eye, skin, and clothing contact. Avoid inhaling dust and ingesting the compound.
>
> If you should spill any of the solid, immediately notify your laboratory instructor.

On the frosted or white circle on a clean, dry 100-mL beaker, write "std" to indicate the $NiSO_4 \cdot 6 H_2O$ standard sample solution. Write "unk" on a second clean, dry 100-mL beaker, which you will use for your $Ni(NH_3)_nCl_2$ sample solution with unknown %Ni²⁺ ion. Clean two 50-mL volumetric flasks and stoppers. Label one flask "std" and the other flask "unk."

Using an analytical balance, weigh on a weighing paper a 0.20 to 0.40 g $NiSO_4 \cdot 6 H_2O$ sample. Transfer the sample to the "std" beaker. Record the mass of $NiSO_4 \cdot 6 H_2O$ to the nearest 0.1 mg on Data Sheet 3.

Using an analytical balance, weigh on a weighing paper a 0.30 to 0.40 g sample of your $Ni(NH_3)_nCl_2$. Transfer the sample to the "unk" beaker. Record the mass of $Ni(NH_3)_nCl_2$ to the nearest 0.1 mg on Data Sheet 3.

Using a graduated cylinder, add 20 mL of distilled water to both samples. Note and record on Data Sheet 3 the color and appearance of each solution. Using separate glass stirring rods, stir each mixture until most of the solid has dissolved. Leave the rods in the beakers to avoid losing any solution adhering to the rods.

> **Note:** Your laboratory instructor will demonstrate the use of volumetric flasks. Prepare the following solutions and transfer them to volumetric flasks in a fume hood.

Measure out 10 mL of concentrated NH_3 solution in a 10-mL graduated cylinder, which need not be dry. Add the NH_3 solution to the solution in the "std" beaker. Then measure another 10 mL of concentrated NH_3 solution, and add it to the solution in the "unk" beaker. Stir each mixture until no solid remains. Note and record on Data Sheet 3 the color and appearance of each solution.

Using a short-stem funnel, transfer the "std" solution into the "std" 50-mL volumetric flask. Rinse the beaker and rod with a **minimum** amount of distilled water from a wash bottle, and pour the rinses into the "std" volumetric flask. Rinse the beaker two more times, using distilled water, but do not allow the volume of solution in the volumetric flask to exceed 50 mL.

Add distilled water to the solution in the flask until the solution level coincides with the junction of the neck and body of the flask. Stopper the flask. Firmly holding the stopper in place, invert the flask 10 times to thoroughly mix the solution. Then fill the flask exactly to the etched mark by adding distilled water from a disposable pipet or medicine dropper. Stopper the flask. Thoroughly mix the solution by inverting the flask at least 25 times, while holding the stopper firmly in place.

Follow the same procedure to transfer your "unk" solution to the "unk" 50-mL volumetric flask. Dilute the "unk" solution to the etched mark, stopper the flask, and thoroughly mix.

Obtain two spectrophotometer cuvettes, and place them in a dry beaker or test tube rack. Clean the cuvettes and rinse them with distilled water. Using a pencil, write "R" on the frosted glass circle near the top of one cuvette for "reference." Mark the other cuvette "S" for "sample." Fill the reference cuvette about three-quarters full with distilled water. Rinse the sample cuvette three times, using a 1–2 mL of your "std" solution each time. Dispose of the rinses into the container provided by your laboratory instructor and labeled "Discarded $NiCl_2/NiSO_4/NH_3$ Solutions." Fill the sample cuvette about three-quarters full with "std" solution. Place the filled cuvettes in a beaker or test tube rack.

> **Note:** If you are using a spectrophotometer other than a Spectronic 20, your laboratory instructor will give you directions for modifying the following procedure.

Set the wavelength on the spectrophotometer to 620 nm. With the cuvette compartment empty and its cover closed, adjust the left-hand knob until the meter reads 0%T. To avoid a parallax error, look directly down at the needle so that you cannot see its reflection in the mirror behind the needle.

Using a tissue, wipe off the water-filled reference cuvette, and insert it into the cuvette compartment. Close the cover. Turn the right-hand knob until the meter reads 100%T. Transfer the cuvette to the beaker or test tube rack. Wipe the filled sample cuvette, and insert it into the compartment. Read the %T. Record this

%T on Data Sheet 3 as the %T of the $[Ni(NH_3)_n]^{2+}$ ion standard solution at 620 nm. Transfer the sample cuvette to the beaker or test tube rack and close the cover.

Adjust the wavelength to 600 nm. Following the above procedure, first with the reference cuvette and then the sample cuvette, obtain and record the %T for the standard solution at 600 nm. Repeat the procedure at wavelength intervals of 20 nm down to 540 nm. From among your five %T readings, determine the approximate λ_{max} for the $[Ni(NH_3)_n]^{2+}$ ion and record it on Data Sheet 3. To more precisely establish λ_{max} for the $[Ni(NH_3)_n]^{2+}$ ion, measure the %T of the standard solution at wavelengths 10 nm less and 10 nm greater than the wavelength you estimate as λ_{max}. Record these additional %T measurements on Data Sheet 3. Select the λ_{max} for the $[Ni(NH_3)_n]^{2+}$ ion and record it on Data Sheet 3.

Empty the sample cuvette into the "Discarded $NiCl_2/NiSO_4/NH_3$ Solutions" container. Rinse the cuvette with distilled water, and then rinse it three times with your "unk" solution, using 1 mL of solution each time. Transfer all rinses to the discard container.

Fill the cuvette about three-quarters full with "unk" solution. Set the spectrophotometer at the analytical wavelength. Check the 0%T setting with the cuvette compartment empty and its cover closed. Check the 100%T setting using the water-filled reference cuvette. Determine the %T of the unknown solution at λ_{max}. Record this %T on Data Sheet 3. Transfer the solutions in your cuvettes to the discard container. Rinse and wash the cuvettes and add any rinses and washings to the discard container.

Finally, check to see whether or not the reference and sample cuvettes are matched at λ_{max}. To do so, rinse and fill the sample cuvette three-quarters full with distilled water. With the wavelength set at λ_{max} and the cuvette compartment empty and closed, adjust the right-hand knob until the meter reads 0%T. Then place the reference cuvette filled with distilled water in the cuvette compartment. Adjust the right-hand knob so that the meter reads 100%T. Finally, measure the %T of the water-filled sample cuvette. Record this %T on Data Sheet 3. If the meter needle goes off the scale to the right of 100%T, the sample cuvette is **more** transparent than the reference cuvette at λ_{max}. To correct for this difference, replace the water-filled sample cuvette in the compartment and adjust the meter to read 100%T. Then reinsert the water-filled reference cuvette and measure its %T. Record this %T on Data Sheet 3.

Transfer the solutions in your volumetric flasks into the appropriate discard container. Rinse the volumetric flasks twice with 10 mL of tap water each time and

twice with 10 mL of distilled water each time. Transfer the rinses into the appropriate discard container. Allow the flasks to drain. Empty the cuvettes into the drain and allow them to dry.

Caution: Wash your hands thoroughly with soap or detergent before leaving the laboratory.

Calculations

I. Synthesizing $Ni(NH_3)_nCl_2$

Record the results of the following calculation on Data Sheet 1.

1. Calculate the mass of $NiCl_2 \cdot 6\ H_2O$ used in the synthesis.

2. Calculate the mass of synthesized $Ni(NH_3)_nCl_2$.

II. Determining the Mass Percent NH_3 in $Ni(NH_3)_nCl_2$ by Titration

Do the following calculations for each determination and record the results on Data Sheet 2.

3. Calculate the mass of $Ni(NH_3)_nCl_2$ sample titrated.

4. Calculate the number of moles of HCl added to the $Ni(NH_3)_nCl_2$ solution.

5. Calculate the number of moles of NaOH required for titration.

6. Calculate the number of moles of NH_3 present in the $Ni(NH_3)_nCl_2$ sample.

7. Calculate the mass percent NH_3 in $Ni(NH_3)_nCl_2$.

8. Calculate the mean mass percent NH_3 in $Ni(NH_3)_nCl_2$.

III. Determining the Mass Percent Ni^{2+} Ion in $Ni(NH_3)_nCl_2$ by Spectrophotometry

Do the following calculations for each determination and record the results on Data Sheet 3.

9. Calculate the mass of $NiSO_4 \cdot 6\ H_2O$ in the standard solution.

10. Calculate the mass of $Ni(NH_3)_nCl_2$ in your unknown solution.

11. Convert all %T readings to A.

12. Calculate the corrected A_s for the $NiSO_4$ solution.

13. Calculate the corrected A_x for the $Ni(NH_3)_nCl_2$ solution.

14. Calculate the mass percent Ni^{2+} ion in $Ni(NH_3)_nCl_2$.

IV. Determining the Empirical Formula and Percent Yield of $Ni(NH_3)_nCl_2$

Record the results of these calculations on Data Sheet 4.

15. Calculate the mass percent Cl^- ion in $Ni(NH_3)_nCl_2$.

16. Determine the empirical formula of $Ni(NH_3)_nCl_2$.

17. Calculate the molar mass of $Ni(NH_3)_nCl_2$.

18. Calculate the number of moles of synthesized $Ni(NH_3)_nCl_2$.

19. Calculate the number of moles of $NiCl_2 \cdot 6\ H_2O$ used in the synthesis.

20. Calculate the percent yield of your synthesized $Ni(NH_3)_nCl_2$ based on the mass of $NiCl_2 \cdot 6\ H_2O$ used in the synthesis.

Post-Laboratory Questions

(Answer the following questions on a separate sheet of paper.)

1. Would the percent yield of $Ni(NH_3)_nCl_2$ be higher, lower, or unchanged if you made the following procedural changes? Briefly explain each answer.

(1) You performed the synthesis reaction at 100 °C instead of at 60 °C.

(2) You dissolved the $NiCl_2 \cdot 6\ H_2O$ in 50 mL of distilled water instead of 10 mL.

(3) You did not wash the $Ni(NH_3)_nCl_2$ crystals with concentrated NH_3 solution before drying them.

2. Would the following procedural changes cause the experimentally determined mass percent NH_3 in $Ni(NH_3)_nCl_2$ to be too high, too low, or unchanged? Briefly explain each answer.

(1) After dissolving a known mass of $Ni(NH_3)_nCl_2$, a student directly titrated the NH_3 with HCl solution, using a mixed bromcresol green–methyl red indicator.

(2) A student added excess standard HCl solution to a known mass of dissolved $Ni(NH_3)_nCl_2$ and back titrated the excess HCl with standard NaOH solution, using phenolphthalein indicator solution.

3. Would the following procedural changes cause the calculated mass percent Ni^{2+} ion in $Ni(NH_3)_nCl_2$ to be too high, too low, or unchanged? Briefly explain each answer.

(1) In Part III of the procedure, a student omitted adding concentrated NH_3 solution to the dissolved $Ni(NH_3)_nCl_2$ sample before analysis using the spectrophotometer, thinking that the additional NH_3 was unnecessary because the Ni^{2+} ion was already complexed as $[Ni(NH_3)_n]^{2+}$ ion.

(2) The student measured the $\%T$ of the standard and unknown solutions at a wavelength that was 20 nm lower than the actual analytical wavelength.

4. (1) Describe the color and appearance changes you observed when you dissolved $NiSO_4 \cdot 6\ H_2O$ in distilled water and then added NH_3 solution.

(2) Describe the color and appearance changes you observed when you dissolved $Ni(NH_3)_nCl_2$ in distilled water and then added NH_3 solution.

(3) Suggest an explanation for the changes you described in (1) and (2).

Data Sheet 1

I. Synthesizing Ni(NH$_3$)$_n$Cl$_2$

mass of weighing paper and NiCl$_2 \cdot$ 6 H$_2$O, g   _____

mass of weighing paper, g   _____

  mass of NiCl$_2 \cdot$ 6 H$_2$O, g   _____

mass of weighing bottle and synthesized Ni(NH$_3$)$_n$Cl$_2$, g   _____

mass of weighing bottle, g   _____

  mass of synthesized Ni(NH$_3$)$_n$Cl$_2$, g   _____

Data Sheet 2

II. Determining the Mass Percent NH$_3$ in Ni(NH$_3$)$_n$Cl$_2$ by Titration

molarity of NaOH solution used, mol L^{-1} _____

molarity of HCl solution used, mol L^{-1} _____

volume of HCl solution added, mL   _____25.00 mL (0.02500 L)_____

  number of moles of HCl added, mol   _____

	determination			
	1	2	3	4
mass of weighing paper plus Ni(NH$_3$)$_n$Cl$_2$, g	_____	_____	_____	_____
mass of weighing paper, g	_____	_____	_____	_____
mass of Ni(NH$_3$)$_n$Cl$_2$ sample, g	_____	_____	_____	_____
final buret reading, mL	_____	_____	_____	_____
initial buret reading, mL	_____	_____	_____	_____
volume of NaOH solution required, mL	_____	_____	_____	_____
number of moles of NaOH required, mol	_____	_____	_____	_____
number of moles of NH$_3$ present, mol	_____	_____	_____	_____
mass percent NH$_3$ in Ni(NH$_3$)$_n$Cl$_2$, %	_____	_____	_____	_____

mean mass percent NH$_3$ in Ni(NH$_3$)$_n$Cl$_2$, %   _____

Data Sheet 3

III. Determining the Mass Percent Ni^{2+} Ion in $Ni(NH_3)_nCl_2$ by Spectrophotometry

molar mass of $NiSO_4 \cdot 6\,H_2O$, g mol^{-1} _____ 262.9 _____

mass of weighing paper and $NiSO_4 \cdot 6\,H_2O$, g   _____

mass of weighing paper, g   _____

    mass of $NiSO_4 \cdot 6\,H_2O$, g   _____

mass of weighing paper and $Ni(NH_3)_nCl_2$, g   _____

mass of weighing paper, g   _____

    mass of $Ni(NH_3)_nCl_2$, g   _____

initial color and appearance of

    $NiSO_4$ solution   _____

    $Ni(NH_3)_nCl_2$ solution   _____

color and appearance of solutions after addition of concentrated NH_3

    $NiSO_4$ solution   _____

    $Ni(NH_3)_nCl_2$ solution   _____

A for the standard $[Ni(NH_3)_n]^{2+}$ Ion Solution using $NiSO_4 \cdot 6\,H_2O$

wavelength, nm	%T of the standard $[Ni(NH_3)_n]^{2+}$ ion solution	A of the standard $[Ni(NH_3)_n]^{2+}$ ion solution
620	_____	_____
600	_____	_____
580	_____	_____
560	_____	_____
540	_____	_____

approximate λ_{max} of the $[Ni(NH_3)_n]^{2+}$ ion, nm   _____

    approximate $\lambda_{max} - 10$ nm, nm _____   _____

    approximate $\lambda_{max} + 10$ nm, nm _____   _____

chosen λ_{max} of the $[Ni(NH_3)_n]^{2+}$ ion, nm   _____

A of the standard $[Ni(NH_3)_n]^{2+}$ solution at λ_{max} (A_s) _____

A of the unknown $[Ni(NH_3)_n]^{2+}$ ion solution

wavelength, nm	%T of the unknown $[Ni(NH_3)_n]^{2+}$ ion solution	A of the unknown $[Ni(NH_3)_n]^{2+}$ ion solution
_____	_____	_____

Correcting Unmatched Cuvettes

	%T	A
sample cuvette using reference cuvette blank	_____	_____
reference cuvette using sample cuvette blank (if necessary)	_____	_____
A_s for the standard $[Ni(NH_3)_n]^{2+}$ ion solution (from determination on Data Sheet 3)		_____
corrected A_s (if necessary)		_____
A_x for the unknown $[Ni(NH_3)_n]^{2+}$ ion solution		_____
corrected A_x (if necessary)		_____
mass % Ni^{2+} ion in $Ni(NH_3)_2Cl_2$, %		_____

Data Sheet 4

IV.   Determining the Empirical Formula and Percent Yield of $Ni(NH_3)_nCl_2$

mass percent Cl^- ion in $Ni(NH_3)_nCl_2$ %	_____
empirical formula for $Ni(NH_3)_nCl_2$	_____
*mass of $Ni(NH_3)_nCl_2$ synthesized, g	_____
molar mass of $Ni(NH_3)_nCl_2$, g mol^{-1}	_____
number of moles of $Ni(NH_3)_nCl_2$ synthesized, mol	_____
*mass of $NiCl_2 \cdot 6\ H_2O$ used in the synthesis, g	_____
molar mass of $NiCl_2 \cdot 6\ H_2O$, g mol^{-1}	237.7
number of moles of $NiCl_2 \cdot 6\ H_2O$ used in the synthesis, mol	_____
percent yield of $Ni(NH_3)_nCl_2$, %	_____

*Reenter the same data for this item as on Data Sheet 1

Pre-Laboratory Assignment

1. Review Experiment 6 for a discussion of the use of a spectrophotometer.

2. (1) Briefly explain why it is necessary to work under a fume hood when using concentrated NH_3 solution.

(2) Briefly explain why it is important to cover containers of concentrated NH_3 solution with a wet paper towel when working with these containers in the open laboratory.

(3) Briefly explain the hazards you should be aware of when working with $NiCl_2 \cdot 6\ H_2O$.

3. Briefly define the following terms as they apply to this experiment.

(1) standard solution

(2) equivalence point

(3) end point

(4) analytical wavelength

4. When zinc chloride ($ZnCl_2$) reacts with NH_3 solution, a white crystalline compound, $Zn(NH_3)_nCl_2$, is produced. A student used the procedure in this module to determine the mass percent of NH_3 in $Zn(NH_3)_nCl_2$. The student added 25.00 mL of $0.2500M$ HCl solution to an aqueous solution containing 0.3145 g of $Zn(NH_3)_nCl_2$. The excess HCl required 25.60 mL of $0.1000M$ NaOH solution for complete titration. Calculate the mass percent of NH_3 in $Zn(NH_3)_nCl_2$, based on the student's data.

5. A student recorded the following percent transmittances for a $0.500M$ chromium(III) ion solution (Cr^{3+}), using a Spectronic 20 spectrophotometer.

wavelength, nm	percent transmittance, %T	absorbance, A
700	64.6	_____
675	52.5	_____
650	24.0	_____
625	7.94	_____
600	3.98	_____
575	3.02	_____
550	7.08	_____
525	15.8	_____
500	32.4	_____
475	34.7	_____
450	37.2	_____

(1)   Calculate the absorbance equivalent to each %T reading. Record these absorbances in the table.

(2)   From the wavelengths listed, determine the analytical wavelength of the Cr^{3+} ion solution and record it below.

analytical wavelength, nm _____

6.   Iron(II) ion salts, such as $FeSO_4 \cdot 7\,H_2O$, react with the organic compound ortho-phenanthroline (phen) to form red coordination compounds with formulas such as $Fe(phen)_nSO_4$. Such compounds have an analytical wavelength of 510 nm.

(1)   A standard solution containing $Fe(phen)_nSO_4$ was made by dissolving 0.0139 g of $FeSO_4 \cdot 7\,H_2O$, with a molar mass of 278.05 g mol^{-1}, in 500.0 mL of water containing excess phen. The solution has a %T of 36.5% at 510 nm. A 500.0-mL solution with unknown $Fe(phen)_nSO_4$ concentration has a %T of 64.3% at 510 nm. What is the molar concentration of $Fe(phen)_nSO_4$ in the unknown solution?

(2)   How many grams of Fe^{2+} ion are there in the unknown solution?